Tectonics

Proceedings of the
27th International Geological Congress

Stratigraphy	Volume 1
Palaeontology	Volume 2
Quaternary Geology and Geomorphology	Volume 3
Sedimentology	Volume 4
Precambrian Geology	Volume 5
Geology of Ocean Basins	Volume 6
Tectonics	Volume 7
Geophysics	Volume 8
Petrology (Igneous and Metamorphic Rocks)	Volume 9
Mineralogy	Volume 10
Geochemistry and Cosmochemistry	Volume 11
Metallogenesis and Mineral Ore Deposits	Volume 12
Oil and Gas Fields	Volume 13
Solid Fuel Mineral Deposits	Volume 14
Non-metallic Mineral Ores	Volume 15
Hydrogeology	Volume 16
Engineering Geology	Volume 17
Remote Sensing	Volume 18
Comparative Planetology	Volume 19
Mathematical Geology and Geological Information	Volume 20
History of Geology	Volume 21
Geological Education	Volume 22
Origin and History of Marginal and Inland Seas	Volume 23

Proceedings of the
27th International Geological Congress
Moscow 4–14 August 1984

Volume 7

TECTONICS

VNU SCIENCE PRESS
Utrecht, The Netherlands
1984

VNU Science Press BV
P.O. Box 2073
3500 GB Utrecht
The Netherlands

First published 1984

ISBN 90–6764–016–6 Volume 7
ISBN 90–6764–009–3 set of 23 volumes

Printed in Great Britain by J. W. Arrowsmith Ltd, Bristol

PREFACE

The leading scientists in the many different fields of geology were invited by the Organizing Committee to present a paper on a specific problem in present-day geological science at the 27th International Geological Congress. The published proceedings of the Congress consist of twenty-three volumes. Each volume is dedicated to a particular aspect of geology. Together the volumes contain all of the contributions presented at the Congress.

The Organizing Committee is pleased to acknowledge the efforts of all of the participating scientists in helping to produce these proceedings.

Professor N. A. BOGDANOV
General-Secretary of the
Organizing Committee

CONTENTS

REGULARITIES OF THE CONTINENTAL CRUST FORMATION IN PHANEROZOIC TECTONIC MAPPING OF CONTINENTS

A.V. PEIVE, A.L. KNIPPER, M.S. MARKOV, A.A. MOSSAKOVSKY, A.S. PERFILIEV, YU.M. PUSHCHAROVSKY, N.A. SHTREIS
Geological Institute of the USSR Academy of Sciences, Moscow, USSR

Leading off the seventies a new trend in theoretical tectonics has been developed at the Geological Institute of the USSR Academy of Sciences. It owes its origin to the reveal of the oceanic crust remains of the geological past in the folded structures of continents, and to study of the structural and material transformations of the crust in the course of geosynclinal development and subsequent tectonic processes responsible for its transformation into the Earth's crust of the continental type.

This trend synthesizes the ideas of mobilism in tectonics, advances in the field of continental geology, the latest data on deep structure of the Earth's geospheres obtained by both the geophysical and the petrological methods, and, finally, the most important results of geological studies of the present-day ocean bottom.

The paramount concern was given to elaboration of two closely related problems. One of them dealt with ophiolites of the continents (their structure, composition and structural position in fold systems), whereas the other - with tectonic evolution of the Earth's crust, its stage character and orientation.

The problem of ophiolites was elucidated through geological and petrological study of the ophiolitic association in various folded areas of Eurasia: in the Urals, Mediterranean, Caucasus, Iran, Sayany, Mongolia, in the USSR North-East, on Cuba, in Italy, the Appalachian Mountains, and in some other places. The materials on rocks of oceanic bottom and ocean islands obtained by special marine expeditions were used for correlation.

These investigations made it possible to state that an ophiolitic complex composed of a peculiar set of magmatic and sedimentary rocks regularly repeats itself

Proceedings of the 27th International Geological Congress, Volume 7, pp. 1–14
TECTONICS

in the same succession in differing by age folded zones: Cenozoic, Mesozoic, Paleozoic and even Late Proterozoic.

The lower part of the ophiolitic complex, the so-called melanocratic basement, is composed of heavily deformed ultrabasites: dunite-harzburgite and lherzolite. Higher up the sequence olivine and olivine-pyroxene gabbro alternate with pyroxenite, wherlites, dunites and anorthosite. This alternation is known under the name of a banded complex which, in its turn, is substituted higher up the sequence by gabbroids containing veins, dykes and bodies of tonalites, diorites and quartzy tonalites. Finally, the uppermost horizons of gabbroids are cut by dykes of basalts, diabases, dolerites and sometimes lamprophyres, frequently transformed into amphibolites which form a so-called complex of parallel dykes.

The upper part of the ophiolitic complex is composed of tholeiitic basalts, spilites, diabases, radiolarites, phtanites, siliceous-terrigenous, pelagic carbonate rocks and black shales correlated to the lowermost parts of the section of eugeosynclinal zones deposits.

It was found out that the ophiolitic complex was not coeval, as suggested previously. Its formation proceeded by several stages that are frequently divided by epochs of tectonic deformations. The studies of relationships of rocks inside the ophiolitic complex showed that its lower part composed of dunites and harzburgites was tectonized and metamorphosed much stronger and more complicatedly than the overlying banded complex. In its turn the upper volcanogenic-sedimentary part of the ophiolitic complex sometimes rests with hiatuses on gabbroids, and even on ultrabasites.

The present-day structural position of ophiolitic complexes in folded belts does not correspond, as a rule, to their primary occurrence in the Earth's crust sequence. In the vast majority of cases, as a result of intense horizontal movements, the ophiolitic complexes of continents are in the tectonically dislocated (allochthonous) state, having been subjected to the processes of tectonic crushing, delamination and piling.

The comparative study of ophiolitic complexes of continents and rocks of the present-day ocean bottom revealed their similarity both by composition and structure of sections. The lower ultrabasitic part of the ophiolitic complexes of continental folded zones can be correlated to rocks of the upper mantle beneath the oceans; the banded complex - to the third layer of the oceanic crust, the pillow tholeiitic basalts and diabases - to the second layer, and jaspers and other sedimentary rocks - to the first layer of the oceanic crust. These comparisons are confirmed by finds of deep-sea oceanic forms in sedimentary rocks of the uppermost parts of continental ophiolitic associations. Consequently, there are good reasons to believe that ophiolitic complexes of the continents are relics, fragments of the ancient oceanic crust on which the eugeosynclines developed. This fundamental conclusion gave rise to the mobilism treatment of paleotectonics of eugeosynclinal areas and served the basis for working out new concepts of regularities of the Earth's continental crust formation. These concepts were shown on the tectonic map of North Eurasia, scale 1:50000000 and some large-scale maps of separate regions (the Urals, Kazakhstan, USSR East) compiled in accordance with quite new principles (1-8).

The problems of tectonic evolution of Earth's crust of continental folded zones, its trend, stage character and the role of geosynclinal process in this evolution appeared as that related to the ophiolite problem. The working out and substantiation of new principle theses stating that all eugeosynclines of the geological past were formed on the oceanic crust, necessitated a historical analysis and study of various geological processes that in the course of development of geosynclinal areas caused transformation of the oceanic crust into the continental one, as all folded systems of continents in the present-day structure are known to be of the continental type of the Earth's crust.

While solving this problem, we were guided by the many years' experience and progress in continental geology which enabled us to elucidate the most important features of geological development of the continental folded zones using the methods of structural-formational analysis, petrology, lithology and historical geology.

Worth attention here are the known evolutional regularities of magmatic and metamorphic processes in folded geosynclinal zones of Phanerozoic; vertical and lateral formational rows in such zones; common peculiarities of development of geosynclinal zones and their transformation into folded systems; essential space and chronological regularities of formation of folded zones and belts.

Simultaneously, we used some concepts of the litosphere plate tectonics, especially a peculiar actualistic approach in distinguishing and interpretation of the major structural elements, geodynamic and sedimentation environments of the geological past.

A great volume of work has been done aimed at revision (from the mobilism position) of the tectonic structure of many folded systems of North Eurasia, the Urals, first of all, the Cacausus and contiguous parts of the Alpine belt, paleozoides of Siberia, Mongolia, mesozoides of the USSR North-East. The analysis of structural-formational complexes from the position of actualism showed that in the present-day complicated nappe-folded structure of differently aged folded systems there are complexes of sedimentary and magmatic rocks similar by composition and structure not only to present-day oceanic sediments, as already mentioned above, but also to rock complexes of island arcs and marginal seas occurring in the zones transitional from ocean to continent. In so doing, the latter within folded zones always prove younger than the oceanic sediments.

Along with such complexes, we diagnosed terrigenous complexes similar to deposits of the continental slope and foot, carbonate shelf deposits characteristic of passive continental margins, as well as extended volcanio-plutonic belts of the orogenic type - analogues of many present-day active continental margins. The development of folded zones in all cases was terminated by epochs of tectonic deformations, mass manifestation of potassium-granitic magmatism and accumulation of continental molasses under orogenic conditions. At the present-day stage such conditions are peculiar to continents only.

Thus, there appeared an idea of the stage character

of the same oriented formation process of the Earth's crust of continents through a successive transformation of the oceanic Earth's crust into a transitional and then into a continental one. Correspondingly, the mountains rock complexes formed at these stages reflect various stages and different degree of a complicated and long process of transformation of the oceanic crust into a continental one, when judged by structural features, composition and metamorphic reworking.

The idea of the stage character of development of the Earth's crust passing through the oceanic transitional and continental stages promoted a new approach to the tectonic zonation of continents relative to the time of formation of the ripe Earth's crust and the time of origination of granite-metamorphic beds in places where the crust has not been formed yet.

Complexes of the oceanic stage incorporate such sedimentary and magmatic formations that were formed in paleobasins with the crust of the oceanic type. They are associated, and frequently underlain by rocks of the melanocratic oceanic basement. It should be emphasized that in the present-day structure of folded belts these complexes, as well as fragments of the melanocratic basement, practically always compose tectonic nappes thrusted over the geological formations of a quite different nature - island arc or epicontinental. The tectonic matching of rock complexes formed under quite different conditions produces an erroneous impression of primary nearness of their accumulation areas. This fact frequently leads researchers into error. Geodynamic conditions of the oceanic stage are characterized by predominance of the general tension on whose background local zones of tectonic piling originated.

This stage is characterized by non-differentiated, or slightly differentiated tholeiite-basaltic volcanics together with radiolarites, jasper, phtanite, micritic limestones and shales attributed to the upper part of the ophiolitic complex. However, they are frequently substituted along the lateral by rocks of a alkaline-basaltic formation, and even by contrasting but essentially sodic volcanogenic association including basalts, albitophyres and quartzy keratophyres, as well as by carbonate reef rocks peculiar to intra-oceanic rises and guyots. These rock complexes are usually distinguished on continents as spilite-diabase, spilite-keratophure, phtanite, carbonate-siliceous formation. The study of continental folded zones showed that intrusion of gabbro-norites, plagiogranites and metamorphism of a green-schist and epidote-amphibolite facies.

Special investigations carried out in the Urals, Alpine belt and in the USSR North-East showed that many paleo-oceanic structures revealed within these folded areas owed their formation to the processes of intraconti-nental gaping fault proceeding on various tectonic surfaces inside the lithosphere. In such cases the oceanic stage is preceded by the riftogenic stage during which coarse-detrital sediments are accumulated. The latter consist of clastomorphic material from separating continental massives and accompanying it subalkaline volcanics. Then, as a newly-formed oceanic basin expands, there appear rocks of an ophiolitic complex. At the same time it is beyond any doubt that in many other cases, especially in such folded belts as paleozoides of Siberia and Central Asia characterized by a long development that commenced as far back as in the Riphean, the oceanic stage of development could have been inherited from the ancient World Ocean.

At the transitional stage, when the Earth's crust is characterized by an exceptionally complicated structure, combination of various types of crust with variable thickness of the basaltic layer, local develop-ment of the granite-metamorphic layer, sharp variations of thickness of the granite-metamorphic cover, com-plexes of sedimentary and magmatic rocks are formed. These differ in composition and structure, and are characterized by considerable lateral changeability.

In the recent structure of the Earth's crust this stage of development is related to structures of island arcs, marginal seas and deep-sea trenches whose geo-dynamic environment, as established now, is determined by predominance of uneven and sporadic tectonic squeezing and piling, especially increasing by the end of the stage. Zones of secondary extension responsible for formation of the marginal seas basins appear on the background of general squeezing.

This is the very stage that specifies tectonic differentiation of paleooceanic areas or their large parts and distinguishing of structure-formational zones with specific vertical rows of formations and their successive transformation into folded systems. This has been already thoroughly studied in the process of development of the geosynclinal theory.

One of the main types of structural-formational zones peculiar to the transitional stage is representative of paleo-island arc structures within which the rock complexes of the oceanic stage are overlain either by thick accumulations of volcanics of the contrasting series or high-alumina basalts complicatedly associated with greywackes, less frequently with limestones. Higher up the sections they are replaced by volcanics of a differentiated calcareous-alkaline series, whose piled and folded accumulations are overlain by lower molassa formations. In deep root parts of such zones the processes of zonal metamorphism and plagiogranitization of volcanogeno-sedimentary rocks take place. They are accompanied by formation of various plagiogneisses and essentially plagioclase crystalline schists and an association with autochthonous and paraautochthonous plagiogranites, granodiorites and plutons of variegated gabbrodiorite-granodioritic composition. These rocks compose lenses, peculiar "embryos" of a newly-formed granite-metamorphic layer that is fixed by geophysical methods at the base of the most number of recent island arcs.

Here, it is likely worthwile to recall that as a result of petrological and geochemical study of metamorphic rocks of Phanerozoic folded zones, as well as Precambrian ones, it was long ago established that the vast majority of metamorphic and magmatic rocks, corresponding by composition and geophysical features properties to the granite-metamorphic layer, had undergone in its history at least two stages of regional granitization of different types: the earlier plagiogranitization and the later potassium granitization. The former proved responsible for formation of various plagiogneisses and plagioclase crystalloschists, plagiogranites and granodiorites, the latter, usually superimposed on the former, resulted in formation of the major ganut of microclines, less frequently orthoclase gneisses and plutons of potassium granites of different types of the granite-alkaline-granite formation.

The same processes also involve microcontinents of a more ancient an older continental crust incorporated in the composition of transitional zones. However, in this case these processes superimpose a quite different substratum.

Another very important type of structure-formational zones corresponds to basins of the marginal seas. Its formational row is represented by tufogenic-siliceous-greywacke, greywacke proper, aspiditic and some other terrigenous formations which contain volcanic rock interbeds of basic and median composition. There occur members and series of wild flysh and olistostromes. Higher up the sections the above formations are substituted by flysh, flyshoid- and then lower down mostly marine molassa. This type of structural formational zones was traditionally distinguished under the name of terrigenous geosynclines.

Formations of the traditional stage are frequently represented by deposits peculiar to shelves, continental slopes and their feet. These are mainly carbonate and terrigenous complexes containing sometimes deep-sea sediments as well. Thus, during the traditional stage of the crust development there occurs a tectonic differentiation of paleooceanic areas and formation of the complicated and characteristic of geosynclinal areas structural-formational zonality. The latter controls the processes of sedimentation, magmatism, metamorphism and folding (tectonic piling of rock masses, including). The folding successively embracing more and more areas, gradually transformed paleooceanic areas into folded zones. Simultaneously, and in direct genetic connection with the abovementioned phenomena, the fundamental process of origination and growth of some segments of the granite-metamorphic layer was proceeding through long evolutionary material and material and structural transformation of rocks of the volcanogeno-sedimentary cover of the so-called basaltic layer, and probably the mantle too. This process was accompanied by higher sharply differentiated heat flows and an inflow and outflow of the matter, as a result of sedimentary and metamorphic differentiation and metasomatosis due to magma intrusions, and especially owing to supply of liquid and gaseous fluids. The results and space-time regularities of this process can be evidenced both by observations of the composition and age of metamorphic rocks in cores of island arcs, and by magmatism and metamorphism manifestations that by age and position in the sections of folded zones correspond to the stage of their development. These are plagiogneisses, other indications of rock plagiogranitization, autochthonous and para-autochthonous plutons of the gabbro-plagiogranite-diorite and diorite-plagiogranite formations, which, together with the lower molassa fixing the time of folded zones formation, are regarded by us as complexes-indices of formation of the granite-metamorphic layer in folded zones.

The analysis of structure and the history of many folded systems showed that formation of the granite-metamorphic layer was a long process proceeding heterochronously in various zones of even one folded area, not speaking of folded belt as a whole. We call such a way of formation of a granite-metamorphic layer characteristic, for instance, of paleozoides of Siberia, Kazakhstan, West Europe, autochthone, comparing it with the allochthone way, when formation of the granite-metamorphic layer resulted from large-scale horizontal dislocations, piling and remobilization of ancient Precambrian continental masses. This is the very way by which the continental crust in linear folded belts (Uralian, Alpine and some others) was formed. Commonly, in the history of any folded area two ways of the continental crust formation co-exist, one of them prevailing.

So, by the end of the transitional stage the Earth's crust within folded systems, when judged by geophysical parameters, acquires features peculiar to a crust of the continental type, i.e. everywhere it consists of three main layers: sedimentary-volcanogenic, granite-metamorphic, and basaltic. Actually, the material composition of these layers, as well as their structure, is characterized by appreciable lateral changeability, considerably differing from that we face with in the present-day continental crust.

First of all, it concerns, as mentioned above, the granite-metamorphic layer of essentially plagiogneissic in the transitional crust, and the granite-gneissic one in the continental crust.

However, there are good grounds to think that it is equally true for the basaltic layer too. In any case, the data available in the present day geology, testify to considerable differences in the rock composition of the basaltic layer of the transitional and continental crust. In the transitional crust it is represented by gabbro-amphibolites heavily tectonized and metamorphosed in green-shale and epidote-amphibolite facies which slightly differ from the basaltic layer of the oceanic crust. In the continental crust, where the geophysical basaltic layer is accessible for geological investigations on Precambrian shield only, it is composed of basic crystalline schists and pyroxene-plagioclase gneisses, or even of more acid gneisses and crystalline schists (when judged by preliminary results of the still going on drilling of the Kola superdeep Hole).

So, it became clear that the material composition of the Earth's crust basic layer considerably differs from that of the transitional and continental crusts. It follows that the true continental crust, peculiar to the present-day continents, requires for its formation additional fundamental transformations that would affect practically the entire section of the crust of the transitional type.

Naturally there arises a question: at what historico-geological moment does this important transformation of material and also structural features of the Earth's crust basic layers take place in the process of its transformation from the transitional into the continental type?

The well studied stage character of development of folded geosynclinal zones of continents can answer this question.

The above transformations proceeded during the global epochs of tectono-magmatic orogenic activity of the Earth's crust with folded zones of continents and adjacent parts of the Precambrian platforms that complete large tectonic cycles - the Alpine, Variscan, Caledonian and others. In the preceding years, especially in the sixties, these global epochs were the subject of animated, and frequently formal discussions: the question was,whether these epochs were an orogenic stage of geogynclinal areas, or they belonged to a quite isolated phenomenon, i.e. repeated orogenesis (deiteroorogenesis), or they should be regarded as manifestations of autonomous of reflected activization, and so on. The work on the tectonic map of North Eurasia showed that there were seven such global epochs in the history of the continent related by us to the completion of formation of new massifs of the continental crust *): 1) Alpine (Oligocene, Miocene, beginning of Pliocene); 2) Late Mesozoic (late Cretaceous); 3) Early Mesozoic (end of Triassic - beginning of Jurassic); 4) late Paleozoic (middle of Carboniferous - early Permian); 5) early-middle Devonian; 6) late Riphean; 7) Karelian (at the end of early Proterozoic).

Great importance of these epochs in the evolution of the continental crust can be explained by the following. These global epochs are characterized by a structural reconstruction of geosynclinal folded areas (alpides, variscides, caledonides, etc.) and the adjacent parts of the ancient platforms that resulted in a new system of structures - arched - clumpy uplifts, inner depressions and marginal down-

*) The epochs of continental crust formation cannot be confused with the epochs of folding or mountain bilding. For instance, the USSR east is characterized by the significant late Jurassic-early Cretaceous epoch of tectonogenesis, but it was not accompanied by growth of continental blocks.

warps unconformably superimposed on structures of folded areas.

During these epochs in the same areas of orogenic activization there occurred simultaneous mass manifestations of potassium granitization (microcliniti-zation, albitization) of preceding development stages, including those of more ancient plagiogneisses. Related granite-alkaline-granitic magmatism was widespread as well. All this testifies to considerable material transformations in the granitic-metamorphic sheet. Noteworthy is that the processes of granitic monotyped magmatism and potassium granitization simultaneously affected within corresponding folded areas the zones with different age of folding and different time of formation of the granite-metamorphic layer. This speaks of synchroneity of the processes of granitization and granite-formation at deep levels of the Earth's crust within vast previously heterogenous folded areas.

These global tectonic epochs coincided with the epochs of mass orogenic surface (so-called subsequent) volcanism whose products formed extended volcanic belts spatially coinciding with areas of tectonic activization: Central Kazakhstan and Altai-Minusinsk-Tuvin and Scotland in the Devonian, Eurasian in the late Paleozoic, Sikhote-Alin and Okhotsk-Chukotsk in the Cretaceous, etc.

Finally, these global epochs were characterized by intense rock formation of which we judge on the basis of thick accumulations of coarse-detrital continental upper molassa of the corresponding age, and which testifies to a sharp increase of the Earth crust thickness within the areas of its manifestation and formation of deep-seated roots necessary for the isostatic uplift.

It is perfectly obvious that all these various forms of manifestation of tectono-magmatic activization coinciding in time and localized within the same folded areas of continents could not be related to fundamental transformations of the Earth's crust deep-seated shells and the mantle, granite-metamorphic, first of all. After these transformations the Earth's crust acquired its continental characteristics. That is why the first manifestations of potassium granitization, alkaline-granitic magmatism, orogenic volcanism and upper continental molassa is regarded by us as an indication of final formation and consolidation of the continental crust. Subsequently, in the process of evolution of the continental crust they can reiterate.

Thus, the process of the continental crust formation in the Phanerozoic is divided into two major stages:

1) Stage of origination and growth of segments of the granite-metamorphic layer through a long evolutionary material and structural transformation of rocks of the volcanogeno-sedimentary deposits and the basaltic layer during the oceanic, but mainly transitional stages of the Earth's crust development;

2) Stage of global, relatively short epochs of strong activization of tectonic movements resulting in tectonic mergence, piling of previously occurring granite-metamorphic layers of various age, separated fragments of ancient continental crusts into monolithic masses - continents, or parts of the continents; in this case, as a result of metamorphic and magmatic processes the growth and increase of the metamorphic layer thickness took place, and its repeated, essentially potassium granitization responsible for material consolidation of the newly-formed continental crust.

All these metamorphic transformations result in shifting the Conrad and Mono discontinuities to other levels corresponding to a new distribution of metamorphic rocks of required compactness, already peculiar to the continental type of the Earth's crust.

The elucidation of regularities of the process of the continental crust formation enables us to approach a new, more profound genetic understanding of the essence of the geosynclinal development, and to determine its place and role in formation of the major Earth's crust layers. From these positions, the geosynclinal development should be treated as a combination of tectonic, sedimentary, magmatic and metamorphic processes causing the structural and material transformation of the oceanic crust into the continental one. During the last years there appears tendency in geology to abandon the term "geosyncline".

Keeping to the new concept of the Earth's crust development that can be formulated as "accretion-structural", the authors consider such a deviation from this term quite rightful and natural. It reflects the characteristic feature of the present-day tectonics: a change of classical concepts of the geosynclinal theory by new ideas of tectonogenesis of the Earth.

REFERENCES

1. Tectonic map of the USSR East and adjacent countries, scale 1:2 5000 000. GUGK, 1979 (in Russian).
2. Tectonic map of Eastern Kazakhstan, scale 1:2 500000. M., GUGK, 1981.
3. Tectonic map of North Eurasia, scale 1:5000000. MINGEO SSSR, 1980.
4. Tectonic map of the Urals, scale 1:1000000, MINGEO SSSR, 1977.
5. Tectonics of Kazakhstan (explanatory note to the tectonic map 2). M., "Nauka", 1982, 138 sheets.
6. Tectonics of continental margins of the Pacific North-West (explanatory note to the tectonic map 1).M.,"Nauka", 1980, 285 sheets.
7. Tectonic of North Eurasia (explanatory note to the tectonic map 3). M., "Nauka", 1980, 220 sheets.
8. Tectonics of the Urals (explanatory note to the tectonic map 4), M., "Nauka", 1977, 120 sheets.

STRUCTURE AND EVOLUTION OF THE EARTH'S CRUST IN SIBERIA ON THE BASIS OF STAGE-BY-STAGE TECTONIC REGIONALIZATION

YANSHIN A.L. (1), BOGOLEPOV K.V., LAVEROV N.P. (2), SURKOV V.S. (3), BASHARIN A.K. (4), BERZIN N.A. (4), BORUKAEV CH.B. (4), VOTAKH O.A. (4), FRADKIN G.S. (4), STAROSELTSEV V.S. (4), CHIKOV B.M. (4), SHCHEGLOV A.P.(5).
1. Institute of Lithosphere, USSR Acad. Sci., Moscow, USSR; 2. Academy of National Economy, Moscow, USSR; 3. Siberian Institute of Geology, Geophysics and Mineral Raw Materials, Novosibirsk, USSR; 4. Institute of Geology and Geophysics, Siberian Branch of the USSR Acad. Sci., Novosibirsk, USSR; 5. Far East Institute of Geology, Vladivostok, USSR.

Vast, more than 10 mln km2 territory of Siberia is extremely diversified in tectonical respect. It comprises the ancient Siberian platform, Paleozoic folded areas of the Central-Asian belt fringing it from the south, Mesozoic folded Verkhoyansk-Kolyma, Taimyr and Mongolo-Okhotsk areas, Mesozoic orogenic belt conjugated with the latter and young West-Siberian platform.

The generalizing tectonical studies persistently emphasized a specificity of Siberia compared to Europe. E.g. the materials on tectonics of Siberia served as a basis of the sagacious idea by V.I. Vernadsky about the Earth's dissymmetry. It is Siberia, where studies identified the Baikal and Mesozoic foldings, which weak manifestations were revealed in West Europe much later.

The structure of some regions in Siberia is studied rather intensively (e.g. West-Siberian plate). This resulted in discovery of very important commercial ore deposits. At the same time, wide special tectonic investigations with an objective to describe the volumetric structure of the whole territory of Siberia have not been carried out. Creation of such three-dimentional model for preparing the fundamental basis for prediction of distribution of commercial ore deposits became the aim of compiling "the Atlas of tectonic maps and key profiles of Siberia".

This huge interdepartmental works, being an important chain in the superprogramme "Siberia" was accomplished

Proceedings of the 27th International Geological Congress, Volume 7, pp. 15–25
TECTONICS

by numerous scientific institutions under the Siberian Branch of the USSR Academy of Sciences, the USSR and RSFSR Ministries of Geology etc. The Atlas includes 7 stage-by-stage and general (summary) tectonical maps as well as maps reflecting recent structure and state of the Earth's crust and 12 key geologo-geophysical profiles. The maps are compiled to be published on a scale of 1:5 000 000, but the authors' models were compiled on more detail scales - from 1:100 000 up to 1:2 500 000. The works have synthesized huge material on geology of Siberia, obtained among other things in a course of specialized investigations along the program "Atlas".

Compilation of the Atlas was based on methodology worked out by IGIG SO of the USSR Academy of Sciences on tectonic zonation by structural-matter properties of mineral masses, modelled in a course of the works. With the aim of detailing the Earth's deep zones a method of stage-by-stage zonation or compilation of maps of megacomplexes was adopted. Much attention during these constructions was given to the system of key profiles.

The megacomplex is a facies series of tectonical complexes (geosyncline, proto- and deuterogenous, plate) with different matter composition, structure and conditions of formation, which were formed during a single tectonical era. The age boundaries of megacomplexes are defined as approximately isochronous in conformity with periods of main tectonical reconstructions of the Earth's crust, manifested on the territory of Siberia. Combination of megacomplexes in space gives a notion about the most general peculiarities of the continental crust structure and main stages and laws of its formation. The succession of megacomplexes creates the chronological framework of the Atlas.

The megacomplexes are subdivided into tectonical complexes. They are large geological bodies - an association of formations, which receive on the map a morphological (geometric) characteristics and tectonical interpretation. Besides traditional geosyncline and plate (platform) tectonical complexes were used also proto - and deuterogenous complexes.

Their identification is new in principle.

The leading recent tectonical concepts accentuate transformation of oceanic crust into continental crust in the geosyncline provinces. The moment of such transformation is fixed by appearance of molassa - formations of protoorogenic class. The later history of the folded area belongs to the continental stage. The tectonical forms typical for it are usually considered less detailed. However the interval between main folding and accumulation of orthoplatform (plate) mantle is frequently very long and complex. Besides protoorogenic complexes, closely connected in space and time with geosynclinal complexes, this time was accompanied by formation of deuterogenous complexes, separated from the above-mentioned ones in space (or separated from them by long interval) - usually by an epoch of structural reconstruction.

The deuterogenous complexes form both extended volcanic belts and isolated from each other bodies of sedimentary and volcanogenic thick formations with essential gradients of their change. They are predominantly composed of continental, rarely epicontinental marine sedimentary and sedimentary-volvanogenic molassa. Among volcanites predominate lavas and pyroclastic alkaline affinity from trachybasalt up to trachyliparite. Widespread are intrusive massifs of granites, syenites, as well as rapakivi, gabbro-anorthosites etc.

The deuterogenous complex overlies the heterochronous folded basement and even the platform mantle. Their separation as an independent structural element helps more completely to reflect lateral and vertical series of tectonical structures, and more deeply to reveal the history of crust formation. The deuterogenesis does not result in formation of new continental crust but in its essential transformation.

Each of the revealed tectonical complexes is described in detail. Special signs show its position by depth, composition and interrelations between its formations and associations, the degree and nature of dislocation of beds and metamorphism, manifestations of ore deposits etc.

The compilation of the Atlas permitted to reflect

completely the structure of each megacomplex and to reveal new feature of general structure and evolution of the Earth's crust in Siberia.

The pre-Riphean formations are shown in a single scheme though by their volume they considerably exceed one megacomplex, because the area of their outcrops is small and the closed territories are not sufficiently studied.

The works on compiling the Atlas made a considerable contribution into studying pre-Riphean deposits in Siberia. In the Aldan complex earlier considered as a single and adopted as a standard of Upper Archean the geologists from the Yakutsk Branch of the USSR Academy of Sciences separate a gneiss complex of intrastructure by a number of properties similar to the most abcient gray gneisses of other shields. The methods of detailed structureal analysis revealed an existence of disconformity, presupposed earlier by geologists, which separates the quartzite-bearing and carbonate-bearing parts of the overlying supercrustal unit. These results permitted to correlate the Aldan shield crystalline complexes with the most ancient formations of other shields in a new way and in a new way to interpret its structure. Among other things, the studies revealed that many of the grabens, superimposed on the crystalline basement and composed of weakly metamorphozed rocks, did not belong to the type of greenstone belts and had Lower Proterozoic age. However these results do not permit at present, to make unambiguous age subdivision of the crystalline basement, overlain by weakly metamorphozed deposits belonging to the Karelian megacomplex. The metamorphic zonation can be presumably considered as superimposed over the structural one while the granulitic facies of metamorphism is typical, in places, also for Lower Proterozoic rocks.

Pre-Riphean deposits, metamorphozed in conditions of mesozone were included into the Karelian megacomplex. On the Siberian platform they are rather well recognized while in the folded areas they are merged with the crystalline basement and practically unseparable from it. The extended belts of Karelides or entire mantle of geosyncline deposits of this age were not discovered on the Siberian platform. The megacomplex occurs locally in the most different structural conditions - in series of grabens, in wide basins of syneclise type (which is typical, by the way, for the Russian platform). The megacomplex is presumably widespread in a form of belt with latitudinal strike only in the Stanovoi province.

Interpretation of the Karelian formations in terms of teaching on geosynclines is difficult task. On the one hand, their formational series correspond to the eu- and miogeosynclines and on the other - overwhelming majority of zones occupy intercontinental position. The role of vast overthrust and nappe overlaps, breaking the order of zones, is not yet clear though their features are undoubtful. For these reasons the zonation of this megacomplex and its basement was based on principle of structuro-formational zones (compared to other maps).

The upper part of the Lower Proterozoic included earlier into the Middle Proterozoic, deserves particular attention. This is an independent structural stage, overlying the underlying deposits with disconformity and represented by association of volcanogenic molassa and high-mature sandstones. In places (the Prebaikal province, etc.) the deposits form typical volcano-plutonic belts, which by analogy with the Anadyr belt can be presumably considered as marginal-continental. The volcanism disappears inside the pracontinent. By the adopted classification the stage belongs to the deuterogenous complexes. It corresponds to the epoch of widespread cratonization, i.e. the process of formation of the Siberian platform basement and temporary consolidation of adjacent areas. As a result, the vast Siberian craton had been formed by the beginning of the Riphean.

The Baikal megacomplex is distinguished in the Riphean. Everywhere it overlies (with few exceptions) the underlying deposits with disconformity. On the Siberian platform the megacomplex is subdivided into two stages. The lower one is predominantly composed of continental quartzite and oligomiktic sandstones and in the marginal parts also of high-aluminiferous shales. It is included into the Lower Riphean. The deposits occupy narrow graben-like troughs, in places extending on the territory of the adjacent blocks. Trustworthy Lower Riphean deposits were established only in the south-eastern part of the platform in the North Prebaikal province, in eastern part of the Anabar massif and on the Yenisei ridge. But they will be possibly correlated with similar by composition rocks recovered by holes drilled on the West-Siberian plate and parts of deposits from the Altai-Sayany province.

The second stage, dated as Middle and Upper Riphean is more widespread. It is composed of abundant carbonate formations, formed in conditions of epicontental seas. Its lower boundary is diachronous. In places, the Upper Riphean directly overlies the platform basement. This part of megacomplex has abundant flysh formations, indicative of basin deepening.

The entire Riphean mantle was not discovered on the Siberian platform and the megacomplex lies on the platform and miogeosyncline troughs, separated by basement protrusions, overlain in places by thin mantle. Such type of structure is typical not only for the Siberian platform but for the major part of the Verkhoyansk- Kolyma province and possibly for a part of the West-Siberian plate. All these areas are united into the Siberian craton.

In the southern part of the craton, the megacomplex gradually increases the thickness in the pericratonic zones, conjugated with the Central-Asian geosynclinal belt. The structure of Riphean deposits within the boundaries of the latter is not yet clear. Some scientists identify here a number of eugeosynclinal zones, separated by ancient massifs, whereas others think that on the territory of this belt predominate non-metamorphosed plate and miogeosynclinal complexes. We support the first viewpoint, though the problems of tectonical interpretation rest on imperfect and poorly substantiated stratigraphic schemes.

Caledonian megacomplex is limited by Vendian (more precisely Yudomian) and Lower Proterozoic boundaries.

On the Siberian platform it is represented by entire, weakly differentiated mantle, composed of carbonate-terrigenous deposits. Completeness of the sequence and abundance of carbonate rocks increase north- and east-northward. On the Taimyr and in the Verkhoyansk province the plate complex is rather quickly replaced for mio-geosyncline complex. The latter fills the troughs, separating relatively elevated blocks.

In the northern marginal part of the Central-Asian belt, the studies identified in some troughs miogeo-synclinal complexes, overlain in places by proto-orogenic ones. Inside the belt, they merge with Riphean deposits into a single unit and compose iso-metric blocks,separated by narrow troughs with increased content of volcanites. Further lies the volcanogenic zone with internal part composed of ophiolic associations and the peripheral part - of island arc associations. The overlying deposits are represented by olistostrome, flyshoid and molassoid formations. In the mountainous Altai and West Sayany, sandy-shale deposits are wide-spread. They are disconformably overlain by terrigenous and carbonate-terrigenous Ordovician and Silurian deposits.

Thus, from the Siberian platform inside the Central-Asian belt the studies observed a successive series of complexes corresponding to recent transitional zones from the ocean to continent. The island arcs occupy the boundary area between large blocks. The recent structure is much complicated by overthrusts and later fault dislocations. As a whole the megacomplex is more differentiated than the Riphean one but their structural connections are undoubtful.

The structure of the Hercynian megacomplex is given on two maps: Middle Paleozoic and Upper Paleozoic-Lower Triassic. On the Siberian platform the lower of them is represented by epicontinental deposits, often red and salt-bearing, occurring in large depressions and syne-clises. Only in the north-eastern part gray formations predominate. In the Altai-Sayany province and Kazakhstan molasses are widespread, including volcanogenic, overlain with platform carbonate deposits.

For the last few years it was found that such units had spcial distribution on the territory of the West-Siberian platform too. The studies identified here a series of large massifs, separated by narrow ophiolitic zones, relatively short, blindly thinning out along the strike. These facts enable us to reconsider a traditional viewpoint on a vast and wide geosyncline belt which passed inthe Hercynian epoch throughout the plate territory. Major part of the massifs have possibly Precambrian if not pre-Riphean basement. The eugeosynclinal complexes occupy only the territory of relatively narrow Ob-Zaisan province its continuation - the Salym branch.

The Upper Oaleozoic (in places together with the Triassic) in all the above-mentioned folded areas is represented by orogenic complex - molassa, frequently volcanogenic. Its structure on the USSR North-East is different. Here it corresponds to thick miogeosynclinal formations of the Verkhoyansk-Kolyma province. Eastwards they give place to typical eugeosynclinal complex occupying the territory earlier belonging to the Kolyma median massif.

The most typical peculiarity of the final stage in formation of megacomplex structure is mass eruption of trapps in the western part of the Siberian craton and Taimyr. In West Siberia the basaltoids occur in large grabens of NS and SE strike.

Thus the megacomplex structure essentially differs towards different sides of the Transasian lineament, which is possibly a boundary between the Atlantic and Pacific segments of the planet.

These differences are more distinct in the structure of Mesozoic megacomplex. In the western part of the Central-Asian belt, few isolated depressions are filled with sedimentary molassa while in its eastern part manifestations of powerful deuterogenesis, accompanied by intensive magmatism are recorded. The lateral series of complexes here is much more variable as the eugeosyncline formation of the Mongolo-Okhotsk province wedged into the deuterogenous area. The deuterogenesis occupies also considerable territory of the Siberian craton and, besides that, most of the Aldan shield. There are abundant alkaline complexes of volcanites, gabbro and granites (including alaskitic). The chain of piedmont troughs trends along boundary with the stable part of Siberian platform.

The West-Siberian plate is overlain throughout by gently dipping plate complex - the main oil-bearing complex of Siberia. From here it passes via the Enisei-Khatanga and Lena-Anabar troughs towards the Lena mouth. This complex is also widespread in the Vilyui syneclise. Here its thickness slightly increases towards the Verkhoyansk-Kolyma province. This is served as a basis for separation of the Perverkhoyansk marginal trough.

Mio- and eugeosyncline complexes are widespread in the Verkhoyansk-Kolyma province. The first ones are typical for the Verkhoyansk zone. For the last few years it became clear that they were intensely folded even in the areas of so-called "zone of gentle dislocations", complicated by overthrusts and along these overthrusts with huge amplitude of replacement are thrusted over the margin of the Siberian platform. The Preverkhoyansk trough either does not exist at all or is completely concealed beneath these overthrusts.

The analysis of structure of the Siberian territory by the data of stage-by-stage regionalization permits to put forward some ideas on evolution of the Earth's crust on this territory.

The continental crust with granodiorite composition was formed already in the Early Archean over at least half of the territory of Siberia and, possibly, over its major part. The Late Archean was marked by shattering of this crust, formation of greenstone belts and some lateral accretion of territory of procontinent, manifested globally. In Siberia this epoch is reflected vaguely due to inadequate knowledge of the stratigraphy of the Archean reworked practically everywhere under the conditions of granulitic facies of metamorphism. At the same time new data on Archean structure essentially altered the earlier notions on the permobile regime of the era.

The Early Proterozoic is characterized by a predominant development of epicontinental (ensialic) troughs of miogeosynclinal type, which may have been partially inherited from greenstone belts but more frequently were superimposed on the crystalline basement. It was marked by general regressive trend of development resulted in wide formation of molassa in the end of the stage. Very peculiar is an epoch of formation of volcano-plutonic belts at the boundary between the Early Proterozoic and Riphean. This time was associated with radical structural reconstruction of continental crust, sealing of ancient structural patterns, stratification, which quite possibly resulted from merging of all sialic masses into a single supercontinent.

The Riphean is marked by the beginning of disintegration of the supercontinent and formation of platforms and geosyncline belts. At the beginning, these processes went inertly. This is expressed in platformal and miogeosynclinal nature of the Lower Riphean throughout this territory. But already in the Middle Riphean the riftogenic troughs, later transformed as a result of axial spreading into eugeosynclines, were formed in the Central-Asian belt. Active spreading throughout the belt territory was started in the Late Riphean and continued up to the Ordovicien in some of its parts (Kazakhstan) and in apophyses (the Urals). The spreading was accompanied by complementary compression in the adjacent zones. Features of this process are recorded by wide distribution of olistostromes, folding and mantle formation in geosyncline belts, manifestations of deuterogenesis in their margins and median massifs. As a result, on the Siberian platform in the south and within the Central-Kazakhstan massif in the north-east more and more folded areas were developed, while the territory of geosynclinal belt, separating them, was reduced. Such a process began possibly at the end of the Middle Riphean and was completed in the Early Paleozoic. The Taimyr and Verkhoyansk-Kolyma geosynclinal provinces accumulated miogeosyncline deposits in relatively quiet conditions.

The Hercynian stage began as new major structural reconstruction. The Ob-Zaisan province underwent intense spreading and the adjacent Riphean-Lower Paleozoic areas and southern part of the Siberian platform were subject to powerful deuterogenesis. These processes resulted in the formation of superimposed depressions with intense magmatism. The end of Devonian was marked by initiation of the Verkhoyansk-Kolyma geosyncline. In the Late Paleozoic the folding covered the Ob-Zaisan zone while formation of Permian volcano-plutonic association and postkinematic alyaskitic granites completed the development of geosynclines in the Central Asia. The territory of Taimyr underwent folding and epiplatformal orogenesis.

A number of syneclises were formed on the Siberian platform in the Mesozoic. The West-Siberian plate was covered by a continuous mantle, and in the Central-Asian belt predominated processes of deuterogenesis particularly intense in its eastern part. Mesozoic folding, which produced slight lateral accreation of the continental crust, was possibly connected complementarily with formation of oceanic basins. Such a connection is presumed in particular between folding in the Taimyr province and the beginning of formation of the Arctic Ocean.

Thus the Post-Early Archean history seems to be not simple accretion of the initial continent by heterochronous folded areas but a process of transformation of the continental crust, its breaking, spreading of sialic blocks and their later convergence. The accretion of the continental crust in the course of about 3.5 mlrd. years was relatively inconsiderable.

In the course of history, the geosynclinal areas undergo evolution. Lower Proterozoic and Lower-Middle Riphean deposits, replaced for the Archean greenstone belt do not have typical eugeosyncline complexes with ophiolite associations. The latter appeared only in the Late Riphean. It is hardly possible to draw a complete parallel between basins with oceanic crust of that time and recent ocean though some similarity is obvious.

Though the essential accretion of the continental crust has not been established, its reworking is traced throughout its history. Beginning from Archean-Proterozoic boundary already, some segments were subject to the processes of deuterogenesis, which manifestations increased possibly in time. They are particularly distinct near the boundary between the Early Proterozoic and Riphean and in the first half of the Mesozoic.

LOOKING FOR CRITERIA OF OROGENIC DIVISIONS.

AUBOUIN Jean.

Université Pierre et Marie Curie, Laboratoire de Géotectonique, Tour 26-0 - 1er Etage, 4 Place Jussieu, 75230 PARIS Cédex 05, FRANCE.

ABSTRACT. - In the light of plate tectonics, the sequence opening-closing of one ocean gives the best definition of one orogenic cycle. So is defined the alpine cycle by the opening and the closing of the Tethys.

The criteria of ocean opening for the beginning of an orogenic cycle is not clear everywhere. For instance, the opening of the Tethys is clear in southwestern Asia, southern Europe and Caribbean : it is the "Tethys of the reconquest", cut across the Pangea in a scissors shape. But in southeastern Asia where was the "Permanent Tethys", part of Panthalassa from which was inhereted the Pacific, there is no clear definition of the beginning of an alpine cycle.

The criteria of ocean closing is the ophiolitic suture. But there are as many ophiolitic sutures as there were oceanic areas to be closed. That has to be understood by reference with the present western Pacific, with its numerous marginal seas. Only the final collision of two main continents determine the end of one orogenic cycle ; each ophiolitic suture define only sub-cycle as a part of the complete cycle.

One orogenic cycle, with its sub-cycles, has only a regional value, related to the opening-closing of a precise ocean ; there is no clear evidence of worldwide orogenic cycle. For instance, alpine cycle is related to

Proceedings of the 27th International
Geological Congress, Volume 7, pp. 27–42
TECTONICS

the opening-closing of the Tethys : in the Americas, the alpine cycle can only be defined in the Caribbean area inherited from the Tethys.

The assemblage of all the continents in a Pangea, as at the end of the Proterozoic and the end of the Paleozoic, determines super-cycles which are universal divisions. In this way, there is only one paleozoic super-cycle, including the caledonian and the variscan cycles ; and the alpine cycle could be part of a super-cycle, the second part of which being, in the future, the possible closing of the Atlantic...

As a conclusion, we can propose : assemblage of all the continents in a Pangea as criteria of orogenic super-cycles of global value ; opening-closing of one ocean as criteria of orogenic cycles of regional value ; ophiolitic sutures as criteria of orogenic sub-cycles of local value.

A LA RECHERCHE DE CRITERES DE COUPURES OROGENIQUES.

La subdivision tectonique des continents est un problème essentiel pour l'établissement des cartes tectoniques continentales.

La solution a pû en paraître simple lorsque la construction des continents paraissait résulter de **cycles orogéniques successifs** d'une durée avoisinant 200 millions d'années ; du moins pour les temps phanérozoïques -cycles calédonien, hercynien, alpin- car pour le Protérozoïque et le Précambrien l'accord est encore loin d'être fait. Elle s'est compliquée lorsque, à ces cycles orogéniques superposés, on ajouta la notion de **tectoniques superposées** au cours d'une même cycle : pour prendre l'exemple du cycle alpin en Europe méridionale (1), on reconnut successivement la superposition de la tectonique alpine, de déformations tarditectoniques puis néotectoniques, le tout précédé d'épisodes paléotectoniques aussi importants. Ainsi, en Méditerranée occidentale, il fut montré que le premier charriage des nappes ophiolitiques les plus internes remontait au Jurassique supérieur (1, 2, 3), et que celles-ci avaient été simplement reprises dans les phases tectoniques ultérieures du Crétacé et du Tertiaire.

On retrouve ici, au plan tectonique, un problème aussi ardu au plan stratigraphique : celui de coupures d'une certaine généralité.

Allons donc **à la recherche de critères des coupures tectoniques.**

Deux ordres de phénomènes paraissent seuls susceptibles de fournir des coupures de premier ordre, étant donné leur importance à l'échelle du globe : selon que l'on considère les océans ou les continents :

- l'ouverture et la fermeture des océans ;
- la séparation et le rassemblement de continents;

1. L'évolution des **chaînes alpines issues de la Téthys** est une illustration du premier critère. Elles naissent : d'une ouverture océanique en ciseaux (4) au travers de la Pangée permienne (5), qui se développe progressivement du Trias à l'Est (Eurasie méridionale) au Jurassique à l'Ouest (Caraïbes) et qui forme la Téthys ; suivie d'une fermeture océanique qui s'échelonne du Crétacé au Tertiaire et d'où naissent les chaînes alpines. Le cycle alpin en Eurasie méridionale et dans les Caraïbes est ainsi encadré entre l'ouverture de la Téthys et la fermeture de celle-ci.

Mais la distinction n'est pas aussi simple : la fermeture de la Téthys est liée à l'ouverture de l'Atlantique qui progresse en ciseaux du Sud au Nord du Jurassique supérieur au Tertiaire. Globalement rabattus vers l'Ouest et le Nord à l'Ouest, vers l'Est et le Nord à l'Est, l'Amérique du Sud d'une part et les continents issus du Gondwana d'autre part entreprennent la fermeture de la Téthys, à partir de la limite Jurassique-Crétacé, en commençant de se rapprocher des continentaux septentrionaux (Amérique du Nord et Eurasie). Ainsi se marque la grande "révolution fini-Jurassique (1, 2, 4) dans l'évolution téthysienne qui passe de l'extension à la compression. Ainsi, le cycle téthysien et le cycle atlantique se superposent-ils au lieu de se succéder. Une complication supplémentaire serait introduite par l'évolution de l'Océan Indien, avec la séparation de l'Inde au Jurassique et de l'Australie au Crétacé.

Si donc l'on conserve la notion de cycle alpin, on voit que c'est **l'ensemble ouverture + fermeture** océanique qui le définit.

Mais un tel critère conduit à définir des **cycles différents** dans leurs limites **pour chaque océan** ; car, comme Téthys et Atlantique le montrent, quand un océan s'ouvre ici un autre se ferme là. L'apparente netteté du "cycle" alpin à l'échelle du globe vient sans doute du fait qu'il est le dernier : ses limites paraîtront moins nettes si, comme on peut le présumer, dans l'avenir l'Atlantique doit entrer à son tour en compression (cf. infra).

2. La formation de la **Pangée** (5) est l'illustration du second critère. Mais on arrive alors à la conclusion que l'ère Primaire ne forme qu'un seul cycle orogénique, et non point deux, successivement calédonien et hercynien, puisque la Pangée ne s'est formée que par le résultat de ces deux orogénèses. Tandis que le cycle

alpin ne serait pas terminé : il faudrait attendre que l'Eurasie méridionale et l'Australie achèvent leur coalescence d'une part, et que l'Atlantique et l'Antarctique se referment d'autre part. D'une certaine manière, l'Atlantique occupe par rapport à la Téthys alpine (Néotéthys), la même situation que la Téthys hercynienne (Paléotéthys) par rapport au Iapetus calédonien.

Sous cet angle, on est amené au fait que, dans la suite ouverture-fermeture, un **cycle orogénique est mieux défini par la fermeture océanique** qui peut se produire en autant d'épisodes (sous-cycles) qu'il y a d'espaces océaniques à fermer. Les cycles calédonien et hercynien (varisque) ne seraient ainsi que des sous-cycles d'un véritable cycle paléozoïque.

3. Les **sutures ophiolitiques** jouent dans cet optique, un rôle essentiel : cicatrices majeures, elles marquent la fermeture océanique.

L'exemple le plus clair est la **suture ophiolitique d'Eurasie méridionale** qui jalonne les collisions successives entre l'Eurasie et les continents issus du Gondwana.

Mais précisément, cette suture n'est simple ni dans le temps ni dans l'espace.

Dans l'espace : à partir de la Méditerranée orientale et vers l'Est, il y a au moins deux sutures ophiolitiques, celle qui correspond au Vardar et celle qui correspond au Pinde (6) ; de là on passe en Asie Mineure, puis en Asie méridionale où les sutures sont plus nombreuses encore : il semble qu'il y en a trois en Afghanistan (7) puis au Tibet (8).

Dans le temps : revenant à la Méditerranée orientale, la suture du Vardar est d'âge Jurassique supérieur, tandis que la suture du Pinde est d'âge Tertiaire inférieur. Allant vers l'Est, on voit apparaître une suture d'âge Crétacé supérieur en Asie Mineure (9), tandis que de plus anciennes, liasiques en Asie méridionale, voire triasiques en Asie du Sud-Est se manifestent (10).

Laquelle de ces sutures est la véritable suture téthysienne ?

C'est ici le problème de la nature réelle de l'océan téthysien qui, dans un premier temps, dans l'optique de la Tectonique des Plaques, fut conçu comme un océan atlantique (au moins jusqu'à la fin du Jurassique).

Mais si l'ouverture "atlantique" de la Téthys est claire en Europe méridionale et aux Caraïbes, dans ce qu'on a pu appeler la "**Téthys de la reconquête**" (4), il n'en est rien en Asie méridionale où se situait la "**Téthys permanente**" dépendant du Pacifique d'où est précisément partie la reconquête. Cette Téthys permanente est bordée au Nord, en marge de l'Eurasie, par une zone de subduction active dès le Trias contre laquelle sont venus s'accoler successivement des éléments continentaux détachés du Gondwana, dont l'Inde n'est que le dernier : ainsi se sont formés les sutures ophiolitiques successives du Tibet, dont la suture du Tsang-Po est la dernière (8). Toute la Téthys orientale était un océan de type Pacifique dès le Trias ; à la différence de la Téthys ocidentale qui fut reconquise comme un océan de type Atlantique.

Mais même au niveau de la Téthys de la Reconquête le **stade atlantique** extensif est limité au Jurassique. La révolution finijurassique (cf. supra) marque précisément le début des subductions d'où naissent arcs insulaires et mers marginales, c'est-à-dire d'un **stade pacifique.** La fermeture de la Téthys occidentale sera aussi celle d'un Pacifique : là aussi, autant de sutures ophiolitiques que d'espaces océaniques à fermer.

Il n'y a ainsi pratiquement nulle part une seule suture ophiolitique téthysienne, mais au contraire une famille de sutures échelonnées dans l'espace et dans le temps, qu'on pourra rassembler dans la mesure où elles sont sub-parallèles et ont une logique apparentée. Tandis que chaque nouvelle ouverture océanique est franchement oblique par rapport aux précédentes, ainsi qu'il en fut pour l'Atlantique depuis la fin du Jurassique, et pour la Téthys à partir du Permien.

En toute rigueur il faut s'efforcer de reconnaître dans ces sutures ophiolitiques multiples celle qui marque l'océan fondamental et celles qui résultent de la fermeture de mers marginales successives. En se fixant pour modèle de référence le Pacifique occidental, cela implique de différencier continents, arcs insulaires à substratum continental (avec plutonisme et volcanisme calco-alcalin), de reconnaître le pendage de(s) subduction(s) (du côté du volcanisme et du plutonisme calco-alcalin), de séparer l'âge de formation de la croûte océanique et l'âge de sa mise en place sous forme de massifs ophiolitiques allochtones. Ce dernier critère est peut être le plus clair : si l'océan fondamental est le plus souvent ancien là où il aborde les zones de subductions, il n'en va pas de même des mers marginales toujours récentes. La situation du Japon illustre ce fait : dans la fosse du Japon, l'océan Pacifique subducté est d'âge jurassique, tandis que la mer du Japon est d'âge tertiaire.

On reconnaîtra donc **la fin d'un cycle orogénique** à la fermeture de l'ensemble des espaces océaniques séparant deux continents qui finissent par entrer en collision, quel que soit le nombre des sutures ophiolitiques qui en résultent.

4. **Le début d'un cycle orogénique est**, au contraire, **l'ouverture océanique**. Ainsi le cycle alpin (téthysien) est-il encadré, en Europe méridionale où il a été défini, par l'ouverture triasique-jurassique de la Téthys et sa fermeture crétacée-tertiaire.

Ainsi est posé le problème de la fracturation continentale.

La plus claire est la **fracturation extensive** dont l'Afrique orientale et la Mer Rouge donnent l'exemple en tant que stades précurseurs d'une ouverture océanique de type atlantique. Elle a donné naissance au concept de "trilogie d'ouverture" (11) sédiments continentaux-évaporites-sédiments marins, qui existe sur les marges de l'Atlantique comme sur les marges de la Téthys aux époques d'ouverture (le Trias, par son nom, évoque cette trilogie dont la référence est, en Europe méridionale, l'ouverture de la Téthys).

Mais il existe une **fracturation compressive** dont l'Asie centrale donne l'exemple à la suite du poinçonnement par le continent indien (12) ; on l'a tout d'abord considérée comme limitée à la déformation du continent (12) ; mais des travaux expérimentaux conduisent à lui rattacher l'ouverture de certaines mers marginales comme la Mer d'Andaman ou la Mer de Chine méridionale (13). Or, bien qu'il ne s'agisse pas de vastes océans, ce sont des espaces de croûte océanique, qui donneront des sutures ophiolitiques lorsqu'ils seront cicatrisés. Une telle cicatrisation est d'ailleurs en cours à Taïwan, aux dépens de la fosse de Manille dans laquelle est subductée la mer de Chine méridionale (15).

Il est bien évident que seule la fracturation extensive à l'échelle de continents peut être considérée à l'origine d'un nouveau cycle orogénique. Mais la fracturation compressive introduit un élément de complication.

Encore faut-il qu'il y ait fracturation continentale. Car la Pangée avait une périphérie océanique où tout indique que la subduction était active : le début du cycle alpin n'est pas clair autour du Pacifique, héritier de la Panthalassa, y compris dans sa partie sud-ouest qui fut la Téthys permanente d'où partit la "reconquête" au travers de la Pangée (4). Que faire, en Asie du Sud-Est, de l'Orogenèse triasique ? un épisode tardif du cycle hercynien ou un épisode précoce du cycle alpin ?

Il n'est pas étonnant que la notion de cycle orogénique soit née en Europe : c'est là que l'ouverture triasique de la "Téthys de la reconquête" scellait le cycle hercynien et ouvrait le cycle alpin. Les choses sont beaucoup moins claires partout ailleurs.

5. Si donc la fracturation continentale extensive est à l'origine des cycles orogéniques au moins locaux, dans quelle mesure est-elle originale, dans quelle mesure hérite-t-elle de traits tectoniques antérieurs ?

Il semble que, fondamentalement, toute ouverture océanique de grande ampleur est nouvelle, indépendante des évènements antérieurs ; comme le montre l'ouverture de l'Atlantique depuis la fin du Jurassique, qui recoupe toutes les structures du Précambrien à l'Alpin ; au point qu'on y a vu la première argumentation pour la mobilité continentale (15). Cela était vrai pour la Téthys qui recoupait obliquement la Pangée au travers du Précambrien en Asie méridionale, de la chaîne hercynienne en Europe du Sud et dans les Caraïbes. Cela est vrai de l'ouverture Afrique orientale-Mer Rouge-Moyen Orient qui, amorcée au travers du Précambrien arabo-africain, se propage maintenant au travers de la ceinture alpine (16).

En vérité, lorsqu'il y a rejeu tectonique d'accidents anciens, bref "héritage tectonique", c'est le plus souvent dans le cas d'accidents compressifs. Ainsi la fracturation de l'Asie centrale par le poinçonnement du continent indien, est-elle largement commandée par la tectonique paléozoïque pré-existante (14) ; ainsi le dessin des chaînes alpines de l'Europe méridionale est-il largement commandé par le réseau des failles tardihercyniennes, réutilisées en coulissage lorsqu'elles sont Ouest-Est comme le front Nord-pyrénéen (17).

Mais dès qu'il s'agit d'ouverture océanique, l'indépendance est la règle, l'héritage l'exception.

La carte tectonique de l'Europe montre clairement l'indépendance des cycles successifs : les édifices calédonien, hercynien, alpin sont obliques les uns par rapport aux autres, au point que les chaînes alpines ont pour avant-pays direct le socle précambrien de la plate-forme russe à partir des Carpathes vers l'Est.

6. Enfin le jeu d'ouverture et fermeture océanique, qui conduit à la formation des chaînes de montagnes, conduit-il à l'accroissement des continents, comme on l'admet généralement ?

S'il n'est pas aisé de donner une réponse pour les périodes les plus anciennes qui ont vu naître la première croûte continentale, il est clair que pour les temps phanérozoïques (et probablement protérozoïques) il n'y a pas eu accroissement des continents mais plutôt conservation.

En effet, considérant le cycle alpin par exemple, où est le matériel océanique ajouté aux continents ? Dans les nappes de charriages océaniques d'une part, les plutons et volcans d'autre part.

Or, les nappes océaniques reposent sur une marge continentale, qu'il s'agisse de la marge d'un continent fondamental comme dans la ceinture téthysienne, ou de fragments insulaires continentaux comme autour du Pacifique, notamment dans les chaînes côtières de type Californie. En toute rigueur, il n'y a pas accroissement du continent mais simple superposition de matériel océanique sur le continent lui-même.

De la même manière, les volcans s'installent sur la marge des continents comme dans les cordillères ouest-américaines, ou en arcs insulaires sur des fragments de socles continentaux comme dans l'ouest Pacifique ; car il ne paraît pas y avoir d'arc insulaire sans fragment continental pour les supporter (18). Ce n'est pas par le volcanisme que s'augmentent les continents.

Par contre, la question reste ouverte pour le plutonisme. Certes, on considère généralement que les intrusions granitiques et granodioritiques se font pendant ou après les grands épisodes de tectonique compresssive : granites et granodiorites traversent des structures qu'elles digèrent et dont elles prennent la place. Mais divers indices existent que la montée des plutons granitiques et granodioritiques correspond en fait à des épisodes d'extension qui suivent les phases compresssives (19) ; ce qui, à la vérité, est plus aisée à comprendre. Si cela se vérifiait, les ceintures granitiques et granodioritiques marqueraient alors une véritable augmentation de la surface des continents.

Dans l'état ctuel des connaissances, on peut donc dire que rien n'indique une augmentation de la surface des continents depuis le Phanérozoïque et peut être le Protérozoïque. Bien au contrire, au fil des ouvertures et fermetures océaniques successives, les continents se conservent en se fragmentant et se rassemblant selon des logiques chaque fois nouvelles. Le destin des deux Amériques est à cet égard exemplaire, qui, détachées séparément de la Pangée, se sont trouvées (mal) soudées par l'Orogenèse alpine.

CONCLUSION.

Au terme de cette analyse, les traits majeurs qui permettent de diviser les contiennts sont les **sutures ophiolitiques** qui marquent la reconstruction des continents aux dépens d'océans subductés, par collision avec d'autres continents ou avec des fragments continentaux. De telles sutures se disposent d'une manière très complexe dont on comprend la logique par référence aux nombreuses mers marginales à fond océanique de l'actuel Ouest-Pacifique.

Rassembler des familles de suture ophiolitique en un ensemble d'ordre supérieur définissant les cycles orogéniques est délicat. S'il est démontré que les masses continentales se rassemblent périodiquement en une Pangée, comme ce fut le cas à la fin du Protérozoïque, et à la fin du Paléozoïque, ce sont ces rassemblements qui devraient marquer les coupures principales qui, par comparaison avec le calendrier stratigraphique traditionnel auraient valeur de "super-cycle".

Le critère d'ouverture océanique, en apparence plus facile à utiliser, n'est pas généralisable à l'échelle du Globe : s'il fut bien clair pour la "Téthys de la reconquête" ouverte dans la Pangée permienne, c'est-à-dire pour les chaînes de l'Europe méridionale et des Caraïbes, il est beaucoup plus flou pour la Téthys permanente en Asie du Sud-Est, et d'une façon général autour du Pacifique. **Un tel critère n'est valable qu'à l'échelle de l'océan qui s'est ouvert** (puis refermé) ; donc de la Téthys de la reconquête pour les chaînes alpines, ou ce qui seraient des chaînes atlantiques, si cet océan devait se fermer dans le futur. Mais il n'y a pas de raison d'extrapoler au-delà des limites de l'océan concerné ; donc de la Téthys pour les chaînes alpines : aux Amériques le cycle alpin n'est clair que dans le domaine caraïbe, dérivé de la Téthys.

La notion de cycle orogénique n'est probablement valable que régionalement. Elle correspondait à un stade de la connaissance qui privilégiait la géologie de l'Europe où, au cours des temps phanérozoïques, des séries d'ouverture-fermeture océaniques ont permis de définir aisément les cycles calédonien, hercynien, alpin. Les connaissances s'étant généralisées à l'échelle du Monde, le calendrier orogénique s'est d'autant compliqué. Sans doute est-il sage, pour l'instant, de distinguer les orogenèses par leur âge, quitte à découvrir dans l'avenir certains synchronismes dans les mouvements de palques. Il sera temps alors de redéfinir des cycles orogéniques. Car vouloir à toute force, dans le Monde entier, faire entrer tous les évènements tectoniques dans les mêmes coupures risque de masquer la réalité au lieu d'aider à sa connaissance.

En conclusion, nous pouvons donc proposer : l'assemblage de tous les continents en Pangée comme critère de super-cycles orogéniques de valeur universelle ; l'ouverture-fermeture d'une océan comme critère de cycles orogéniques de valeur régionale ; les sutures ophiolitiques comme critère de sous-cycles orogéniques de valeur locale.

(1) AUBOUIN J. (1973). - Des tectoniques superposées et de leur signification par rapport aux modèles géophysiques : l'exemple des Dinarides ; paléotectonique, tectonique, tarditectonique, néotectonique.
Bull.Soc.géol.Fr., (7), XV, n°5-6, p.426-460.

(2) DERCOURT J. (1970). - L'expansion océanique actuelle et fossile : ses implications géotectoniques.
Bull.Soc.géol.Fr., (7), XXIV, p.229-242.

(3) MERCIER J. (1973). - Etude géologique des zones internes des Hellénides en Macédoine entrale (Grèce).
Ann.géol.pays helléniques, t.XXIV.

(4) AUBOUIN J., BLANCHET R., STEPHAN J.F. et TARDY M. (1977). - Téthys (Mésogée) et Atlantique : données de la géologie.
C.R.Acad.Sc., Paris, sér.D, t.285, p.1025-1028.

(5) DIETZ R.S. and HOLDEN J.C. (1970). - Reconstruction of Pangea : Breakup and dispersion of continents, Permian to Present.
J.Geophys.Res., 75, p.4939-4956.

(6) BONNEAU M. (1982). - Evolution géodynamique de l'arc égéen depuis le Jurassique supérieur jusqu'au Miocène.
Bull.Soc.géol.Fr., (7), XXIV, p.229-242.

(7) TAPPONIER P., MATTAUER M., PROUST F. and CASSAIGNEAU C. (1981). - Mesozoic ophiolites sutures and large scale tectonic movement in Afganistan.
Earth.Planet.Sc.Letter, 52, p.355-371.

(8) TAPPONIER P., MERCIER J.L. et al. (1981). - Tibetan side of the India-Eurasia collision.
Nature, 294, p.405-410.

(9) RICOU L.E., MARCOUX J. and WHITECHRUCH H. (1983). -The mesozoic organisation of Taurides : one or several oceanic basins ?
Journ.Geol.Soc.London, in press.

(10) AUDLEY-CHARLES M.G., LATREILLE M. and NAYOAN G.A.S. (1980). - Comparison of orogenic features of the Tethyan mountain belt from Himalaya to the East Asian Arcs.
In Geology of the alpine chains born of the Tethys, J. AUBOUIN, J. DEBELMAS, M. LATREILLE ed., 26th I.G.C. Paris, p.155-165, BRGM, Orléans.

(11) LE PICHON X. and HAYES D.E. (1971). - Marginal offsets, fractures zones and the early opening of the South Atlantic.
J.Geophys.Res., 76, p.6283-6293.

(12) TAPPONIER P. and MOLNAR P. (1976). - Slip line field theory and large scale continental tectonics.
Nature, 264, n°5584, p.319-324.

(13) TAPPONIER P., PELTZER G., LE DAIN A.Y., ARMIJO R. and COBBOLD P. (1982). - Propagating extrusion tectonics in Asia : New Insights from simple experiments with plasticine.
Geology, 10, p.611-616.

4) BOWIN C., LU R.S., LEE C.S. and SHCOUTEN H.(1978). -Plate convergence and Accretion in Taïwan-Luzon Region.
A.A.P.G., 62, 9, p.1645-1672.

(15) WEGENER A. (1915-22). - Die Entstehung der Kontinente und Ozeane.
Braunschweig ed., 1ère, 2ème, 3ème édition, 1 vol.

(16) BIJU-DUVAL B. (1976). - Commentaires de la carte géologique et structurale des bassins tertiaires du domaine méditerranéen.
Revue de l'Inst.Franç.Pétrole, vol.XXIX, n°5, Technip éd., Paris.

BIJU-DUVAL B. (1976). - Esquisse photogéologique du domaine méditerranéen au 1/2.500.000.
1 carte, éd. IFP-CNEXO.

(17) LE PICHON X., BONNIN J., FRANCHETEAU J. et SIBUET J.C. (1971). - Une hypothèse d'évolution tectonique du golfe de Gascogne.
In Histoire structurale du golfe de Gascogne, vol.II, p.VI, 11, p.1-44, Technip éd. Paris.

(18) AUBOUIN J. and BLANCHET R. (1981). - Subduction and Tectonics : discussion of the results of the IPOD program in active margins.
In : Geology of continental margins, R. BLANCHET et L. MONTADERT éd., 26th I.G.C. Paris, Coll.C3, **Océanologica Acta**, p.283-294.

(19) PITCHER W.S. and BUSSELL M.A. (1977). - Structural control of batholiths emplacement in Peru : a review.
Journ.Geol.Soc.London, 133, 3, p.249-256.

A NEW TYPE OF ACTIVE MARGINS : THE CONVERGENT-EXTENSIONAL MARGINS, AS EXAMPLIFIED BY MIDDLE AMERICA TRENCH OFF GUATEMALA.

AUBOUIN Jean
BOURGOIS Jacques
AZEMA Jacques

Université Pierre et Marie Curie, Département de Géotectonique, Tour 26-0 - 1er Etage, 4 Place Jussieu, 75230 PARIS Cédex 05, FRANCE.

ABSTRACT. - Middle America Trench off Guatemala was drilled during Legs 67 and 84. During Leg 67 the basaltic basement of the oceanic Cocos plate and of the trench itself were drilled through down a relatively thin sedimentary cover of Pleistocene to lower Miocene age. The lower-mid and upper slope of the inner wall were drilled during Leg 84. At the penetration sites, basement was found to consist of ophiolitic rocks covered by sediments of Early Eocene (upslope and midslope) to late Cretaceous (downslope). Petrologic studies of the mafic and ultramafic rocks suggest that they could represent the offshore prolongation of the pre-Campanian ophiolitic complexes cropping out in Central America.

As a consequence the DSDP/IPOD Guatemalan transect across the Middle America Trench has defined a non-accretionary convergent margin ; since late Cretaceous time, subduction has not resulted in the development of an accreted complex. So, landward seismic reflectors are either tectonic contacts in the basement or layers in the post late Cretaceous sediments.

On the other hand, seismic and Seabeam data give evidences of normal faulting : on the seaward wall and in the trench, reactivating a primary structural network

Proceedings of the 27th International
Geological Congress, Volume 7, pp. 43–63
TECTONICS

inherited of the East Pacific rise rifting ; on the landard wall as collapse accompanying the downgoing of the oceanic Cocos plate into subduction.

Thus, we propose Guatemala margin as the model of a new type of Active margins : the "Convergent Extensional margins" withtout accretion. It is opposed to "Convergent Compresional margins" where accretionary process is working.

We suggest the active margins that exhibit a broad continental shelf (i.e. = Cordilleran margins) belong to the "Convergent-Extensional" type.

INTRODUCTION.

Two major types of continental margins are distinguished in the classical view of plate tectonics, Passive and Active. Passive margins develop as a result of rifting and spreading along an ocean ridge and are marked by extensional and occasional transcurrent structure within a single plate. Active margins develop as a result of the subduction of an oceanic plate beneath a continental or island arc and often marked by compressional and transcurrent structure along the boundary between two plates.

Compressional stress is emphasized along active margins and the evidene cited for compression are earthquake focal mechanisms and the global plate motion scheme. Global plate motion certainly indicates plate convergence but compressional stress requires friction between plates across the subduction zone and if the friction is low, the compressional stress will be correspondingly little. Thus convergence does not necessitate compression.

The concept of the accretionary prism is related to the view of compressional stress and considerable friction between plates (fig. 1). A succession of wedges accumulates at the front of the continental margin in proportion to the amont of the seafloor subducted and the corresponding amount of sediment scrapped off. A gradual uplift and rotation of the imbricate stack results from continued operation of this process and tectonic slices of oceanic material are superposed in a reverse stratigraphic order with the older at the top and the younger at the bottom of the prism (Seely, 1979).

As the concept of an accretionary prism evolved, marine seismic data were cited to support the general model and landward dipping reflections (fig. 2B) were interpreted as reflections from fault planes of the imbricated ocean sediment in the Middle America (Seely and others, 1974 ; Moore, Watkins and others, 1979 ; Moore and others, 1981), Marianna (Karig, 1971), Japan (Von Huene, Uyeda, 1981) trenches.

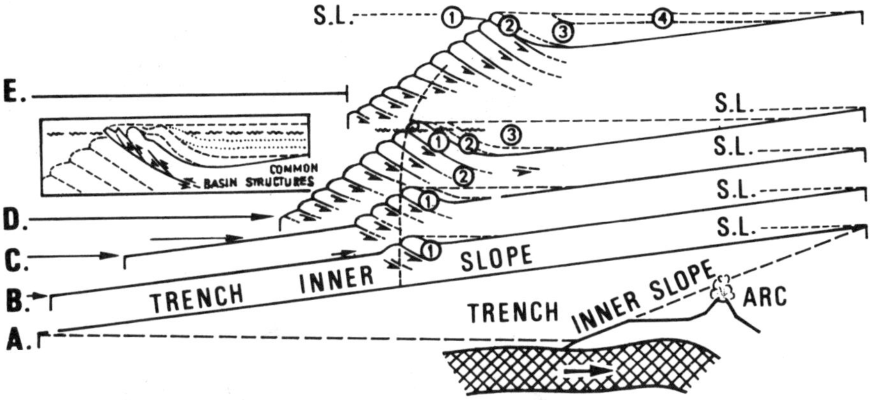

Fig.1.-The accretionary prism model (after Seely, 1974). Note the general uplift of the margin and the reverse stratigraphic order of accretionary units.

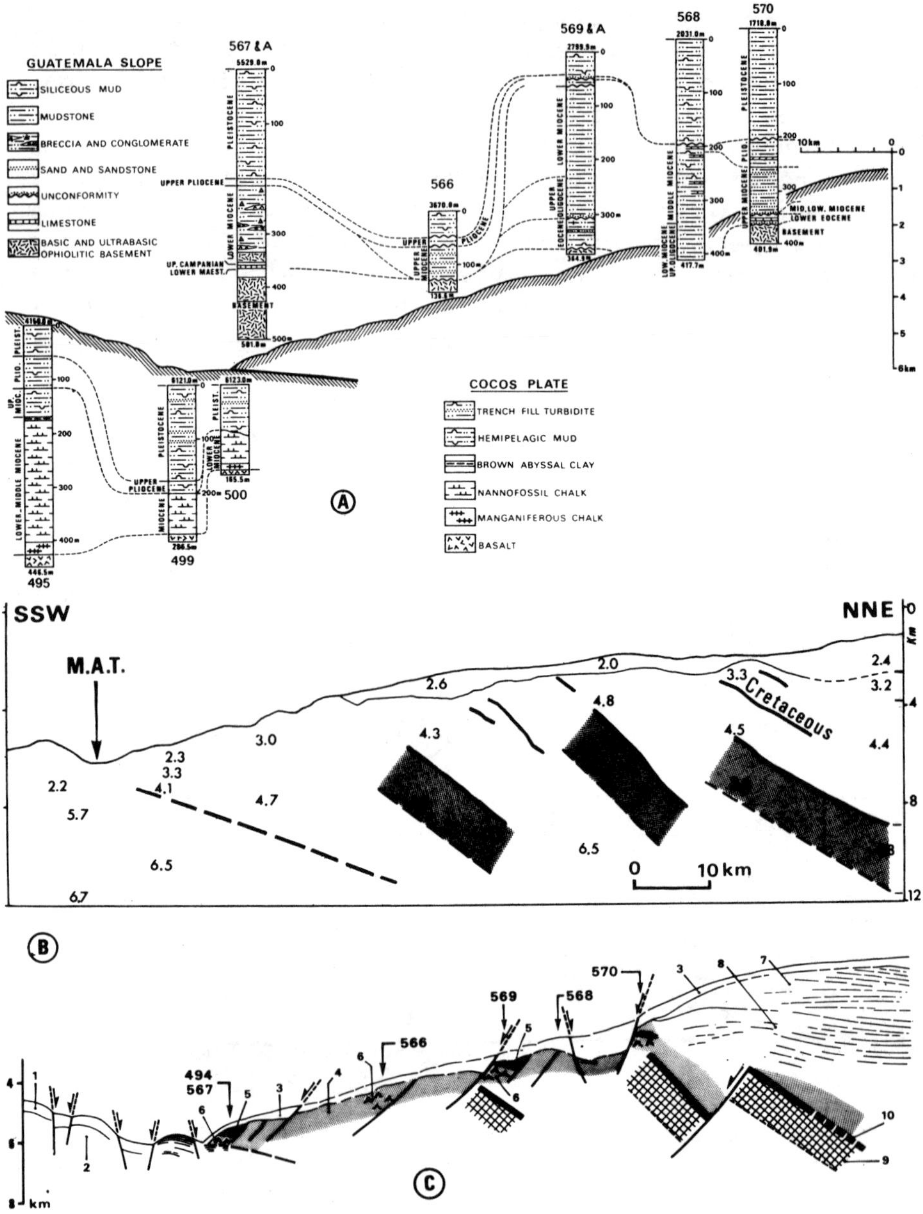

GUATEMALA SLOPE
SILICEOUS MUD
MUDSTONE
BRECCIA AND CONGLOMERATE
SAND AND SANDSTONE
UNCONFORMITY
LIMESTONE
BASIC AND ULTRABASIC OPHIOLITIC BASEMENT
567 & A
566
569 & A
568
570
COCOS PLATE
TRENCH FILL TURBIDITE
HEMIPELAGIC MUD
BROWN ABYSSAL CLAY
NANNOFOSSIL CHALK
MANGANIFEROUS CHALK
BASALT
495
499
500
A
SSW
NNE
M.A.T.
Cretaceous
0
10 km
B
570
569
568
566
494
567
C
km

Fig. 2. - Data from the Mid-America trench off Guatemala.

2A. Simplified stratigraphic columns summarizing the lithostratigraphy and biostratigraphy of Legs 67 and 84 drilling sites (after von Huene, Aubouin and others, 1980 ; Aubouin, von Huene and others, 1982).

2B. Summaries of seismic reflection and refraction studies showing velocity structure and major landward dipping reflectors on the Leg 67 and 84 transects.

2C. Interpretation of the structure of the Guatemalan margin. The major landward dipping reflectors are interpreted as thrust fault bringing pre-Campanian ophiolitic slabs above Upper Jurassic - Upper Cretaceous oceanic Nicoya complex as on land in Costa Rica. 1. Miocene to Pleistocene sedimentary cover of Cocos plate ; 2. Basalts ; 3. Pliocene to Pleistocene slope deposits ; 4. Acoustic basement ; 5. Upper Senonian to Miocene sediments and 6. Ophiolitic rocks ; 7. Oligocene and Miocene forearc sediments ; 8. Paleocene and Eocene forearc sediments ; 9. Interpreted as a Nicoya Complex remnant from magnetic and gravity data ; 10. Major reflectors interpreted as a pre-Campanian thrust fault ; from on-shore data it could be the same as the overthrusting of the Santa Elena Peridotite on the Nicoya complex.

The combined study of seismic reflection records and Deep Sea Drilling Project (DSDP) cores from active margins, however, showed that the reflections are not necessarily from fault planes (Moore and others, 1981) and the interpretation of other active margin geophysical features was also questioned. The DSDP results show that the interpretations of geophysical data without supplemental geological data are very incertain. Landward dipping reflections cannot only be bedding planes, but also older tectonic features, or artifacts of the geophysical technique ; sedimentation is very complex, and tectonic histories contain both compressional and extensional episodes. Without geological information beneath the slope deposits, which have generally been of atleast Miocene age, the interpretations of geophysical data only in accord with the classical accretionary model is likely to be in error. The DSDP active margin transects of drill holes indicated clearly that seismic reflection records, especially single channel records, cannot be interpreted reliably without geological (temporal and environmental) control.

Only on the Barbados margin (DSDP Leg 78A) did drilling give direct evidence of accretion (Biju Duval, Moore and others, 1981). The top (upper Miocene to Quaternary) of the upper Cretaceous to Quaternary oceanic sediment sequence is repeated and although the contact with the main subduction zone was not drilled, disturbances in the stratigraphic order were recovered and high pore pressures (300 psi) were measured. The very nearness of the main subduction zone was thus indicated in addition of the indications from seismic reflection records.

Drilling on the Mexican margin (DSDP Leg 66) of the Middle America Trench (Acapulco Trench) provided indirect arguments for accretion (Moore, Watkins and others, 1979). Coarse grained sediment was drilled on the margin, ranging from upper Miocene to Quaternary. Quaternary sand was firest recovered in the trench (Site 486) and then older (upper Miocene) sand was drilled upslope (Site 492), and Pliocene sand was recovered at a site inbetween (midslope, Site 491). This disposition was interpreted as resulting from tectonic slices in reverse stratigraphic order (Moore, Watkins and others, 1982), as in the accretionary model. Clearly, this is only one interpretation, which implies that coarse grained sediments are always confined to trench deposits. However sand could also be trapped and ponded in basins on the margin, as shown in many places around the Pacific (e.g. Peru-Chile trench (Thornburg and Kulm, 1981 ; Schweller and others, 1981 ; Shepherd and Moberly, 1981), and see below).

Various drilling results can only be explained with models other than the Seely (1974) accretionary model. Instead of uplift during subduction, subsidence was observed on the Japan Trench margin (Leg 57, Langseth and others, 1981 ; von Huene and others, 1982), Marianna Trench margin (Leg 60, Hussong and Uyeda, 1981). Instead of very young sediment at the front of the continental slope, old material was recovered in the Middle America Trench off Guatemala (upper Cretaceous and Eocene, Leg 67), and was subsequently recovered all along a cross section of the slope (Leg 84). The Leg 84 results are essential (fig. 2A and 3). In four holes the same pre-early Eocene (and probably pre-Campaniean) ophiolitic basement was recovered. If the Campanian-Maestrichtian limestone were recovered in place at the very front of slope (Site 494 and 567), a Cretaceous unconformity was also recovered in two holes. So, the ophiolitic basement of the Guatemalan Trench landward slope belongs to the substratum of Central America because similar ophiolites crop out in Costa Rica. There, pre-Campanian ophiolites of the Santa Elena peninsula are overthrusted (Azéma et Tournon, 1980 ; Tournon et Azéma, 1980 ; Bourgois et al., 1982) on the Nicoya complex of upper Jurassic-Certaceous age (Dengo, 1962, 1973 ; Barr et Escalante, 1969 ; De Boer, 1979 ; Schmidt-Effing, 1979, 1980 ; Azéma and other, 1979 ; Kuijpers, 1980 ; Azéma and others, 1982). And, as a consequence, no net accretion has occured in the Middle America Trench off Guatemala, at least since the very beginning of the Eocene (around 60 MY B.P.) if not since the pre-Campanian (around 70 MY B.P.). The ophiolitic complex was drilled at the front of the slope to approximately 10-20 m above the subduction zone where high pore pressures (350 psi) caused collapse of the hole and abandonment of drilling.

In the Miocene to Quaternary sediment of the trench landward slope, sand was recovered, demonstrating that sandy facies are not restricted to the trench. The Miocene sequence at Site 568 in the midslope area is thicker and coarser than the presently ponded turbidites filling the trench axis.

Thus the combined results of Leg 67 and 84 demonstrate that : 1) landward dipping reflections in seismic records are ancient discontinuities and not features formed during the present tectonic regime (fig. 2C) ; 2) there has been no net accretion of the sediment entering the Middle America Trench off Guatemala since early Eocene time ; and 3) the assumption that sandy sediment is a facies deposited only in the trench and not on the slope is violated in a gross way and sandy facies cannot be used as a paleoenvironmental indicator.

GUATEMALA TRENCH, END-MEMBER MODEL.

The Middle America Trench off Guatemala is as well studied as any in the World. It is the only trench drilled to igneous oceanic crust (Sites 499 and 500 of Leg 67) and the only landward trench slope drilled to basement in 4 places (Sites 566, 567, 569 and 570 of Leg 84) ; the Middle America Trench is the first, and so far, the only trench to have been precisely surveyed by SEABEAM (Aubouin and others, 1981 ; Aubouin and others, 1982) and deep tow (Moore and others, 1982). These datas allow us to propose a Guatemala Trench model as a model strongly supported by facts(fig. 3).

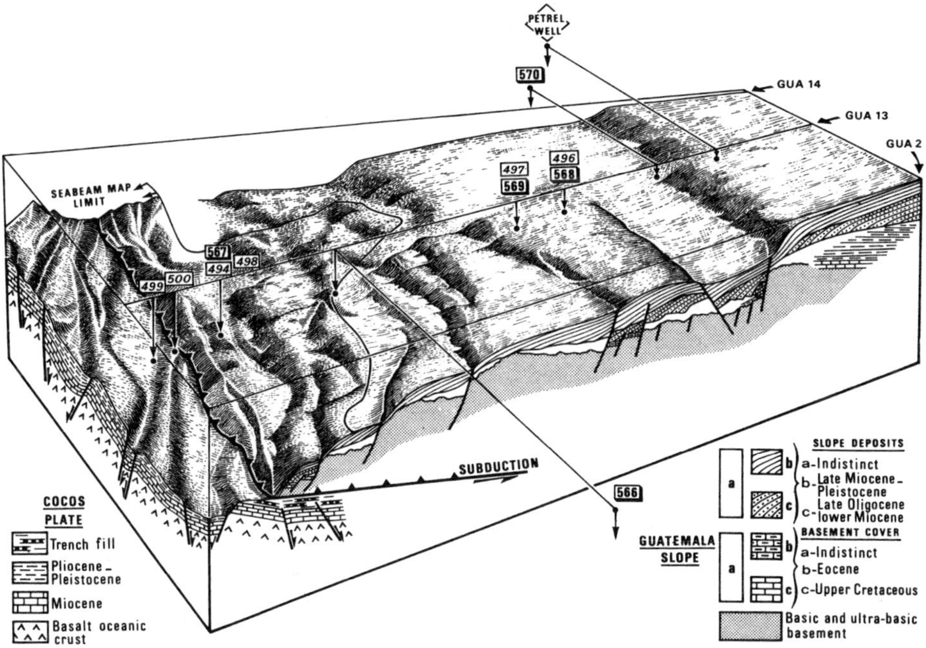

Fig. 3. - Bloc diagram showing the convergent zone between Cocos plate and Caribbean plate in Mid-America trench off Guatemala (after Aubouin, von Huene and others, 1982).

The Cocos plate has a horst and graben structure inherited from sea floor spreading at East Pacific rise, as shown by Cyamex diving (Rangin, Francheteau, 1981). The horst and graben enter the trench obliquely, dividing it into a succession of diamond shaped basins separated by bottle-necks corresponding, respectively, to graben and horst of the ocean plate (Aubouin and others, 1982). Beyond the trench, the Cocos plate sinks into the subduction zone beneath the continental margin.

All these features appear on the seismic reflection records (Ladd and others, 1982) ; the SEABEAM (fig. 4) and the deep tow (More and others, 1982) surveys. All features are confirmed by drilling (see above, fig. 2A). The oceanic crust of the Cocos plate that was drilled at the foot of the continental slope (Leg 67, Site 500), is 150 m higher than in the trench itself (Site 499), (fig. 2A and 3) and shows clear extensional features, specifically normal faults in the cores of undisturbed Miocene chalk.

Thus, as the Cocos plate flexes down into the subduction zone, the tensile stress from plate bending causes an extensional horst and graben structure which breaks along the original plate grain inherited from the spreading ridge, rather than breaking parallel to the axis of flexure. This is a rather normal trench phenomenon. However the less common features along the IPOD transect, are the normal faults on the landward slope on the trench (fig. 2 and 3). The landward normal faulted slope is documented by the SEABEAM survey (fig. 4) and the seismic reflection record GUA-13 (Ladd and others, 1982). Thus along this part of Guatemala, the trench has a morphology of a graben (fig. 3).

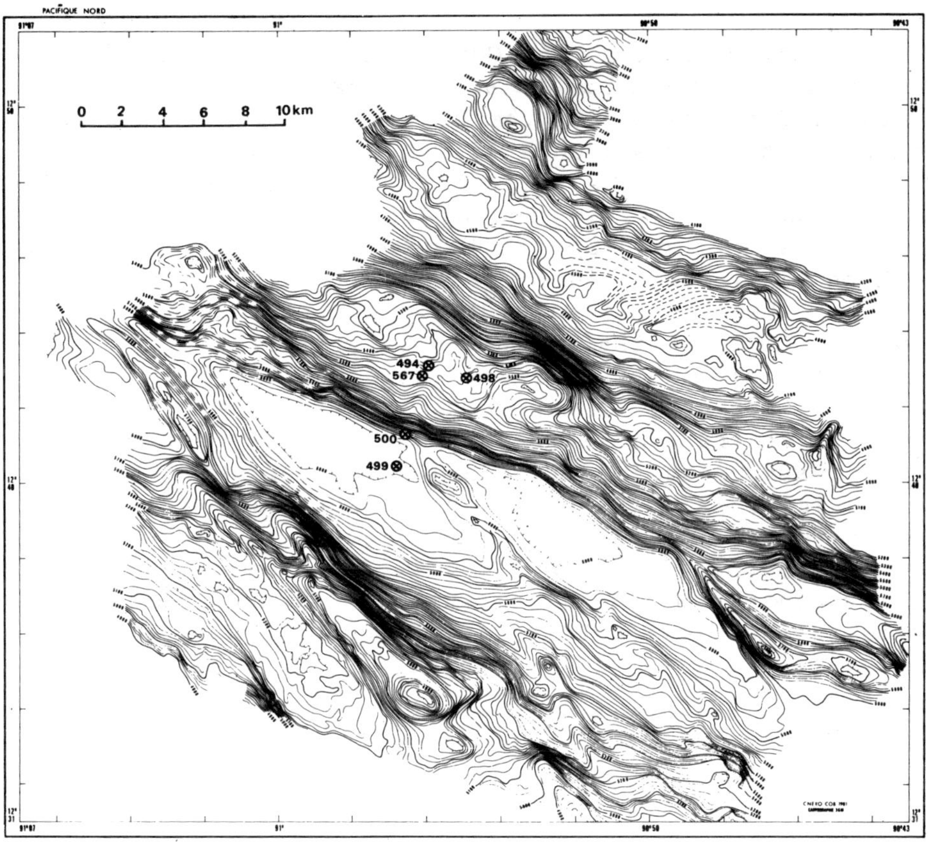

Fig. 4. - SEABEAM survey of the Leg 67/84 area in the Middle America Trench after Aubouin and others, 1982. Note the horst and graben structure of the Cocos plate. As a consequence, the diamond shape basins and bottle necks in the trench are formed ; the position of site 500 on the southwest side of a horstlike structure entering in the trench is shown. The linear boundary of the continental slope at the trench and the normal fault morphology on the continental slope are well displayed.

So, we propose the Guatemala trench as a model of "Convergent-Extensional active margin" (CE active margin). The collapse of the trench landward slope can be explained as a consequence of the sinking of the oceanic Cocos plate into the subduction zone. Perhaps the ocean plate subsidence was acclerated slightly there by causing a loss of support beneath the landward plate : perhaps sinking was caused by thinning of the continental crust or by thermal subsidence (Langseth and others, 1981). Whatever the fundamental cause, collapse is obviously aided by pore-pressure at near lithostatic levels as measured during failure of the Hole at Site 567 just above the top of the subduction zone. This elevated pore pressure essentially causes the upper plate to ride with very little friction above the subducting lower plate (von Huene and Lee, in press). Certainly, the subduction of all ocean floor sediment and accomodation of the ocean floor relief requires extensive decoupling of plates across the Guatemalan subduction zone.

DISCUSSION.

Could the Guatemala model be related to its special position (fig. 5) South of the sinistral Polochic-Motagua strike-slip zone (that is to say, extensional to the South were Caribbean plate is going east, and compressional to the North) ? Or is it a general model of a CE active margin as we proposed above ?

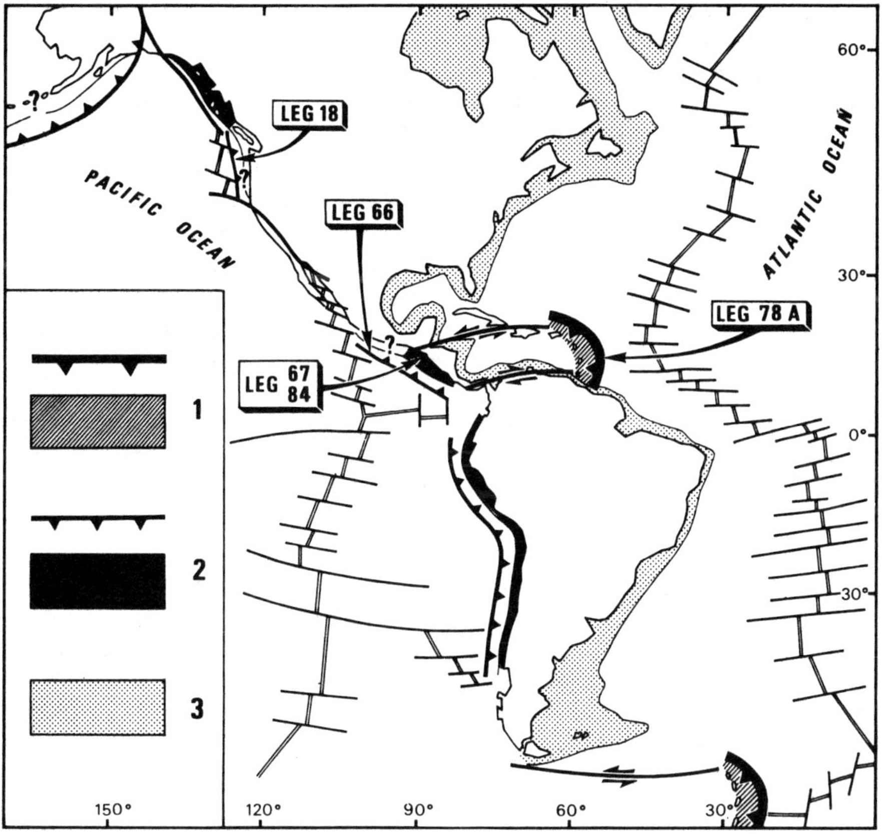

Fig. 5.- Convergent extensional margins in the Eastern Pacific.
1. Convergent compressive margins (i.e. : C.C. margins) ; 2. Convergent extensional margins (i.e. : C.E. margins) ; 3. Passive margins.

One must note that on one hand Barbados, which could be the model of classical Convergent-Compressional active margin, involves a thick trench-fill sequence of terrigenous sediments partly derived from the Orinoco and Amazone submarine deltas, which differs on the other hand from the Guatemala trench where the fill is only 300 m thick. The thickness of oceanic sediment involved in subduction could be one of the main factors in determining accretion or no accretion. But rate of convergence, age of the oceanic crust, etc. are others. Anyway, it appears that not more than the top of the sedimentary trench fill is accreted in Barbados trench. Accretion of oceanic crust by some form of the classical model (Dickinson & Seely, 1979) seems very difficult.

Quite a few examples of extensional structure have been shown on convergent margins. Superficial slumping is common on trench slopes where tectonism causes oversteepening and subsequent failure.

However fundamental normal faulting is known (fig. 5) in the Aleutian area off Kodiak Island (Fisher and von Huene, 1980), the Marianas (Hussong and Uyeda, 1981), Peru-Chili (Sheperd and Moberly, 1981 ; Thornburg and Kulm, 1981) and Japan (Langseth and others, 1981 ; von Huene and Uyeda, 1981 ; von Huene and others, 1982). Generaly the amount of data in these areas is insufficient to evaluate the role of these faults in the develoment of the margins whether it be a passive or an active one.

As a first approach, we suggest that the active margins that exhibits a broad continental shelf, as the Cordilleran margins (fig. 5), belong to the "Convergent extensional" type.

Nevertheless this discussion is only a first approach to the problem of CE margins -Japan and Marianna margins are island arcs margins- but we think that extension is now too well known to be readily dismissed as a special case or of a shallow mass movement of superficial origin.

ACKNOWLEDGEMENTS.

The authors have greatly profited from the extensive discussions and the critical reviews of Roland von Huene who made helpful comments on a draft of the manuscript. We acknowledge him very much.

Aubouin J., von Huene R., Azéma J., Coulbourn W.T., Cowan D.S., Curiale J.A., Dengo C.A., Faas R.W., Harrison W., Hesse R., Ladd J.W., Muzilev N., Shiki T., Thompson P.R. et Westberg J. (1979). - Premiers résultats des forages profonds dans le Pacifique au niveau de la fosse du Guatemala (fosse d'Amérique Centrale) (Leg 67 du "Deep Sea Drilling Project" : mai-juin 1979).
C.R.Acad.Sc., Paris, série D, t.289, p.1215-1220.

Aubouin J., Stephan J.F., Renard V., Roump J., Londsale P. (1981). - Subduction of the Cocos plate in the mid America Trench.
Nature, vol.294, p.146-150.

Aubouin J., Stephan J.F., Roump J. and Renard V.,- (1982). -The Middle America Trench as an example of a subduction zone. **Tectonophysics**, 86,- p.113-132.

Aubouin J., von Huene R., Baltuck M., Arnott R., Bourgois J., Filewicz M., Helm R., Kvenvolden K., Lienert B., McDonald T., McDougall K., Ogawa Y., Taylor E. and Winsborough B. (1982). - Subduction sans accretion : la marge pacifique du Guatemala ; premiers résultats du Leg 84 du Deep Sea Drilling Project (Janvier-Février 1982).

C.R.Acad.Sc., Paris, série II, t.294, p.803-812.

Aubouin J., von Huene R., Baltuck M., Arnott R., Bourgois J., Filewicz M., Helm R., Kvenvolden K., Lienert B., McDonald T., McDougall K., Ogawa Y., Taylor E. and Winsborough B. (1982). - Leg 84 of the Deep Sea Drilling Project ; Subduction without accretion : Middle America Trench off Guatemala. **Nature**, vol.297, p.458-460.

Azéma J., Sornay J. et Tournon J. (1979). - Découverte d'Albien supérieur à Ammonites dans le matériel volcano-sédimentaire du complexe de Nicoya (province de Guanacaste, Costa Rica). **C.R.Acad.Sc.**, Paris, série D, 3, p.129-131.

Azéma J. et Tournon J. (1980). - La péninsule de Santa Elena, Costa Rica : un massif ultrabasique charrié en marge pacifique de l'Amérique Centrale. **C.R.Acad.Sc.**, Paris, série II, t.290, p.9-12.

Azéma J., Origlia I., Tournon J. et De Wever P. (1982). -Nouvelles données sur la présence de Crétacé moyen au sein des formations volcano-sédimentaires de l'autochtone relatif de la péninsule de Santa Elena (Costa Rica, Amérique Centrale). **9ème Réunion Annuelle des Sciences de la Terre**, Paris, 1982, Société Géologique de France éd., p.22.

Barr K.W. y Escalante G. (1969). - Contribucion al esclarecimiento de la edad del complejo de Nicoya, Costa Rica. Publication Geologica ICAITI, Guatemala, 2, p.43-47.

Biju Duval B., Moore J.C., Blackinton G., Bergen J.A., Claypool G.E., Cowan D.S., Guerra R.T., Hemleben C.H.J., Marlow M.S., Natland J.H., Pudsey C.J., Renz G.W., Tardy M., Willis M.E., Wilson D. and Wright A.A. (1981). - Premiers résultats des

forages IPOD implantés lors de la croisière 78A du Glomar Challenger au NE de la ride de la Barbade (arc des Petites Antilles) : tectonique frontale d'un prisme d'accrétion.
C.R.Acad.Sc., Paris, série II, t.293, p.621-628.

Bourgois J., Azéma J., Tournon J., Bellon H., Calle B., Parra E., Toussaint J.F., Glaçon, G., Feinberg H., De Wever P. et Origlia I. (1982). - Ages et structures des complexes basiques et ultrabasiques de la façade pacifique entre 3°N et 12°N (Colombie, Panama et Costa Rica).
Bull.Soc.géol.Fr., 7ème série, t.23, n°3,-p.545-554.

De Boer J. (1979). - The outer arc of the Costa Rica orogen (oceanic basement complexes of the Nicoya and Santa Elena peninsulas).
Tectonophysics, 56, p.221-259.

Dengo G. (1962). - Estudio geologico de la region de Guanacaste, Costa Rica.
Instituto Geografico Nacional, Costa Rica, 113 p.

Dengo G. (1973). - Estructura geologica, historia tectonic y morfologia de America Central.
Instituto Centroamericano de Investigacion y Tecnologia Industrial (ICAITI), Mexico, Buenos Aires (Centro Regional de Ayuda Tecnica), 2a ed., 52 p.

Dickinson W.R. and Seely D.R. (1979). - Structure and stratigraphy of forearc regions.
Am.Ass.Petrol.Geol.Bull., vol.63, n°1, p.2-31.

Fisher M.A. (1979). - Structure and tectonic setting of the continental shelf southwest of Kodiak Island, Alaska.
AAPG Bull., 63, p.301-310.

Hussong D.M. and Uyeda S. (1981). - Tectonics in the Mariana arc : results of recent studies, including D.S.D.P. Leg 60.
In : R.Blanchet et L.Montadert, Géologie des marges continentales, 26ème Int.Geol.Cong., Paris, Oceanologica acta, n°sp., p.203-212.

Karig D.E. (1971). - Structural history of the Mariana Island arc system.
Geol.Soc.Amer.Bull., 82, p.203-212.

Kuijpers E. (1980). - The geologic history of the Nicoya ophiolite complex, Costa Rica, and its geotectonic significnce.
Tectonophysics, 68, p.233-255.

Ladd J.W., Ibrahim A.K., McMillen K.J., Latham G.V. and von Huene R.E. (1982). - Interpretation of seismic reflection data of the Middle America Trench offshore Guatemala.
In : Aubouin J., von Huene R., Init.Repts.DSDP 67 : Washington D.C. (U.S. Govt. Printing Office),- p.675-689.

Langseth M.G., von Huene R., Nasu N. and Odada H.- (1981). -Subsidence of the Japan Trench forearc region of Northern Honshu. **In** : R.Blanchet et- L.Montadert, Geologie des marges continentales, 26ème Int.Geol.Congr., Paris, Océanologica acta, n°sp., p.173-179.

Moore J.C., Watkins J.C., Shipley T.H., Bachman S.B., Beghtel F.W., Butt A., Didyk B.M., Leggett J.K., Lundberg N., McMillen K.J., Niitsuma N., Shepard L.E., Stephan J.F. and Stradner H. (1979). - Progressive accretion in the middle America Trench, Southern Mexico.
Nature, 281, p.638-642.

Moore J.C., Watkins J.S. and Shipley T.H. (1981). -Summary of accretionary processes, Deep Sea Drilling Project, Leg 66 : offscraping, underplating and deformation of the slope apron. **In** : J.S.Watkins, J.C.Moore, Init.Rept.DSDP, 66, Washington D.C., (U.S. Govt. Printing Office), p.825-836.

Moore J.C., Watkins J.C., McMillen K.J., Bachman S.B., Leggett J.K., Lundberg N., Shipley T.H., Stephan J.F., Beghtel F.W., Butt A., Didyk B.M., Niitsuma N., Shepard L.E. and Stradner H. (1982). - Facies belts of the Middle America Trench and forearc region, Southern Mexico : results from Leg 66 D.S.D.P. **In** : J.K.Leggettt ed., Trench forearc Geology, Geol.Soc.London, p.77-94.

Moore G.F., Lonsdale P. and von Huene R. (1982).- Near-bottom observations of the Middle America Trench off Guatemala. **In** : Aubouin J., von Huene R., Init.Rept.DSDP, 67, Washington D.C., (U.S. Govt. Printing Office), p.707-718.

Rangin C. et Francheteau J. (1981). - Fine scale morphological and structural analysis of the East Pacific Rise, 21°N (Rita project). **In** : X.Le Pichon, J.Debyser, F.Vine éd., Géologie des Océans, 26ème Int.Geol.Congr., Paris, Oceanologica acta, n°sp., p.15-24.

Schmidt-Effing R. (1979). - Alter und genese des Nicoya komplexes, einer ozeanischen palaokruste (Oberjura bis Eozan) im Sudlichen Zentralamerika. **Geologische Rundschau**, Stuttgart, 68, p.457-494.

Schmidt-Effing R. (1980). - Radiolarien der Mittel-Kreide aus dem Santa Elena macizo von Costa Rica. **Neue Jahrbuch fur Geologie und Palaontologie**

Abhandlungen, Stuttgart, 160, p.241-257.

Schweller W.J., Kulm L.D. and Prince R.A. (1981). -Tectonics structure and sedimentary framework of the Peru-Chile Trench.
In : L.D.Kulm, J.Dymond, E.J.Dasch, D.M.Hussong éd., Nazca plate : crustal formation and Andean convergence. **Geol.Soc.Amer.**, Memoir 154,-p.323-349.

Seely D.R. (1979). - The evolution of structural highs bordering major forearc basins.
In : Geological and geophysical investigation of continental margins, edited by J.S.Watkins and L. Montadert, **Am.Assoc.Petrol.Geol.Mem.**, 29, Tulsa, Oklahoma, p.245-260.

Seely D.R., Vail P.R. and Walton G.G. (1974). - Trench slope model.
In : Burk and Drake ed., The Geology of continental margins, New York, Springer Verlag,-p.249-260.

Shepherd G.L. and Moberly R. (1981). - Coastal structure of the continental margin, Northwest Peru and southwest Ecuador.
In : L.D.Kulm, J.Dymond, E.J.Dasch, D.M.Hussong ed., Nazca Plate : Crustal formation and Andean convergence. **Geol.Soc.Amer.**, Memoir 154,-p.351-391.

Thornburg T. and Kulm L.D. (1981). - Sedimentary basins of the Peru continental margin : structure, stratigraphy and Cenozoic tectonics from 6°S to 16°S latitude.
In : L.D.Kulm, J.Dymond, E.J.Dasch, D.M.Hussong ed., Nazca Plate Crustal formation and Andean convergence. **Geol.Soc.Amer.**, Memoir 154,-p.393-422.

Tournon J. et Azéma J. (1980). - Sobre la estructura y la petrologia del macizo ultrabasico de Santa Elena (provincia de Guanacaste, Costa Rica). Instituto Geografico Nacional, Informe Semestral, San Jose, Costa Rica, Enero a Junio, p.17-34.

von Huene R. and Uyeda S. (1981). - A summary of results from the IPOD transects across the Japan, Mariana, and Middle America convergent margins. **In** : R.Blanchet et L.Montadert, Géologie des marges continentales, 26ème Int.Geol.Congr., Paris, Oceanologica acta, n°sp., p.233-239.

von Huene R., Aubouin J., Azéma J., Blackinton G., Carter J.A., Coulbourn W.T., Cowan D.S., Curiale J.A., Dengo C.A., Faas R.W., Harrison W., Hesse R., Hussong D.M., Ladd J.N., Muzylov N., Shiki T., Thompson P.R. and Westberg J. (1980). - Leg 67 : The Deep Sea Drilling Project Mid-America Trench transect off Guatemala. **Geol.Soc.Amer.Bull.**, vol.91, p.421-432.

von Huene R., Langseth M., Nasu N. and Okada O. (1982). -A summary of Cenozoic tectonic history along the IPOD Japan Trench transect. **Geol.Soc.Amer.Bull.**, v.93, p.829-846.

von Huene R. and Lee H. (in press). - The possible significance of pore fluid pressures in subduction zones. **In** : Watkins, Drake, éd., **Am.Assoc.Pet.Geol-.Memoir.**

"CONTINENT-OCEAN" TRANSITION ZONES IN THE GEODYNAMIC SYSTEM OF THE ARCTIC OCEAN

POGREBITSKY YU.E.
VNIIOkeangeologia, Leningrad, USSR

Geological evidence available nowadays on the Arctic Ocean enables to characterize main features of morphology, tectonic structure and kinematics of the formation of its transition zones towards the surrounding continents. Distinct spatial and temporal position of these zones in the developing geodynamic system of the Arctic Ocean allows us to introduce a genetic factor into their typization.

MORPHOLOGY

The transition zone between the abyssal and the shelf is usually characterized by two sides of the relief: continental slope and continental rise. For the Arctic Ocean, in spite of a rather gentle inclination of these sides, their intersections (relief ribs) are, as a rule, distinct, but wavy curved along the strike due to tectonic and sedimentation processes. In this relation an intersection of surfaces of the shelf and continental slope, so-called shelf edge (or continental shoulder), is highly significant. This main rib marking the upper edge of continental plate usually bifurcates, a more contrasting break in profile generally plunging from 150-200 m to 500-750 m. Comparison of transverse sections shows that in this case there develops, instead of shelf edge, an additional side having a greater inclination than the shelf, but lesser steepness than the continental slope. The configuration of the chamfered edge or the "border facet" (straight, crescent convex oceanwards, crescent convex landwards) emphasizes morphostructural features of the slope which are associated with transverse uplifts and subsidences of continental margin. Along the border facets the Arctic Ocean transition zones can be divided along the strike into separate morphostructural links. According to the subdivision 21 links have been recognized on the basis of

Proceedings of the 27th International
Geological Congress, Volume 7, pp. 65–79
TECTONICS

GEBCO chart (Fig. 1). In length, 75% links recognized, except for two links 200 km long each (Nos 1, 5) and three links 500 to 600 km long (Nos 2, 11, 12), are characterized by persistent dimensions within the range of 300 to 350 km. The width and steepness of the continental slope do not depend on the extent of links and average about 60 km and 3-4°, respectively. Thick alluvial fans (links 1, 3, 13) where the width of the slope reaches 150-200 km and gradient does not exceed 1.5° are an exception. If thick alluvial fans exist, the transition zone has no border facet.

From the character of relief and metric data the morphostructures of the "abyssal - shelf" transition zone are profoundly different from the "abyssal - relict continent" boundaries; the relict continent is represented by bathyal steps - Voring, Yermak, Morris Jessup, Chukchi Plateau and Lomonosov Ridge. No border facet is typical of them; their confining slopes are straightened, continental rises are reduced, slopes reach 8-10°. Morphologically, along the edge of British-Greenland riegels the transition zone is analogous to the "abyssal - relict continent" boundary.

As a whole, the Arctic Ocean abyssal basins (Fig. 1) and the transition zones expressed in the relief (without zones situated along the boundary of British-Greenland riegels) appear to be roughly equal in area with 3-5% difference. Similar relationships are typical of sea bottom areas having spreading structure (Norwegian-Greenland Sea and Eurasia Basin) and the adjacent transition zones including the Lomonosov Ridge slope. Correspondingly, the abyssal non-spreading Amerasia Basin bottom and its surrounding transition zones are equal in area as well.

Thus, morphologically the "continent - ocean" transition in the Arctic Ocean geodynamic system is characterized by:

1) division into links of the single morphostructural zone caused by regular undulations of continental slope edges;

2) independence of its surface structure and the extent of development from the genesis and history of the ocean floor formation.

TECTONIC STRUCTURE

From geophysical data the transition zone under study corresponds to a wedge-like crustal block about 150 km wide and 28-30 km thick along the shelf edge and about 10 km along the contact with the oceanic plate; the block is confined by the floor surface and Moho, generalized inclinations of which measure 1.5° and 6°, respectively. The continental slope (app. 3°) and its corresponding steep Moho side (up to 10°) are inscribed into the middle part of the hypothetical wedge (Fig. 2). Regional gravimetry suggests that the wedge-like block is isostatically compensated. Seismic activity is observed as a rule in local areas associated with intersections of longitudinal and latitudinal faults.

Three structural stages are recognized in the wedge-like block structure: 1) lower stage, represented by folded basement; 2) intermediate stage, represented by platform cover remnants; 3) upper coilogenic stage, caused by downwarping and represented by continental margin basins (5). In the transition zone the folded basement surface combines linear dislocated sedimentary sequences and inliers of granite-gneissic crustal layer, both related to the Baikalian and Caledonian orogenic belts and traced there by geophysical anomalies from land. The intermediate stage is chiefly represented by Early-Middle Paleozoic carbonate formations of epi-Caledonian and epi-Baikalian plates. The upper, coilogenic stage includes two structural-formational complexes: lower, Permian-Mesozoic preoceanic complex represented by volcano-terrigenous formations which fill in the basins and troughs; and upper, Cenozoic synoceanic complex represented in the transition zone by essential marine terrigenous sediments which fill in troughs and grabens or form covers over the transition zone surface.

The inner crustal structure within the wedge-like block is controlled by longitudinal, relative to the slope, vertical dislocations. Their development is related to the formation of coilogenic stage and, predominantly, to its upper complex which is obviously illustrated by the occurrence of Cenozoic sequences base. The system of dislocations includes continental foredeeps, marginal uplifts and a zone of flexures and faults which couples them (Fig. 2).

Considerable difficulties experienced in carrying out continuous seismic reflection profiling in the Arctic Ocean do not allow presently to systematize parameters of these structures and describe the relationships between Cenozoic dislocations and basement structures in plan. As exemplified by the Norwegian-Greenland Sea and Canada Basin, they can either cut Mesozoic troughs, grabens and horsts at low angles, or inherit their strike.

As a whole, the following brief evidence on the above structures can be assumed.

Continental foredeeps (or foredeeps of oceanic plates):

- are situated in place of continental rise and their axial zones roughly coincide with nonanomalous marginal zones $(\Delta T)_a$ of abyssal plates;
- occur along the continental rise of all the continental blocks, except for the Eurasia flange of Lomonosov Ridge in its polar part where, against calculations of magnetic layer depth (2), seismic data do not support the presence of a foredeep;
- are characterized by lineation and great extent, the latter exceeds that of the conjugate links of continental slope; in the axial zone the thickness of sediments reaches 4-5 km;
- seismic section of the continental limb is characterized by numerous refractors suggesting the presence of formations of the poorly folded lower terrigenous complex, intermediate stage and, probably, folded basement (1, 3);
- oceanic limb turning to gentle swell consists, as suggested from seismic data, of oceanic layers I, II and III.

Flexure-fault zone of the continental slope:

- has a common forming mechanism and is divided into the same links both from the structural and morphological features;
- each link has several own (not traced through links) longitudinal steps, synsedimentary relative to the upper

complex of sediments; the steps are confined by normal faults dipping towards abyssal plains and curved in plan similar to the border facet;

- step level is downwarped in most cases;
- numerous low-velocity refractors pitching towards the abyssals and the presence of high-velocity horizons (V_p = 5.5-6.5 km/sec) corresponding to the "granitic" layer have been recognized from seismic data in the sedimentary cover;
- all the three structural stages have been recognized, as a rule, from velocity characteristics, the intermediate stage and the lower Permian-Mesozoic complex of the upper stage locally represented in the form of erosion remnants;
- the thickness of the upper Cenozoic complex of sediments generally does nog exceed 2.5 km and is characterized by small landslide faults and related local plications of strata (3, 9).

Marginal uplifts:

- form, along the transition zone edge, a chain of separate structures corresponding in length to the morphostructural links of the continental slope;
- are represented by buried narrow linear horsts or by block-arched uplifts of islands;
- according to seismic data, the buried horsts generally situated beneath the lower edge of the border facet consist of folded and metamorphic basement formations overlapped by Cenozoic complex deposits;
- in the overlapping Cenozoic deposits the buried horsts occur as synsedimentary growing anticlines;
- block-arched uplifts of islands (Spitsbergen, Franz Josef Land, Severnaya Zemlya) situated along the shelf edge beyond the border facet include formations of all the three stages, except for Upper Cenozoic formations.

Longitudinal dislocations of the transition zones are complicated by transverse uplifts and subsidences which determine its subdivision into morphostructural links. Subsidences between block-arched uplifts of islands are supplemented by trenches (Svyataya Anna, Franz Victoria, Voronin Trenches, M'Clure Strait, Amundsen Gulf) well defined in relief. The trenches stretch into the shelf marking a recent stage in the destruction of continental margin.

The buried grabens of the Laptev Sea shelf and the Naris Strait separating Greenland from the Canadian Arctic Archipelago are of particular interest among transverse structures breaking the transition zones. The Canadian Arctic Archipelago is associated with a deep fault along which horizontal regmatic displacements or horizontal throws of large lithospheric blocks take place (4).

The Cenozoic sedimentary basin which is forming within the Laptev Sea is located along the extent of the Eurasia Basin and can be considered as its continental limb. The basin base is represented by an ancient massif overlain by a cover similar to that of Siberian platform cover. The superimposed complex of Cenozoic sediments (including Danian deposits of the Upper Cretaceous) overlies the Laptev Massif by a thin (up to 1 km) cover and fills trenches reaching 3-4 km in thickness. Geophysical data suggest the presence of a central trench and two marginal - Taimyr and Novosibirsk - trenches. These tectonically active structures of the Laptev Basin are coupled with active ocean zones; the central trench continues the rift fault system of the Gakkel mid-ocean ridge, and the marginal trenches are situated on the continuation of the continental foredeeps (5). A close relation to the oceanic structures is emphasized by the distribution of earthquake epicentres. The central trench of the Laptev Basin on the deep seismic profile corssing the Buorkhaya Gulf is characterized by the presence of a "pillow" of mantle of lower density (V_p = 7.5 km/sec) located at a depth of 30 km. a crustal thickness totalling 35 km in the area. But "spreading" magnetic field anomalies do not pass from the ocean into the Laptev Basin.

Tectonic features taken together suggest that the "continent-ocean" transition zone in the Arctic Ocean is an epi-platform mobile belt with a miogeosynclinal mode of development. Its structural evolution is closely related to Cenozoic descending movements forming an abyssal plain of the ocean as a single depression. Heterogenous transverse dislocations breaking the integrity of the belt suggest that the space of exchange processes between the ocean and continent extends beyond the transition zone.

KINEMATICS OF MOTIONS

Continuous seismic reflection profiling performed where ice conditions permit suggests that the Cenozoic sedimentary complex developed in place of continental slope is divided at least into three formations differing in structure and litho-acoustic parameters. The lower formation is persistent in thickness and consists probably of silty and psammitic strata rhythmically interbedded. The middle formation has a maximum thickness among Cenozoic sediments at the slope base and thins out both ocean- and shelfward. It is characterized by a common obscure bedding with frequent intraformational fractures and deformation of strata with local overlapping. Genetically, this is probably a turbidite formation. The upper formation with thin-bedded rhythmic structure is persistent in thickness and mantles marked roughness of the floor. Bottom sampling shows that the formation consists chiefly of clayey marine and glaciomarine Pliocene-Quaternary deposits. The age of the lower and middle formations can be determined only tentatively. For the entire Arctic Basin, by correlating from a change in environment with logs of oceanic layer I of the Norwegian-Greenland Sea and with data available on the Amerasia sector shelves, the lower formation can be attributed to the pre-Oligocene epoch of general shallow-water sedimentation, while the middle formation - to Oligocene-Miocene time of pelagic sedimentation in the ocean. Possible errors in dating of the formation boundary cannot affect a principal three-stage scheme of the development of transition zone:

stage I (pre-Oligocene) - quiet downwarping in the general system of descending motions embracing the future abyssal plains and vast areas of surrounding continents;

stage II (Oligocene-Miocene) - active motions along

the edge of subsiding abyssals;

stage III (Pliocene-Quaternary) - quiet downwarping in the general system of descending motions, embracing abyssal, marked in morphostructure, and continental slopes.

So, the continental slope zone structurally separated as a "continent-ocean" transition during Oligocene-Miocene time; this is in good agreement with data in favour of the fact that in the Arctic Basin the sediments of oceanic layer I became most pelagic in the Miocene (10).

Relict continental blocks (Voring, Yermak, Chukchi Plateau, Lomonosov Ridge) form a peculiar submarine terrace regularly downwarped from the continental margin toward the basin centre, from -750 m to 1500 m. It can be assumed that the terrace marks a general downwarping of the Earth's crust in place of the ocean during the Cenozoic with no superimposed Oligocene-Miocene collapse. In this case the amplitude of the latter for the Arctic Ocean is estimated to be 2.5-3 km, and the rate of subsidence is about 10 cm per 1000 years.

POSITION IN THE GEODYNAMIC SYSTEM

The geodynamic system can be generally determined as a space of interrelated geological processes relaxing energetic disturbance in the Earth's body (6). Eventually these processes manifest themselves by different material movement, i.e. heterogenous mass movements. The major mass movement ever occurred on the Earth's surface is solid flow which can be considered, to a first approximation, as balanced with endogenic relief-forming mass movement. From the solid flow point of view, the Arctic Ocean geodynamic system covers an entire basin and vast areas of surrounding continents including dividing ridges. This geodynamic system is clearly expressed in morphostructure: oceanic basins appear to be situated in the centre of the grandiose Arctic Geodepression characterized by a closeness and regular inclination of lithospheric surface from the boundary divides towards the oceanic bed. According to relief and exogenic processes, three subconcentric zones may be recognized in the Arctic Geodepression at the present stage of its evolution. The outer zone is characterized by ascending motions and development of denudation processes. It is built up of a chain of ridges and highlands and can be named the boundary orogenic belt. The intermediate zone is represented chiefly by accumulation and denudation-accumulation plains and marginal sea shelves where descending motions are predominant. By general inclination typical of this zone from the boundary orogenic belt towards the centre of the depression it can be recognized as a continental centricline. The inner zone or abyssal core consisting of deep basins and separating seamounts is an area of conjugate descedning and horizontal (spreading) motions of the crust and an area of pelagic deposition. Paleogeographic reconstructions of the Arctic Geodepression as a whole and datings of magmatic activity outbursts in the boundary orogenic belt and megacyclic sedimentation in the continental centriclinal basins allow to draw the following main conclusions concerning the structural evolution of the geodepression (6).

The Arctic Geodepression singled out on the Earth's surface near the Early/Late Permian.

The development of morphostructures of the boundary orogenic belt is similar to multistage recurrent growth either in situ or with some displacement towards the centre of the depression.

Sedimentary basins of the continental centricline undergo multiphase recurrent subsidence; their depocentres have remained virtually unmoved.

Active movements in orogens are recorded earlier than in adjacent basins: a wave of activity generated in the orogenic framing moved from the periphery towards the centre of the geodepression.

The development of the geodepression core is characterized by two main stages: Permian-Cretaceous, preoceanic, and Cenozoic, including Danian, oceanic stages. At the first stage the geodepression core contained arched mantle-crustal tumuli (geotumors) from which terrigenous ablation took place towards the present shelves (7). At the end of the Cretaceous the geotumors underwent rifting inversion and oceanic basins with spreading structure of the bed (Norwegian-Greenland Sea and Eurasia Basin) formed in their place in the Cenozoic. The Amerasia Basin formed on the down-warping limb of geotumor. Its separation as a basin possessing oceanic depths took place in the Cenozoic (Oligocene-Miocene) due to sediment starvation of the intensively subsiding core of Arctic Geodepression.

Compositionally, the development of Arctic Geodepression leads to a simultaneous formation of neogenic femic lithosphere in the boundary orogenic belt (continental massif), a wide transition zone of relict lithosphere situated in between. Its reworking results in the formation of assembly of small (daughter) geodynamic systems (epicontinental sedimentary basins and their orogenic barriers) repeating bipolar direction of the process as a whole.

The transition zone of continental slope holds a unique position. In essence, it is the most active border of destructive transformations of relict lithosphere, the former being situated in the "exo-contact" of the forming massif of new femic lithosphere. From the standpoint of dynamic relations, the transition zone of continental slope is complemented by that of the near-orogenic slope situated in the "exocontact" of forming massif of new sialic lithosphere. In the process of development of geo-dynamic system the transition zones come closer and an area of relict lithosphere clutched in between reduces. The migration and formation of transition zones depend on the interaction of expanding bipolar lithospheric massifs which is pulse-wavy in nature: a pulse of active ascending motions in boundary orogens is replaced with time by a pulse of active subsidence of the ocean floor (Danian-Paleocene tectonic activity of the boundary orogenic belt - Oligocene-Miocene collapse of the ocean; neotectonic activity of orogens - future phase of active disintegrating transformations around the periphery of deep basins). From this point of view, the transition zones of continental slope observed in the Arctic Ocean were caused by the first phase of collapse.

CONCLUSION

By morphotectonic patterns the "continent-ocean" transition zones in the Arctic Ocean belong to the Atlantic type. Ideas that this type is determined by passive reworking of the continental plate edge by autonomous processes of gravitational sliding is erroneous in principle. The structure of the transition zones is the result of active interaction of forming oceanic and continental lithospheric massifs separated by a wide belt of relict lithosphere; gravitational sliding plays a minor part there.

A pulse-wave interaction of the forming lithospheric massifs allows to introduce an additional classification criterion according to which the "continent-ocean" transition zones in the Arctic Ocean can be assigned to the subtype of collapse phase I. In contrast, subtypes of collapse phases II and III can be recognized among the transition zones of the North and Equatorial Atlantic. It is precisely this fact that explains an almost complete truncation and subsidence to the sea floor of continental marginal basins of America, Africa and Western Europe. The known hypotheses of monotonous deepening of the ocean as the neogenic lithosphere cools in time from the t $^{1/2}$ law (8) or any other aperiodic law do not satisfy available geological data.

REFERENCES

1. Geologic framework of the Alaskan continental terrace in the Chukchi and Beaufort Seas. - Grantz A., Holmes M.L. and Kososki B.A. - Canada's Continental Margins. CSPG, Calg., Canada, 1975, p. 669-700.
2. Karasik, A.M. Basic specifics of the history of development and the structure of the Arctic basin bottom according to aeromagnetic data. In: Marine Geology, Sedimentology, Sedimentary Petrography and Geology of the Ocean. Academy of Sciences of the USSR; Ministry of Geology of the USSR, National

Committee of Geologists of the Soviet Union, Leningrad, 1980, p. 178-193. (Trans. (1981): NASA GSEC-07).

3. Marine geophysical survey on the Northern Svalbard Margin - Sundvor E., Gidskehaug A., Myhre A., Eldholm O. - The Norwegian Petroleum Directorate's Scientific Report No. 5, Bergen, 1978, p. 1-25.
4. Monahan D. and Johnson G.L. Physiography of Nares Strait: importance to the origin of the Wegener Fault. Meddelelser om Grønland, Geoscience 8, 1982, p. 53-64.
5. Novoe v geologii antarkticheskikh morei SSSR. - Gramberg I.S., Lopatin B.G., Pogrebitsky Yu.E. - Vestnik AN SSSR, No. 2, 1981, s. 84-90.
6. Pogrebitsky Yu.E. Geodinamicheskaya sistema Severnogo Ledovitogo okeana i ee strukturnaya evolutsia. "Sovetskaya geologia". 1976, No. 12, s. 3-22.
7. Ronkina Z.Z. and Vishnevskaya T.N. Glaukofan v osadochnykh tolshchakh zapadnoi chasti Sovetskoi Arktiki. "Sovetskaya geologia", 1982, No. 7, s. 90-93.
8. Sorokhtin O.G. Globalnaya evolutsia Zemli. Moskva, Nauka, 1974, 182 s.
9. Sundvor E. and Nysaether E. Geological outline of the Norwegian continental margin between 60° and 68° North. Canada's Continental Margins. CSPG, Calg., Canada, 1975, p. 267-281.
10. Udintsev G.B. (ed.) Resultaty glubokovodnogo burenia v Atlanticheskom okeane v 38-om reise "Glomar Challenger". Litologia i petrografia. Moskva, Nauka, 1978, 192 s.

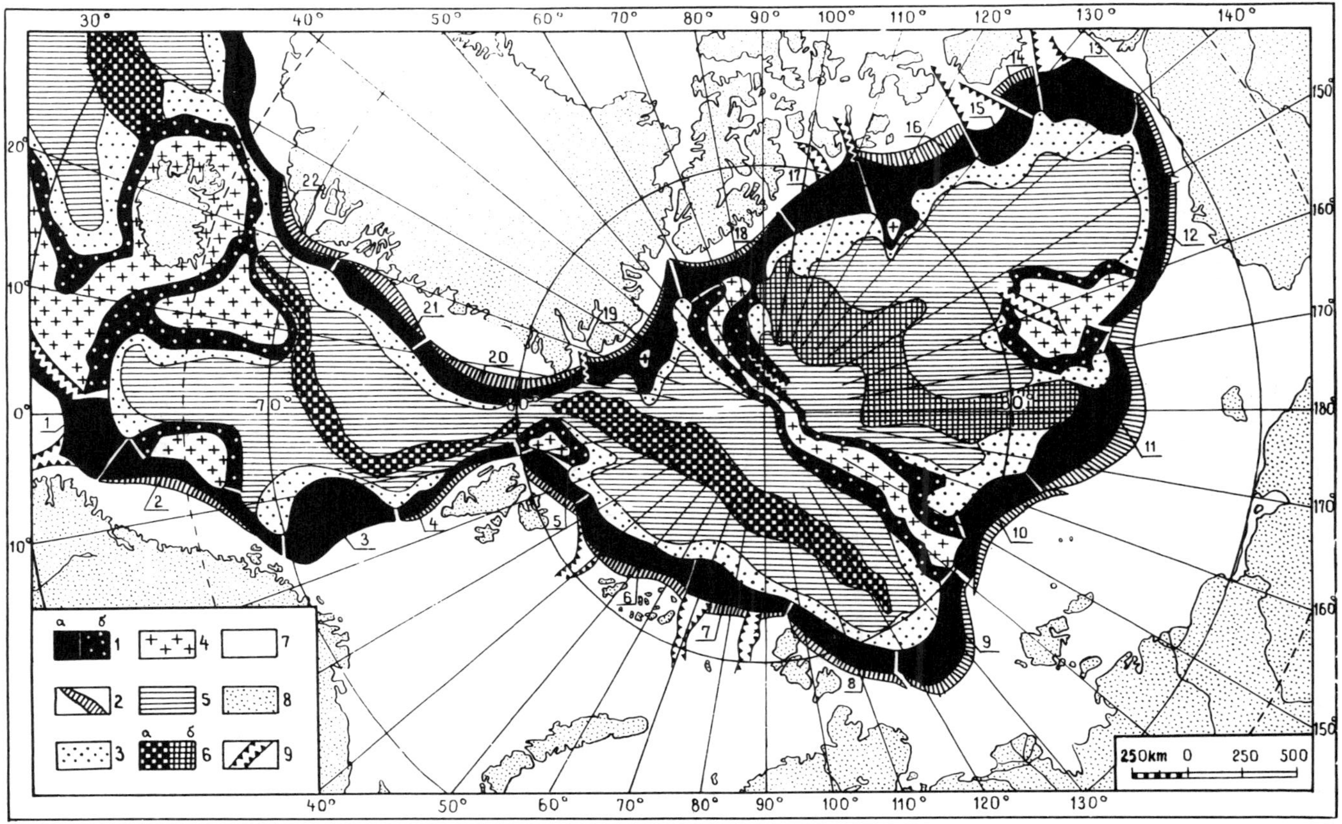

FIGURE 1. Caption overleaf.

FIGURE 1. Morphostructural zoning of transitions.
1-3) Components of transition zones: 1 - continental slope (a - along the shelf edge, b - along the edge of bathyal plateux and riegels); 2 - border facet; 3 - continental rise. 4) Bathyal plateaux. 5) Abyssal plains. 6) Abyssal mountains (a - rift mid-ocean ridges, b - arched block uplifts). 7) Shelf. 8) Land. 9) Continental marginal trenches. Numbers (1-21) on the map indicate links of the "abyssal-shelf" transition (see text).

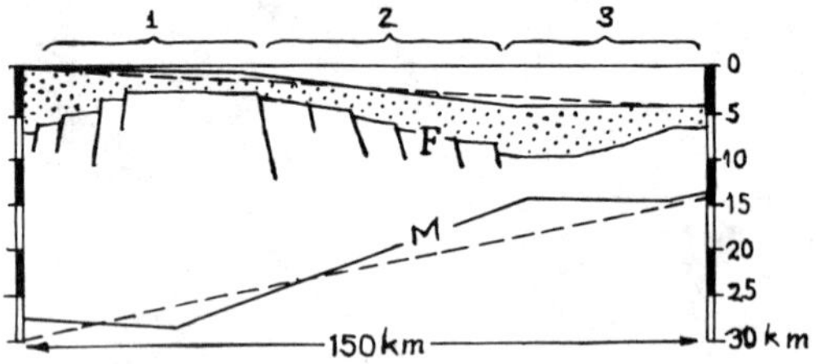

FIGURE 2. Crustal section of the transition zone.
1) Continental marginal uplift. 2) Flexure-fault zone. 3) Continental foredeep. Generalized block boundaries are shown by dashed lines. F - basement boundary. M - Moho discontinuity. Sedimentary cover is dotted.

FIGURE 3. Transition zones in the Arctic Ocean geodynamic system.
1-3) Basic dynamic zones of the Arctic Geodepression: 1 - abyssal core; 2 - continental centricline (areas: a - stable or rising, b - actively subsiding); 3 - boundary orogenic belt (dividing line is shown by dots). 4) Axes of daughter orogens. 5) Transition zones (continental and mountain slopes). 6) Conventional boundary of orogens. 7) North Atlantic abyssal. 8) North Pacific abyssal. 9) Continental areas beyond the Arctic Geodepression (areas: a - stable or rising, b - actively subsiding).

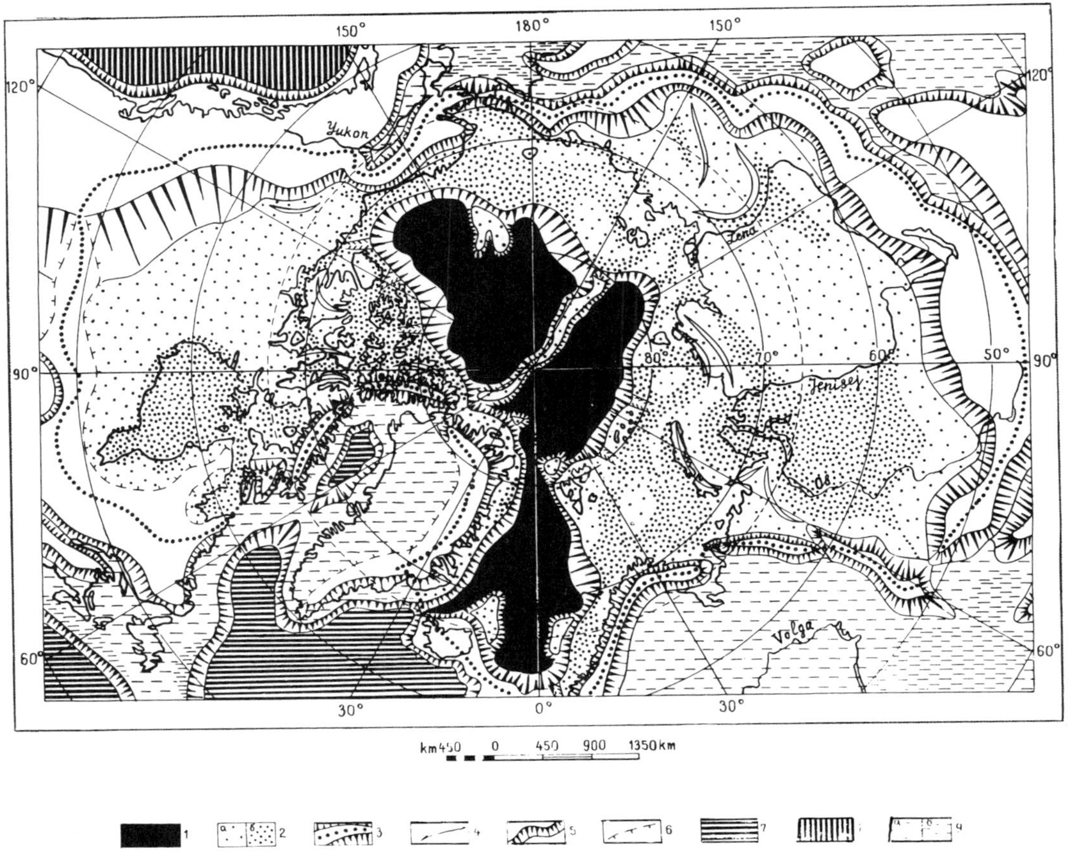

FIGURE 3. Caption on opposite page.

GEODYNAMICS AND LATE CENOZOIC EVOLUTION OF THE ASIA/PACIFIC TRANSITIONAL ZONE

B.V. BARANOV, L.P. ZONENSHAIN
Institute of Oceanology Academy of Science of USSR, Moscow, USSR

INTRODUCTION

The Pacific floor is bounded by trenches along the whole Asia/Pacific transitional zone. The western Pacific margin significantly differs from its eastern margin near Americas, where oceanic plates are subducted directly under the continents. On the western margin, a broad zone of 2000 to 3000 km wide, comprising volcanic island arcs and back-arc basins is situated between the Pacific ocean and Asiatic continent. Though this difference was noted long ago, no satisfactory explanation has been found yet.

The Pacific plate is moving towards Asia at rate 9,3 cm/year near the Kuril islands (1). It has not any common boundary with the Eurasian plate. Small plates, the Sea of Okhotsk, Amurian, and Philippine plates, are located between the Pacific and Eurasia plates. The small plates share an interaction of the two large plates.

The geodynamic features of the western Pacific margin are shown in Fig. 1. They are described in numerous publications (2, 3, 4, 5, 6, 7, 8). In this paper we shall try to throw some light on such features which earlier remained in the shadow.

RECENT GEODYNAMICS

The Pacific plate is sinking westward under trenches along Benioff zones. Maximum depths of these zones up to 500 to 600 km are recorded under the Kuril, Japan, Izu-Bonin and Mariana arcs (9). Meanwhile, earthquake hypocenters no deeper than 200 km are registered under the eastern Aleutian and Ryukyu trenches and 50 to 80 km only under south-western Japan/Nankai trench/ and the western Aleutian arc. In case Japan small depths of

Proceedings of the 27th International
Geological Congress, Volume 7, pp. 81–91
TECTONICS

Benioff zone is interpreted as the beginning of underthrusting of the Philippine plate under Japan (9), and, in case of the western Aleutian arc, as the transformation of underthrusting to strike slip movement (10). Present volcanism is absent along these segments of the island arc. Angles of Benioff zone inclination vary from 45 to 60° for the Aleutian, Kuril and Japan arcs to almost vertical for the Izu-Bonin and Mariana arcs.

Magnitude of shallow earthquakes is changed from the north to the south, from the Aleutian arc to the Mariana one. According to Kanamori (11), great earthquakes with $M \geqslant 7{,}7$ point to the coupling, and absent of great earthquakes to decoupling between underthrusting and overriding plates, that is, the different degree of mechanical interactions between them. The Aleutian, Kuril and Japan arcs are characterized by high seismicity and frequency of earthquakes with magnitude more than 7,7. It means that the underthrusting Pacific plate is coupled with overriding plate. High energy earthquakes are absent in the Izu-Bonin, Mariana and Ryukyu arcs indicating plate decoupling.

Plate cohesion, that is, plate coupling or decoupling, is proved to be of great importance for island arc geodynamics. In case of coupling, for example, in the Kuril and Japan arcs, compressional conditions predominate within the whole body of an island arc, and compressional axes, according to focal mechanism solutions (13, 14) are oriented across the island arc strike. In case of decoupling (the Ryukyu and Mariana arcs) extension across an island arc is inferred from focal mechanism solutions and from deep-sea drilling data even in fore-arc regions (15). Degree of coupling seems to be reflected in island arc magmatism. When plates are decoupled a possibility appears for deriving shallow-chamber tholeiitic basalts in fore-arc region.

Island arcs are broken by transverse faults into a series of blocks ("keyboard") of 100 to 200 km long (16, 17). Earthquakes with $M \geqslant 7{,}7$ occur repeatedly within each block (or within a series of adjacent blocks) approximately during each 100 year period

(18, 19, 20) and in different blocks at different time. This can be explained by oceanward shifting of the block during some great earthquake and then its returning back toward the arc due to its coupling with an underthrusting plate (21). As the underthrusting rate is about several cm/year and oceanward block shift is equal to several meters, the time which is necessary to return the block in its previous position is about 100 years. It well agrees with the frequency of great earthquakes.

Recent data indicate the existence of an additional seismic zone along the island arc rear. This zone separates the island arc of 150 to 200 km wide from the marginal basin floor. Its geodynamics varies within island arc and is determined by interaction between the three units: an underthrusting plate, overriding plate and an island arc situated in-between. Two main types of such interaction exist: compression and extention.

Extention occurs when decoupling takes place in the fore-arc region (22). An island arc is splitted off from an overriding plate and this process is accompanied by back-arc rifting and eventually by generation of an oceanic crust of the inter-arc basin. Now, extention is typical of the Okinawa trough and Mariana inter-arc basin. An extentional centre in the Okinawa trough coincides with the axial graben system (23). Thê central part of the Mariana inter-arc basin is similar to an axial zone of an oceanic ridge with slow spreading (24). This extention began 7 to 8 my ago.

The relationship of back-arc basin generation with plate decoupling is well correlated with two models of marginal basin formations: 1. with the model of oceanward arc migration due to subsidence of old and cold lithosphere under its own weight into the mantle (25) and 2. with the model of retreating of an overriding plate from a subduction zone anchored in the deeper mantle (22, 26). Both models are well supported by focal mechanism solutions and structural data indicating extention in the Okinawa trough and Mariana basin, but they failed to explain relatively quick changes of extension to compression.

Compression in back-arc regions occurs when plates are coupled (22). It is concentrated along the boundary of an island arc with an inactive marginal basin. Compression can be seen structurally either in reverse faults or in strikeslip faults. Reverse faults are known in the rear of the Kuril and Japan arcs where the Pacific plate is underthrusting orthogonally to the arc strike. The strike-slip fault pattern is typical of the Aleutian arc due to oblique subduction. Compressional conditions appeared in the rear of the Japan arc comparatively recently, 5 to 7 my ago following extensional conditions (27). The change in tectonic environment was accompanied by the corresponding change in directions of faults, dyke swarms and volcanic chains.

It should be emphasized that the marginal basin opening is not a continuous process, but goes through discrete time episodes each of several my long. A period of the basin opening is followed by compression when an island arc is overthrusted on to marginal basin floor. Extentional and compressional episodes can be repeated several times during the evolution of the same island arc/marginal basin system.

A structural pattern of the inner part of the transitional zone between the arc-trench system and the Asia continent is determined by interaction of the small Sea of Okhotsk, Amurian and Philippine plates and by interaction of these small plates with the large Eurasian and North American plates (Fig. 1). Now the Eurasian and North American plates are moving apart north of their pole of rotation at 61,9°N and 143,1°E, and converging south of the pole.Accordingly, rift-type grabens are located in the Chersky Ridge along the northern segment of the Eurasia/North America plate boundary. Generally, strike-slip faults occur south of the pole position (26). The Sea of Okhotsk plate is placed in "scissors" between the North American and Eurasian plates and convergence of these two plates induces the clockwise rotation of the Sea of Okhotsk plate around the pole at 50,4°N and 145,3°E near the western boundary of the plate (28). According to Okhotsk plate motion, we have recent thrusts in southern Sakhalin and dextral strike-slip displacements along the faults in northern Sakhalin. The Amurian plate is rotated relative to the Eurasian plate around the pole 56,9°N and 117,4°E. So, to the west of the pole position, plates are moving apart and grabens of Baikal Rift zone are originated (26).

OLIGOCENE PALEOGEODYNAMIC RECONSTRUCTION (FIG. 2)

The recent pattern of the western Pacific volcanic arcs and marginal seas began to form in the Late Oligocene, when the Kuril-Kamchatka, Japan, Izu-Bonin and Mariana arcs were originated. At that time some volcanic arc already existed in place of the Bowers Ridge in the Bering Sea. In the Oligocene the Aleutian arc has separated the Mesozoic floor of the Bering Sea from the Pacific ocean. The oceanic floor of the Okhotsk and Japan Seas and Komandor Basin of the Bering Sea was accreted at the time interval from 38 to 20 my ago. Direct evidence for the floor age are known only for the Komandor Basin where tholeiitic basalts were drilled with the isotopic age equal to 29,6 my (29).

The age of the Japan and Kuril basin floor can be inferred only from indirect evidence. Seismic reflection data show a sedimentary thickness of about 3000 m. If the minimum present accumulation rate established for the upper 300 m in the Japan Sea equals to 0.1 to 0.2 mm per year is taken the age of the floor is no more than 30 my old.

Judging from the magnetic anomaly pattern in the Caroline Basin and the Western Philippine Sea (30), the spreading axis existed in the Western Pacific at the early Tertiary. It served as the accretional boundary between the Pacific plate, to the north, and Caroline plate, to the south. The spreading centre coincided with the present Central Philippine Fault in the Western Philippine basin and was linked through a transform fault with the Caroline basin spreading center. It can be supposed that the Caroline plate floor which was much younger than the surrounding Pacific floor of the Mesozoic was created due to back-arc spreading in the rear of the Melanesian arc prograding southwards. The spreading axis became extinct in the Late Oligocene, and later, the Mariana-Jap-Palau arc-interarc basin system divided the previously single Western Philippine and Caroline basins.

CONCLUSIONS

The present geodynamic of the Asia/Pacific transitional zone is mainly related with interactions between the underthrusting Pacific plate, overriding small plates and island arc blocks. Conditions of plate decoupling and coupling are intermittent along the transitional zone and during its evolution, and accordingly extentional and compressional conditions gave way to one another. This is a reason for development of strongly different and contrasting geodynamic feature, such as trenches, arc and back-arc basins. The Asia continent and the Pacific floor are constantly converged during the Cenozoic and a strip of ocean floor of no less than 2000 km wide was consumed under the West Pacific trenches since the Oligocene (Fig. 2). On the whole, despite of plate convergence, Asia was moving apart from the Western Pacific subduction system and this, probably, has been and is now accounted for marginal sea opening.

Origination of the present-day Western Pacific island arcs and marginal basins in the Oligocene coincides with the reorganization in plate motion due to Eurasia/India collision and with the change of the Pacific plate motion from northward to west-west-northward as it is clearly seen from the bend of the Hawaiian hot spot track.

REFERENCES

1. Galushkin Yu.I., Ushakov S.A. Instantaneous kinematic of relative motion of the lithospheric plates of the Earth. In: Structure of the lithospheric plates. Moscow, 1979, p. 27-53.
2. Gnibidenko G.C. Tectonics of the back-arc floor of the Far East. Moscow, Nauka, 1979, 160 p.

3. Sychev P.M. Deep and surface tectonic processes of the north-western Pacific mobile belt. Moscow, Nauka, 1979, 208 p.
4. Structure of the crust and upper mantle in the Asia/Pacific transitional zone. Novosibirsk, Nauka, 1976, 367 p.
5. Erlikh E.H. Recent structure and quaternary volcanism of the West Pacific belt. Novosibirsk, Nauka, 1973, 243 p.
6. The geophysics of the Pacific ocean floor and its margin. Ed. by Sutton, G.H., Manghnani M.H. and Moberly R., Am. Geophys. Union, Washington, 1976.
7. Island arcs, deep-sea trenches and back-arc basins. Ed. by Talwani M. and Pitman III W.C., Am. Geophys. Union, Washington, 1977.
8. Geodynamics of the Western Pacific. Ed. by Uyeda S., Murphy B.W. and Kobaysni K., Tokyo, 1979.
9. Isacks B.L. and Barazangi M. Geometry of Benioff zones: lateral segmentation and downwards bending of the subducted lithosphere. In: Island arcs, deep-sea trenches and back-arc basins. Am. Geophys. Union, Washington, 1977, p. 99-114.
10. Kanamori H. Tectonic implication of the 1944 Tonankai and the 1946 Nankaido earthquakes. Phys. Earth Planet. Interiors, 1972, v. 5, p. 129-139.
11. Cormier V. Tectonics near the junction of the Aleutian and Kurile-Kamchatka arcs and the mechanism for middle Tertiary magmatism in the Kamchatka basin. Geol. Soc. Am. Bull., 1975, v. 86, p. 443-459.
12. Kanamori H. Seismic and aseismic slip along subduction zones and their tectonic implication. In: Island arcs, deep-sea trenches and back-arc basins. Am. Geophys. Union, Washington, 1977, p. 163-174.
13. Ichikawa M. Reanalyses of mechanism of earthquakes which occurred in and near Japan and statistical studies on the nodal plane solutions obtained, 1926-1968. Geophys. Mag., 1971, v. 35, p. 207-273.
14. Stauder W. and Mualchin L. Fault motion in the larger earthquakes of the Kurile-Kamchatka arc and of the Kurile-Kamchatka corner. Journ. Geophys. Res., 1976, v. 81, p. 297.

15. Hussong D.M. and Uyeda S. Tectonic processes and the history of the Mariana arc: a synthesis of the results of Deep Sea Drilling Project Leg 60. In: Initial Report of the Deep Sea Drilling Project, vol. LX, 1982, p. 909-925.
16. Mogi K. Relationship between the occurrence of great erathquakes and tectonic structures. Bull. Earthquake Inst. Tokyo Univ., 1969, v. 47, p. 429-451.
17. Stauder W. Fault motion and spatially bounded character of earthquakes in the Amchitka Pass and Delarof Island. Journ. Geophys. Res., 1972, v. 77, p. 2072-2080.
18. Fedotov S.A. Earthquake prediction for Kurile-Kamchatka zone. In: Problem of tsunami. Moscow, Nauka, 1968, p. 121-132.
19. Utsu T. Space-time pattern of large earthquakes occurring off the Pacific coast of the Japanese Islands. Journ. Phys. Earth, 1974, v. 22, p. 325.
20. Kelleher J., Sykes L.R. and Oliver J. Possible criteria for predicting earthquake locations and their application to major plate boundaries of the Pacific and the Caribbean. Journ. Geophys. Res., 1973, v. 78, p. 2547-2586.
21. Lobkovsky L.I. and Baranov B.V. About generation tsunami in subduction zones. In: Generation and propagation of tsunami. Moscow, Institute of Oceanology, 1982.
22. Uyeda S. and Kanamori H. Back-arc opening and the mode of subduction. Journ. Geophys. Res., 1979, v. 84., p. 1049-1061.
23. Lee C., Shor G.G., Bibee L.D., Lu R.S. and Hilde T.W.C. Okinawa trough: origin of a back-arc basin. Marine Geology, 1980, v. 35, p. 219-241.
24. Bibee L.D., Shor G.G. and Lu R.S. Inter-arc spreading in the Mariana trough. Marine Geology, 1980, v. 35, p. 183-197.
25. Molnar P. and Atwater T. Interarc spreading and cordillerian tectonics as alternates related to the age of subducted oceanic lithosphere. Earth Planet. Sci. Letter, 1978, v. 41, p. 330-340.
26. Zonenshain L.P. and Savostin L.A. Introduction to

geodynamics. Moscow, Nedra, 1979, 310 pl.
27. Nakamura K. and Uyeda S. Stress gradient in arc-back-arc regions and plate subduction. Journ. Geophys. Res. 1980, v. 85, p. 6419-6428.
28. Savostin L.A., Verjbitskay A.I. and Baranov B.V. Recent plate tectonics of the Sea of Okhotsk region. Dokl. Akad. Nauk USSR, 1982. v. 266, N4.
29. Stewart R.J.,Natland J.H. and Glassley W.R. Petrology of volcanic rocks recovered on DSDP Leg 19 from the North Pacific ocean and the Bering Sea. In: Initial Report of the Deep See Drilling Project, v. XIX, 1973, p. 615-628.
30. Watts A.B., Weissel J.K. and Larson R.L. Sea-floor spreading in marginal basins of the Western Pacific. Tectono-physics, 1977, v. 37, p. 167-181.

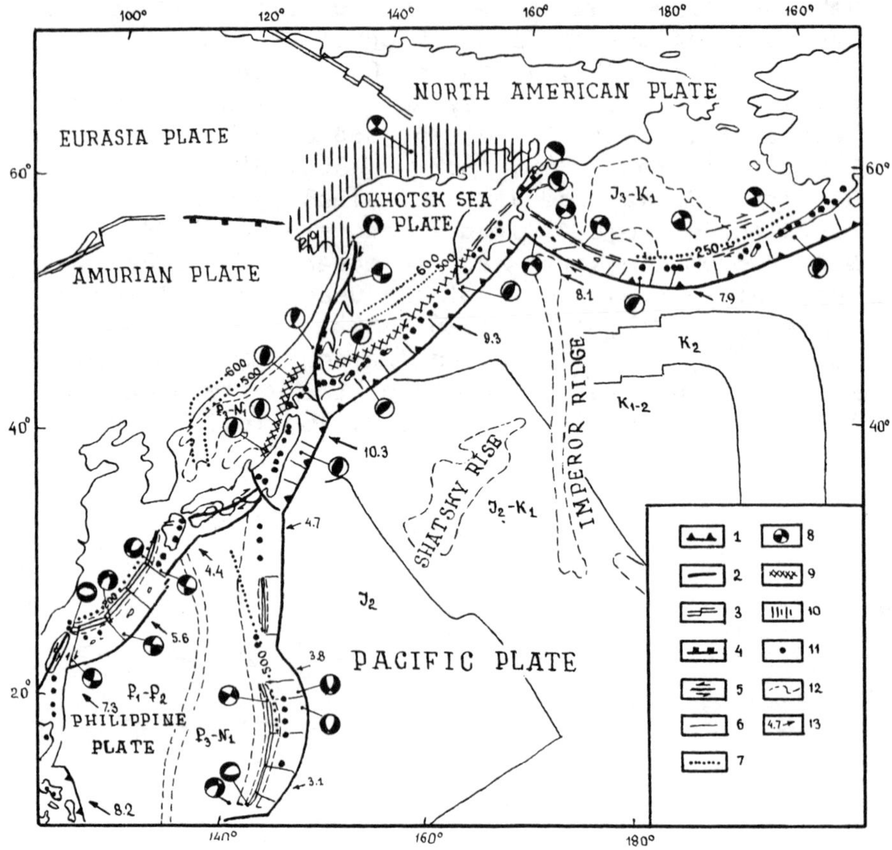

FIGURE 1. Recent geodynamics of the transitional zone of the north-western Pacific.
1-2 - consumption boundaries: 1. - with coupling, 2 - with decoupling; 3 - accredtion boundaries; 4 - collission boundaries; 5 - transform faults; 6 - faults, mainly, strike-slip; 7 - depths of Benioff zones, km; 8 - focal mechanism; 9 - additional seismic zone in the rear part of the island arc; 10 - zone of diffuse seismicity on continent; 11 - active volcanoes; 12 - deep basins in the marginal sea and rises in the ocean; 13 - relative motion vector and rate in cm/year.

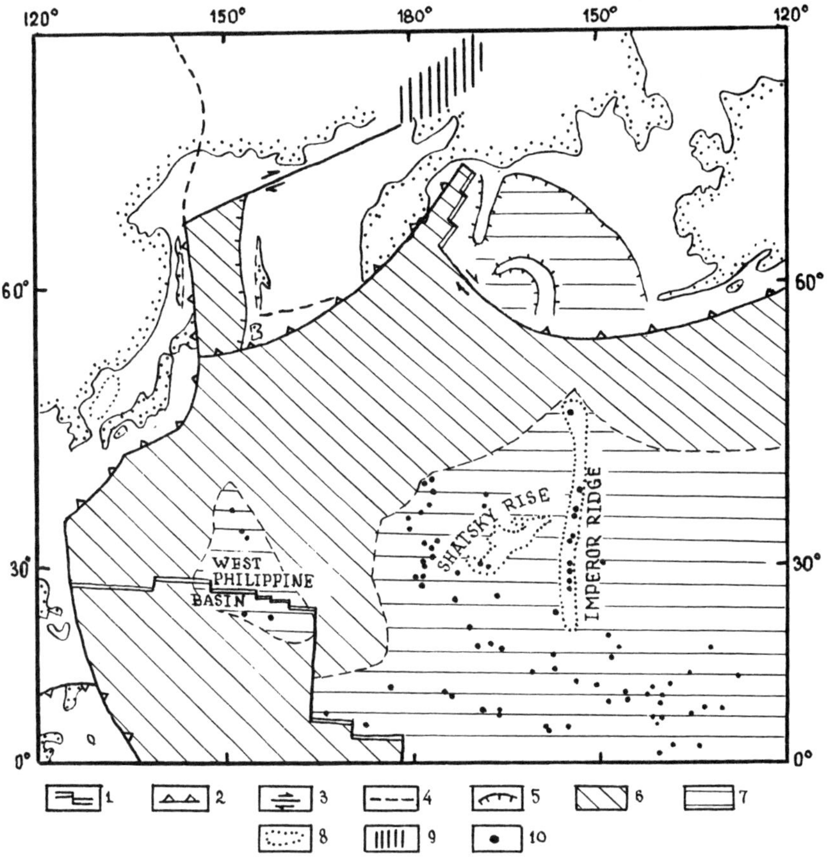

FIGURE 2. Paleogeodynamic reconstruction of the lithospheric plates of the north-western Pacific in the Oligocene. It is based on the reconstruction by L.P. Zonenshain and L.A. Savostin (26). 1-4 - plate boundaries: 1 - accretion, 2 - consumption, 3 - strike-slip, 4 - unknown; 5 - boundary between the oceanic and continental type of the earth crust; 6-7 - oceanic crust of the Pacific ocean: 6 - consumed now in subduction zones, 7 - nonconsumed now in subduction zones; 8 - ridges and rises into Pacific ocean and marginal seas; 9 - deformation zone on continent; 10 - site of deep sea drilling.

ABOUT SOME ASPECTS -SO CALLED CLASSICAL- OF THE PERI-PACIFIC TECTONICS.

AUBOUIN Jean

Université Pierre et Marie Curie, Laboratoire de Géotectonique, Tour 26-0 - 1er Etage, 4 Place Jussieu, 75230 PARIS Cédex 05, FRANCE.

ABSTRACT. - Some Classical aspects of the peripacific tetonics are discussed here.

The present pacific dissymetry with Cordilleras on the East side and Islands Arcs on the west side, could have been inverse during Mesozoic. The actuel marginal seas are not older than the Tertiary : and, for instance, during the Jurassic there was a franciscan marginal sea on the East whereas Japan was part of the asiatic continent.

If the Andes Cordillera -between Huancabamba and Bariloche transverse zones- seems to correspond to a continuous subduction during Mesozoic and Tertiary, in other parts subduction ended into collision with islands arcs at different times. These collision belong either to "Taïwan type" when the island arc is overthrusted onto the continent or to the "Californian type" when the island arc is underthrusted beneath the continental margin.

The tectonic belts older than the present subduction zone show generally a continental vergency as in the "Periaustralian Ophiolitic Crown" from New Guinea to northern New Zeeland through New Caledonia, in the Phillipina-Taïwan belt, in the Hokkaïdo-Sakhalin belt (in a certain way there is a "Periasiatic Ophiolitic Belt"

Proceedings of the 27th International
Geological Congress, Volume 7, pp. 93–107
TECTONICS

with continental vergency). Even in western North America there are evidences of an eastern continental vergency for the ophiolitic thrusts during Nevadian orogeny. As a consequence, collision of Taïwan type seems the main tectonic process and marginal seas such as Tasman sea or south China sea are the main model for paleogeographic interpretations.

The big strike-slip motions conducting to "collages" as proposed in western North America have to be examined in the light of their transverse components. What it the part of "charriages" in these collages which are, anyway, examples of collision ?

In some parts as in Northern Chili or North-western Mexico, the mesozoic basins, as the present Coast, cut obliquely the paleozoic tectonic belts. Were these regions the margins of a new Pacific born in Triasic-Jurassic time as southernmost Atlantic for the first or westernmost Tethys for the second ?

SUR QUELQUES ASPECTS, SOIT DISANT CLASSIQUES, DE LA TECTONIQUE PERIPACIFIQUE.

La ceinture orogénique péri-pacifique forme, avec la ceinture téthysienne, l'un des deux ensembles orogéniques récents du Monde ; si l'on excepte le dispositif qui entoure la Mer de la Scotia entre l'Amérique du Sud et la péninsule antarctique.

Si la ceinture téthysienne est le lieu d'élection de la collision entre les continents, la ceinture péri-pacifique donne l'exemple de la subduction ; c'est d'ailleurs là que ce phénomène majeur a été découvert ; c'est là qu'il a été précisé par différentes campagnes de sismique ou même de forage dans le cadre du Programme International de Forage Océanique Profond (IPOD). C'est donc une ceinture exemplaire à bien des égards.

1. DE LA DISSYMETRIE DU PACIFIQUE.

Le trait le plus apparent est la dissymétrie du Pacifique : Cordillères ouest-américaines à l'Est ; Arcs insulaires péri-asiatiques et péri-australiens à l'Ouest, avec leurs mers marginales associées.

1a. Il est clair que cette dissymétrie tient à une subduction beaucoup plus active au front des Amériques qu'au long de l'Asie : la croûte océanique qui fait face au Japon est jurassique, tandis qu'il n'y a plus de croûte océanique plus ancienne que le Tertiare face aux Amériques (et même en certains endroits la croûte est sub-actuelle comme au large du Canada entre la ride Juan de Fuca et le continent). Le fait est encore plus marqué pour l'Amérique du Nord que pour l'Amérique du Sud : l'Amérique du Nord s'est avancée jusqu'à subducter la ride est-pacifique elle-même dont la tectonique extensive entaille le bord nord-américain depuis 3,5 millions d'années, de l'ouverture du golfe de Basse Californie au complexe de failles de San Andreas.

Ce fait qui avait été déjà intrigué A. WEGENER (1), réclame encore une explication définitive. Une interprétation plausible est celle de l'"avance tectonique" des continents américains (2), liée à l'ouverture de l'Atlantique, avance diachronique, très marquée pour l'Amérique du Nord enfin séparée de l'Eurasie au Tertiaire, par rapport à l'Amérique du Sud déjà séparée du Gondwana au Crétacé inférieur.

1b. Or aucune des mers marginales de l'Ouest Pacifique n'est antérieure au Tertiaire, si l'on excepte la mer de Tasman née à l'extrême fin du Crétacé (Campanien-Maestrichtien) (3).

On peut donc se poser la question de la pérénnité de l'actuelle dissymétrie : qu'en était-il avant le Tertiaire, au Mésozoïque ou du moins à certaines époques du Mésozoïque ?

Tout d'abord, pour restituer la paléogéographie mésozoïque dans l'ouest Pacifique, il faut rapporter les actuels arcs insulaires à la marge continentale dont ils sont détachés ; ce qui reconstitue une marge continentale franche. On a trop pris l'habitude de considérer que les ceintures orogéniques, métamorphiques, plutoniques, volcaniques du Japon, par exemple, se sont formées à toutes les époques dans une position d'arcs insulaires. Or, pour prendre les célèbres ceintures métamorphiques de Sanbagawa et de Ryoke d'âge Jurassique supérieur-Crétacé (4), il faut penser qu'alors le Japon était jointif avec le continent asiatique qui forme d'ailleurs son substratum (la zone de Hida est constitutée de terrains précambriens-paléozoïques apparentés à ceux de Corée). Le dispositif était celui de l'actuelle Californie, avec un Coast Range japonais à ophiolites et schistes bleus, et une Cordillère chinoise traversée de granites dont les plus occidentaux se rencontrent jusqu'à Pékin.

Ensuite, dans l'Est Pacifique, la Californie n'a pas toujours été la marge du continent américain : bien des arguments montrent que les ensembles franciscains sont issus d'une mer marginale mésozoîque dont les terrains ont été charriés sur un bloc continental entré en collision avec la marge continentale américaine à la faveur de la subduction californienne aujourd'hui éteinte (5). L'image Jurassique-Crétacé de la Californie est ainsi plus proche de l'actuel ensemble Asie-Mer du Japon-Japon que de la Californie actuelle.

Si donc on se place à la fin du Jurassique, moment où se sont édifiées des ceintures métamorphiques de haute pression-basse température tant au Japon qu'en Californie, la dissymétrie du Pacifique était probablement inverse de l'actuelle : Cordillère et Coast Range étaient à l'Ouest en Chine et au Japon, arcs insulaires et mers marginales à l'Est en Amérique du Nord.

On ne saurait généraliser cette remarque : il est bien certain, par exemple, que la subduction au large de la Cordillère des Andes était, au cours du Secondaire et du Tertiaire, dans la situation qu'elle occupe actuellement à peu de chose près : rien n'indique l'existence de mers marginales dans l'histoire des Andes. Du moins de la transversale de Huancabamba au Nord à la transversale de Bariloche au Sud ; car, dans les Andes septentrionales de Colombie au Nord, et les Andes méridionales de Patagonie au Sud, le dispositif de mers marginales se rétablit, d'où sont nées des nappes ophiolitiques charriées sur le continent sud-américain (6).

2. DES COLLISIONS DANS LA CEINTURE PERI-PACIFIQUE.

Le Pacifique est généralament considéré comme le site exemplaire des subductions actuelles et passées.

Si l'en est bien ainsi aujourd'hui et si dans le passé les subductions furent incontestablement actives, les collisions n'ont pas manqué. La plus remarquable est celle de l'Australie avec l'Indonésie, dont la limite se place actuellement au niveau de l'île de Sumba, à l'Ouest de laquelle se poursuit la subduction de l'Océan Indien, tandis qu'à l'Est la collision avec le continent australien est réalisée (7). La collision de Taïwan avec le continent asiatique est non moins remarquable, en prolongation de la subduction de la fosse de Manille (8).

Et, si la subduction andine parait avoir été continue, c'est une collision avec une élément continental qui est probablement responsable de la formation du Coast Range de Californie (5).

Le Pacifique donne ainsi l'exemple de deux types de collision (9) : la "**collision à la taïwanaise**", où l'élément insulaire passe par dessus la marge du continent principale ; la "**la collision à la californienne**", où l'élément insulaire passe en dessous de la marge du continent principal.

L'importance de ces collisions peut être beaucoup plus grande qu'on ne l'admet généralement : par exemple, la suture ophiolitique du Japon septentrional (Hokkaïdo), qui pase à Sakhaline et en Sibérie orientale, marque probablement la suture fondamentale entre l'Eursie et l'Amérique (10) (à laquelle appartiendrait ainsi la Sibérie orientale).

3. DES VERGENCES TECTONIQUES.

En extrapolation du modèle de la subduction, il est généralement admis que les édifices tectoniques péri-pacifiques ont une vergence océanique : la référence est trouvée dans le Coast Range de Californie ou le Japon méridional, dont les unités tectoniques chevauchent effectivement vers l'Océan. Ainsi les chaînes péri-pacifiques sont-elles interprétes comme dérivées de plus ou moins vastes prismes d'accrétion (11).

Cependant, il est loin d'en être toujours ainsi au cours du temps. Ainsi, la Couronne ophiolitique péri-australienne (12), qui va de la Nouvelle Guinée à la Nouvelle Zélande par la Nouvelle Calédonie, est-elle charriée vers le continent australien, de l'Eocène au

Miocène inférieur. Ainsi, les nappes de Taïwan avec leurs ophiolites sont-elles charriées vers le continent asiatique, du Miocène au Pliocène (13). Ainsi, les nappes ophiolitiques du Japon septentrional (Hokkaïdo) et de Sakhaline sont-elles charriées vers le continent asiatique (9).

Comme il y a une **Couronne ophiolitique péri-australienne** charriée vers l'Australie, il semble y avoir une **Ceinture ophiolitique péri-asiatique** charriée vers l'Asie (9) ; sans que ces ceintures correspondent à un édifice unique mis en place lors de mêmes phases tectoniques ; mais la fréquene de telles structures est suffisante pour être notée. Ce sont les structures liées aux subductions les plus récentes qui sont déversées vers le Pacifique.

De la même manière, dans le Coast Range californien et son prolongement septentrional, de nombreux indices existent d'une tectonique précoce à vergence continentale (14) : les nappes névadiennes mises en place à la fin du Jurassique sont charriées vers le continent ; ce sont des cisaillements ultérieurs qui les ont repris en vastes charriages vers l'Ouest. Là encore, seules les structures (relativement) récentes ont une vergence pacifique.

Sans qu'on puisse généraliser encore, il semble bien, cependant, que le **dispositif principal est le charriage océanique sur la marge continentale**, et non pas l'inverse. Cela tient peut être au fait que le destin normal de l'océan principal est de retourner au manteau supérieur dans les zones de subduction, sans y laisser de traces, comme le modèle de la subduction actuelle dans la fosse d'Amérique centrale en donne l'exemple (15). Tandis que les édifices tectoniques résulteraient toujours d'une collision avec un obstacle continental -arc insulaire, fragment de continent-qui, le plus généralement, chevauche sur la marge du continent principal : la **"col-**

lision à la taïwanaise" serait ainsi le modèle principal de collision, la "collision à la californienne" n'étant qu'un accident relativement rare.

Quant à la "**collision à la téthysienne**" qui est celle de vastes continents, elle montre toujours des nappes ophiolitiques charriées sur l'un des continents (16). Il y aurait donc, du Pacifique à la Téthys, une certaine unité dans la vergence continentale des structures que les apparences actuelles avaient quelque peu masquée.

4. DES COULISSAGES.

Sur l'exemple de la faille de San Andreas, de vastes coulissages ont été conçus autour du Pacifique, dont la démonstration a été variable. Récemment, les mesures de paléoinclinaisons magnétiques (17) ont apporté un regain d'arguments à de tels coulissages ; l'Ouest de l'Amérique du Nord peut apparaître comme résultant de "collages" successifs de lames coulissées du Sud vers le Nord (18). Bien qu'il reste beaucoup à démontrer, notamment sur l'ampleur de ces coulissages, le faisceau d'arguments parait convaincant.

Ces coulissages se comprennent dans le cadre du Pacifique mésozoïque dans lequel la plaque de Kula avait un mouvement de direction Nord-Nord Ouest, ainsi sub-parallèle aux marges américaines. Par contre, depuis la limite Eocène moyen-Eocène supérieur (40 MA), moment où la ride est-pacifique a pris une position annonciatrice de l'actuelle, les mouvements de la plaque pacifique, plus proche d'une direction Ouest-Est, rendent problématiques de tels coulissages. Ce n'est que récemment, lorsque le bord Sud-Ouest du continent nord-américain a rencontré la ride est-pacifique que les coulissages se sont de nouveau produit le long du système de failles de San Andreas, qui peut apparaître comme transformant de la ride est-pacifique à la ride Juan de Fuca (19).

Si cela est vrai, il s'ensuit que de tels déplacements se sont produits aussi dans l'Ouest Pacifique au Mésozoïque : déjà, des indices en ce sens ont été recueillis au Japon méridional (20) ; et l'on pourrait considérer que la suture ophiolitique du Japon septentrional (Hokkaïdo), pratiquement Nord-Sud, a une composante en coulissage très imporante.

Mais beaucoup reste à faire dans ce domaine : car, si les paléolatitudes peuvent être déterminées, il n'en est pas de même des paléolongitudes qui restent inconnues. On peut rechercher l'origine des blocs collés à la marge nord-américaine soit en Amérique du Sud par une coulissage pur ou presque pur, soit dans le Pacifique sud-ouest par un mouvement d'extension dont le coulissage n'est plus qu'une composante mineure (21).

Et de toute façon, **quelle que soit l'importance des coulissages, des collages, des charriages qui en résultent, tout se résout, en fin de subduction, par des collisions**.

5. DE L'ANCIENNETE DU PACIFIQUE.

C'est une idée classique que le Pacifique est un océan ancien sinon permanent. Certes, on n'y trouve pas de fonds océaniques plus anciens que le Jurassique, comme d'ailleurs dans les autres océans (22) ; mais ceci ne saurait être considéré comme un argument, car on peut admettre que la croûte océanique plus ancienne a été subductée sous les marges continentales du Pacifiques.

Cependant, le passage du Paléozoïque au Mésozoïque ne se fait pas partout en continuité : il y a une discordance post-hercynienne en beaucoup d'endroits d'Amérique du Sud et d'Amérique du Nord. Localement, les limites du Pacifique triasique sont sécantes par rapport aux structures hercyniennes des continents américains, tout comme les limites de l'océan atlantique. Ainsi en est-il, au niveau du Chili septentrional, où la chaîne hercynienne, qui traverse le continent sud-américain depuis la province de Buenos Aires jusqu'au Pacifique, se trouve recoupée à plus de 45° par les limites triasiques (23). Ainsi en est-il, au niveau Mexique, où les structures des Appalaches décalées dans les Monts Ouachita et le rift du Rio Grande, butent perpendiculairement sur l'édifice des Sierra Madre comme sur la côte pacifique actuelle (24). Ces deux exemples montrent que, au moins localement, le Pacifique parait naître d'une rupture continentale.

Il faut, certes, tenir compte des coulissages qui se sont produits au cours du Mésozoïque et qui ont débité la marge continentale en lanières. On pourrait cependant voir dans ces régions les marges d'un nouveau Pacifique, s'ajoutant à l'ancien, né d'une distension triasico-liasique, comme la Téthys occidentale pour le Nord-Ouest du Mexique ou comme l'Atlantique austral pour l'exemple du Nord du Chili. Il s'agirait d'une marge téthysienne originelle dans le premier cas, tandis que l'Amérique australe ne serait qu'une "pièce rapportée" au continent sud-américain par l'orogenèse hercynienne.

CONCLUSIONS.

Il apparait donc que bien des traits les plus classiques du Pacifique, doivent être replacés dans le cadre spécifique de l'époque actuelle ; ou du moins du Tertiaire. Ainsi en est-il : de la dissymétrie du

Pacifique, peut être inverse au cours du Mésozoïque ; des vergences tectoniques plus fréquemment vers le continent que vers l'océan ; des subductions, qui bien souvent furent bloquées par les collisions dont sont nées les ceintures ophiolitiques péripacifiques.

En ceci -par exemple, pour les vergence tectoniques des nappes ophiolitiques- le Pacifique parait moins original qu'on ne pourrait le croire par rapport à la Téthys ; laquelle devait avoir des caractères pacifiques avant que la collision de l'Eurasie et des continents issus du Gondwana ne les masque ; et vice et versa.

D'ailleurs, le Sud-Ouest Pacifique est l'endroit où la Téthys rejoint le Pacifique dont elle est née (25) : on ne s'étonnera pas d'y voir les nappes ophiolitiques y chevaucher vers les continents, comme il est de règle dans la Téthys elle-même.

(1) WEGENER A. (1915-22). - Die Entstehung der Kontinente un Ozeane.
Braunschweig éd., 1ère, 2ème, 3ème éditions, 1 vol.

(2) CONEY P.J. (1970). - The geotectonic cycle and the new global tectonics.
Geol.Soc.America Bull., 81, p.739-748.

(3) MOLNAR P., ATWATER T., MAMMERICKX J. and SOUTH S.M. (1974). -Magnetic anomalies, Bathymetry and the tectonic evolution of the south Pacific since the Late Cretaceous.
Geophys.J.R.Astron.Soc., 40, p.383-420.

(4) MIYASHIRO A. (1961). - Evolution of metamorphic belts.
Jour. of Petrology, vol.2, p.277-311.

(5) HSU K.J. (1971). - Franciscan melange as a model for eugeosynclinal sedimentation and underthrusting tectonic.
J.Geophys.Res., 76, p.1162-1170.

(6) AUBOUIN J. (1975). - Réflexion sur les bordures pacifiques : l'exemple des Cordillères sud-américaines.
C.R.Acad.Sc., Paris, sér.D, t.280, p.99-102.

DALZIEL I.W.D. and PALMER K.F. (1979). -Progressive deformation and orogenic uplift at the southern extremity of the Andes.
Geol.Soc.America Bull., Part I, v.90, n°3, p.259-280.

BOURGOIS J., CALLE B., TOURNON J. and TOUSSAINT J.F. (1982). - The Andean ophiolitic megastructures on the Buga-Buenaventura transverse (Western Cordillera - Colombia).
Tectonophysics, 82, p.207-229.

(7) AUDLEY-CHARLES M.G., LATREILLE M. and NAYOAN G.A.S (1980). - Comparison of orogenic features of the Tethyan mountain belt from Himalaya to the East Asian Arcs.
In : Geology of the alpine chains born of the Tethys, J. AUBOUIN, J. DEBELMAS, M. LATREILLE ed., 26th I.G.C. Paris, p.155-165, BRGM, Orléans.

(8) BOWIN C., LU R.S., LEE C.S. and SCHOUTEN H. (1978). -Plate convergence and Accretion in Taïwan-Luzon Region.
A.A.P.G., 62, 9, p.1645-1672.

(9) AUBOUIN J. (1982). - Vergence océanique et vergence continentale dans l'Ouest Pacifique : subductions, mers marginales, orogènes anciens. **C.R.Acad.Sc.**, Paris, sér.II, t.294, p.285-290.

(10) CADET J.P., JOLIVET L., LALEVEE F. and AUBOUIN J. (1983). - Collision of Eurasian and North American plates in Northern Japan. **Eur.Union.Geos.**, 2nd meet., Strasbourg.

(11) KARIG D.E. (1971). - Origine and developpement of marginal basins in the Western Pacific. **J.Geophys.Research**, 76, p.2542-2561.

DICKINSON W.R. and SEELY D.R. (1979). - Structure and stratigraphy of forearc regions. **A.A.P.G.**, vol.63, n°1, p.2-31.

SEELY D.R. (1979). - The evolution of structural highs bordering major Forearc Basins. **In** : WATKINS, MONTADERT, WOOD-DICKERSON, Geological and Geophysical investigations of continental margins. **A.A.P.G. Memoir**, n°29, p.245-260.

(12) AUBOUIN J., MATTAUER M. et ALLEGRE C. (1977). - La couronne ophiolitique périaustralienne : un charriage océanique représentatif des stades précoces de l'évolution alpine. **C.R.Acad.Sc.**, Paris, sér.D, t.285, p.953-956.

(13) HO C.S. (1979). - Geologic and Tectonic framework of Taïwan. **Mem.geol.Soc.China**, Taipei, n°3, p.57-72.

(14) ROURE F. (1982). - Mise en évidence d'une tectonique majeure du Jurassique supérieur (phase névadienne) dans le Nord-Est de l'Orégon

(secteur d'Hutington).
C.R.Acad.Sc., Paris, sér.II, t.294, p.921-925.

MATTAUER M., COLLOT B. and Van den DRIESSCHE J. (1983). - Alpine model for the internal metamorphic zones of the North Amerian Cordillera.
Geology, vol.11, 1, p.11-15.

(15) AUBOUIN J., BOURGOIS J., Von HUENE R. et AZEMA J. (1982). - La marge pacifique du Guatemala : un modèle de marge extensive en domaine convergent.
C.R.Acad.Sc., Paris, sér.II, t.295, p.607-614.

(16) AUBOUIN J. (1980). - De la tectonique des plaques à la genèse des chaînes de montagnes.
In : Livre Jubilaire de la Société Géologique de France, **Mém.h.-sér.Soc.géol.Fr.**, n°10,- p.163-180.

(17) IRVING E. (1979). - Paleopoles and Paleolatitudes of North America and speculations about displaced terrains.
Can.Journ.Earth Sci., vol.16, p.669-694.

(18) DAVIS G.A., MONGER J.W.H. and BURCHFLIED B.C. (1979) . - Mesozoic construction of the Cordilleran "collage", central British Columbia to Central California.
In : Howell and Mc Dougall ed., Pacific coast Paleogeography symposium, SEPM, vol.2, p.1-31.

(19) ATWATER T. (1970). - Implications of plate tectonics for the Cenozoic tectonics of western North America.
Geol.Soc.America Bull., 81, p.3513-3536.

(20) FAURE M. (1982). - Phase précoce Ouest-Est dans la zone de Sanbagawa dans la partie orientale de

Shikoku (Japon SW).
C.R.Acad.Sc., Paris, sér.II, t.295, p.505-510.

(21) NUR A and BEN AVRAHAM Z. (1977). - Lost Pacific continent.
Nature, 270, p.41-43.

BEN-AVRAHAM Z., NUR A., JONES D. and COX A. (1981). -Continental Accretion : from oceanic plateaus to allochtonous terranes.
Science, 213, p.47-54.

(22) PITMAN W.C., LARSON R.L. and HERRON E.M. (1974). -The age of oceans basins.
Geol.Soc.America., Map and chart MC6.

(23) AUBOUIN J., BORRELLO A.V., CECIONI G., CHARRIER R., CHOTIN P., FRUTOS J., THIELE R. et VICENTE J.C. (1973). - Esquisse paléogéographique et structurale des Andes méridionales.
Rev.Géogr.Phys.Géol.Dyn., (2), XV, fasc.1-2, p.11-72.

(24) CORDOBA D.A., TARDY M., CARFANTAN J.C., CAMPA M.F. et RANGIN C. (1980). - Le Mexique mésogéen et le passage au système cordillerain de type Californie.
In : J. AUBOUIN, J. DEBELMAS, M. LATREILLE, Les chaînes alpines issues de la Téthys, 26ème Cong-.Géol.Intern.Paris, Coll. C5, BRGM éd.,p.18-29.

(25) AUBOUIN J., BLANCHET R., STEPHAN J.F. et TARDY M. (1977). - Téthys (Mésogée) et Atlantique : données de la géologie.
C.R.Acad.Sc., Paris, D, t.285, p.1025-1028.

TECTONICS OF INNER-CONTINENTAL FOLDBELTS

ZONENSHAIN L.P.
Institute of Oceanology, USSR Academy of Sci., Moscow, USSR

Foldbelts are formed along consumption plate boundaries on the place of previous oceans and their margins due to progressive convergence and eventual collision of continents. They can be divided into those (1) where continental collision is already finished, examples are the Paleozoic Central Asiatic, Uralian, Appalachian-Caledonian belts, Mesozoic Verkhoyansk-Kolymian, Indo-Sinian belts; (2) where collision is continuing now, as in the Alpine-Hymalayan belt, Carribean Region, Indonesia; (3) where collision has not yet began, as in island arcs. This classification is somewhat conventional as it reflects an evolutionary succession supposing each inner-continental foldbelt goes through corresponding stages in its development from the marginal oceanic position to closure of an ocean and continental collision. Inner-continental foldbelts only will be considered below. Three belts which are known to the author from personal investigations are choosen as examples: the Uralian, Central Asiatic and Verkhoyansk-Kolymian belts.

THE URALIAN BELT

The mobilistic interpretations of the evolution and structure of the Urals are given in numerous publications (Hamilton, 1970; Ivanov et al., 1972; Peive et al., 1977; Rujentsev, 1976; Abdulin et al., 1977; Perfil'ev,1979; Puchkov, 1979; Korinevsky, 1983; Zonenshain et al., 1983). The uralian foldbelt has N-S orientation being 2000 km long and up to 400-500 km wide in the south where almost a whole section of the belt from the East European to Kazakhstan continent can be seen (Fig. 1). The belt consists of the metamorphic basement, or Pre-Uralides, formed before Ordovician and overlying allochtonous Uralidian

Proceedings of the 27th International
Geological Congress, Volume 7, pp. 109–130
TECTONICS

terrain aging from Ordovician to Upper Paleozoic. The Pre-Uralides correspond to the basement of continents and microcontinents the collision of which has created the foldbelt. They are outcropped in anticlinoria. The Uralides fill synclinoria which are mainly allochtonous synforms. As a rule, they consist of ophiolites below and island arc volcanics above indicating the Uralian structures were born from a Paleozoic ocean. Folding was completed in the latest Permian-Early Triassic.

The East European passive margin constitutes the western border of the belt. The sedimentary cover begins from Ordovician on the east and from Middle Devonian on the west and includes shelf limestone and continent derived clastics up to Lower Permian. Shelf sediments are changed eastwards by deep-water shales and cherts accumulated in continental rise environment (Puchkov, 1979). Bimodal volcanic series appear among Ordovician in the eastern part of the passive margin. They evidence for rifting and continental break-up when a large continental block, or Mugodjarian microcontinent, was splitted off from Europe giving way for the Uralian ocean opening (Fig. 2).

The thrust front is situated inside the passive margin domain. The Uralian foredeep filled by Permian molasse is stretched along the thrust front.

Outer fold zones of the Urals include the para-autochtonous, deformed and thrusted basement and cover of the European passive margin (Bashkirian anticlinorium and Zilairian synclinorium) and overlying allochtonous terrains of the Uralides (Sakmarian and Kraka allochtons). The allochtons are chaotical zones within which oceanic (ophiolite) and island arc complexes are mixed. Ophiolites are mainly of the Lower Silurian age. Island arc calc-alkaline volcanics belong to Silurian. First motion of nappes and emplacement of ophiolite occurred in the late Early Devonian when ophiolitoclastic clistostroma was formed (Rujentsev, 1976; Korinevsky, 1983). Ophiolite obduction seemed to be in back arc environment as a result of origination of a short-living underthrusting zone dipping from the continent under marginal basin oceanic floor. Next movements of thrusts are recorded in Late Carboniferous. Clastics derived from the east, i.e. from the Urals, reached firstly the eastern part of the passive margin in Late Devonian.

Inner zones of the Urals include (from west to east): (1) the Magnitogorsky synclinorium and its southern continuation in the West Mugodjarian synclinorium, (2) East Uralian and Mugodjarian anticlinoria, (3) East Uralian and Irgizan synclinoria, (4) Trans-Uralian anticlinorium, and (5) Oktyabrsko-Denisovsk zone (synclinorium). Ophiolites in the base of the Uralides have somewhere the Silurian age as in the marginal allochtons and in some places their Middle Devonian age is proved. Three complexes different in age can be distinguished among overlying island arc series: the Lower Devonian/Eifelian (Irendyk Suite), Givetian/Frasnian (Ulutau and Coltuban Suites) and Tournaisian/Lower Visean (Beresovskaya Suite) ones. Each of them marks the existence of an independent volcanic arc. Island arc complexes of different age distributed in separated strips being older on the west and younger on the east. All the three complexes overly one another in the easternmost East Uralian synclinorium. Each older (western) arc complex is overlain by deep water chert and turbidites derived from the younger, more eastern arc. The west Pacific pattern can be suggested for the evolution of the Urals in Devonian and Early Carboniferous with a Benioff zone dipping under East Europe, with arc splitting, interarc basin opening, arc advacement oceanwards and subsiding of remnant arcs left behind. Indeed, tholeiitic basalts with ocean affinities and bimodal basalt-rhyolitic series occur being contemporaneous with or slightly older than arc volcanics. They evidence for back arc spreading. The ocean floor of a back arc basin is reconstructed with certainty in the West Mugodjarian synclinorium where it was formed during a very short time interval of Early Givetian, no more than during 5 m.a. The ancient oceanic floor of the back arc basin consists here of layered gabbro, sheeted dykes and pillow lavas. A special study allowed to reconstruct a fossil spreading pattern, very much similar to the present one, with the spreading rate of 5 cm/year and lava extrusion depth of 3000 m (Zonenshain et al., 1983).

Volcanic activity was stopped in Early Visean and calm shallow water carbonate sedimentation took place in Late Visean, Namurian and Bashkirian. The cessation of island arc volcanism appeared to be related with movement of the Mugodjarian microcontinent from the east into the subduction zone. As a result of micro-continent/arc collision and continental crust under-thrusting, granite-gneissic domes began to rise from the mobilized subsided crust giving initiation for anticlinoria growth and high temperature metamorphism around. An absolute age of metamorphism equal to 370-310 m.a. (Milovsky et al., 1977) coincides with the time of cessation of arc volcanism.

A radical reorganization occurred in the middle of Visean after microcontinent/arc collision. A Benioff zone was overflipped eastwards (in present day coordinates) to the Kazakhstan margin and changed its inclination dipping, compared to the previous zone, in the opposite direction under the Kazakhstan continent. A new volcanic arc, i.e. the Valeryanovsk arc, was originated. It was active during the Late Visean, Namurian and, partly, Bashkirian time. The Oktyabrsko-Denisovsk zone situated between the Valeryanovsk arc and Mugodjarian anticlinorium consists of tectonic slices with Devonian and Early Carboniferous ophiolites and clastics, corresponding a forearc subduction wedge and subsequent collision suture.

The oceanic crust of the Uralian paleoocean was completely consumed under the Valeryanovsky arc before Late Carboniferous and then a direct Kazakhstan/East European continental collision began. As the latest Benioff zone was inclined under the Kazakhstan continent, the whole structure of the Uralian foldbelt acquired vergence toward the East European margin with a general movement of nappes in the same direction.

Fig. 2 shows the evolution of the Urals in Paleozoic.

The Uralian paleoocean existed in Ordovician, Silurian and Devonian being nearly 2000 km wide. Accretion of a new oceanic crust and its consumption in trenches occurred simultaneously at that time. This process continued about 100 m.a. Then, from the Middle Devonian time, closure of the ocean began with consumption of its crust during 70-80 m.a. Continental collision lasted from Late Carboniferous to Early Triassic, from 320 to 230 m.a., i.e. during 90 m.a.

Two main features of the Uralian evolution should be emphasized: (1) its development along the island arc pattern, on the place of island arcs and marginal basins, and (2) very long duration of a continental collision stage. Due to the first feature volcanics and related chalcophile mineralization are abundant in the Urals. Due to duration of continental collision, all primary structural roughnesses in counters of collided continents and island arcs were forcedly smoothed and the Uralian belt acquired a distinct linear shape.

CENTRAL ASIATIC BELT

The Central Asiatic belt is situated between the Siberian and Chinese platforms (Fig. 3) including Late Precambrian and Paleozoic foldzones of Sayan Mts, Altai, East Kazakhstan and Mongolia. The belt is 4000 km long and 1000 km wide. Considered below is the longitudinal section of the belt from the South Siberian margin near the Baikal Lake southwards through the Khangai Upland of Central Mongolia, the Mongolian Altai and Gobi Altai Mts, Gobi desert to the northern margin of the Chinese platform (Zonenshain, 1972, 1973; Dergunov et al., 1971; Mossakovsky, 1970).

Though the Central Asiatic belt has the same age as the Uralian belt it differs strongly having rather a mosaic pattern instead of the linear shape. Paleozoic foldzones are intermittent with Precambrian massifs of angular outlines. Other distinction from the Urals is an abundance of granite batholiths and continental volcanics.

Ophiolites are widely spread within the belt indicating the existence of a large Asiatic paleoocean in Late Precambrian and Paleozoic. The age of ophiolites is changed from 900-800 m.a. in the northern part of the belt to 500-400 m.a. in the southern part, i.e. duration of life of the ocean was nearly 500 m.a.

The cross-section of the belt includes the next isopic zones (see Fig. 3B):

(1) The East Sayan massif formed before Vendian and joined the Siberian continent;

(2) The Djida and Ider zones containing Vendian/Early Cambrian oceanic and island arc sequences deformed before Late Cambrian;

(3) The North-Mongolian zone where the Precambrian basement is outcropped;

(4) The Khangai-Khentei zone filled by thick clastic turbidites of the Devonian and Carboniferous age deformed in the end of Paleozoic;

(5) The Bayan-Khongor zone composed of Late Pre-Cambrian ophiolites laying allochtonously over an ancient sialic basement;

(6) The Central Mongolian zone - the Proterozoic massif covered by shallow water coarse Ordovician sediments;

(7) The Lake zone with widely spread Vendian/early Cambrian ophiolites and island arc volcanics deformed during Cambrian;

(8) The discontinuous strip of small ancient massifs along the northern slope of the Mongolian and Gobi Altai Mts.;

(9) The South-Mongolian zone which consists of the Middle Paleozoic, from Ordovician to Lower Carboniferous, sequence deformed by the Hercynian orogeny before Late Carboniferous; ophiolites are numerous in this zone being mostly dismembered and included into melanges; island arc volcanics predominated in Devonian and Early Carboniferous; the structure is characterized by chaotical stacking of tectonic slices putting together complexes initially originated in different environment, the structure resembles, in a whole, a forearc subductional melange;

(10) The South Gobi zone, an ancient crystalline massif covered by shallow water Vendian/Cambrian sediments;

(11) The Inner Mongolian zone consisting of two subzones; the northern one contains a thick Carboniferous and Permian clastic sequence including olistostromas, and the southern one is distinguished by Late Paleozoic ophiolites and island arc volcanics; these Late Paleozoic rocks mark a suture of the Paleo-Tethys ocean; they were deformed in Early Triassic;

(12) The Chinese platform.

Lare granitic batholiths of the Lower Paleozoic age occupy the northern part of the belt and the Upper Paleozoic volcanic belt of the Andean type is stretched in the south. Alkaline plutonic and extrusive bodies occur on some places being manifestation of hot spot magmatism during different time intervals from Devonian to Late Jurassic.

The most probable plate tectonic interpretation of the evolution of the Central Asiatic belt can be drawn in such a way (Fig. 4).

Crust generation in the northern Asiatic paleoocean was completed no later than in the Earliest Cambrian as all known ophiolitic complexes are overlain by sediments with Lower Cambrian Archaeocyatha. At least two oceanic basins existed between the Siberian continent and the Central Mongolian microcontinent on the place of the Djida-Ider and Lake zones. The width of the basins can not be estimated. The width of the whole Asiatic paleoocean was, judging from global reconstructions, no less than 3000 km. The widespread Middle Cambrian, Salairian orogeny gives evidence for closure of the northern basin near the Siberian margin as a result of collision of microcontinents with Siberia. An active continental margin of the Nevadan type was formed with a subduction zone dipping under the Siberian continent. Granodiorite-plagiogranitic bodies were melted above the subduction zone.

During Ordovician and Silurian (from 500 to 400 m.a.) the Siberian continent's framing was of the passive margin type. A new oceanic basin, the South Mongolian basin, as a part of the new originated Paleo-Tethys ocean was actively opened at that time on the place of the South Mongolian zone for which Ordovician and Silurian ophiolites are the characteristic feature.

Two events took place in Early Devonian: firstly, an island arc was originated near the Central Mongolian margin on the Paleo-Tethys oceanic floor, and secondly, the North Mongolian continental crust was broken up and blocks were moved apart behind the arc to create a marginal basin in between, on the place of the Khangai-Khentei zone which has been filled by clastics derived from adjacent continents.

The volcanic arc and related subduction zone were active during Devonian and Early Carboniferous. They consumed the South Mongolian oceanic crust until the South Gobi microcontinent which was previously torn away from the Chinese continent reached the trench and stopped the subduction. Collision of the South Gobi microcontinent with the island arc and subsequently with the Siberian/North Mongolian continental margin induced orogeny which was accompanied by thrusting, high grade metamorphism and granite intrusions. Collision had a short duration, no more than 20 m.a. In Late Carboniferous a subduction zone was overflipped on the opposite edge of the South Gobi microcontinent thus joining it to Siberia.

As a result, an active continental margin of the Andean type marked by the subaerial calk-alkaline volcanics was formed along the southern periphery of the Siberian continent in Late Paleozoic. The Paleo-Tethys floor was consumed under the active margin. Eventually, the ocean was closed in Late Permian and Chinese/Siberian collision began. It lasted no more than 10-15 m.a. Folding of the Inner Mongolian and Khangai-Khentei zones was related with the collision. No oceanic space remained in the Central Asia in Mesozoic.

Thus, the evolution of the Central Asiatic belt is characterized by: (1) involving large sialic massifs in orogeny, (2) change in time of island arc conditions by active continental margins of the Nevadan and Andean types, (3) comparatively short duration of continent/island arc and continent/continent collisions, (4) hot spot magmatism superimposed on different former structures. Appearance of granitic batholiths and subaerial volcano-plutonic belts is related to the first feature. Due to short collision intervals, many zones, especially ancient massifs, preserve their configuration and therefore the belt has the mosaic pattern. Moving the region over hot spots led to magmatic activization with alkaline magmatism and rare metal mineralization.

THE VERKHOYANSK-KOLYMIAN BELT

The Verkhoyansk-Kolymian belt is stretched as a sinuous arc from the Siberian Arctic coast to the Sea of Okhotsk over the distance of 2500 km being about 1000 km wide (Fig. 5). A summary description of the belt is given in numerous works (Tilman, 1973; Krasny, 1973; Natapov et al., 1977; Parfenov et al., 1981; Fujita, 1978). The belt consists of (1) the Verkhoyansk foldzone, (2) the magmatic arc of the Chersky Ridge, (3) the Kolymo-Omolon massif, (4) the South-Anui suture zone, (5) the Chukotka foldzone.

The Verkhoyansk zone is composed of a thick (up to 10-12 km) sedimentary sequence beginning from Vendian and ending in Upper Jurassic. No structural unconformities are present within the sequence but a significant reorganization took place in late Early Carboniferous: shelf limestones sometimes with evaporites predominating earlier were followed by clastic, also shallow water sediments deposited during Late Paleozoic and Early Mesozoic. Clastics were derived from uplifted basement blocks of the Siberian platform. This sequence in a whole presents a sedimentary wedge of the Siberian passive margin with thickness and marine conditions increasing offshore. The change of carbonate sedimentation to clastic one can be related either with increased relief or with the Siberia's drift from the aridic to humidic climate belt when large rivers appeared. Very long existence of the Siberian passive margin during 500 m.a. witnesses for a large adjacent ocean. Folding began in the eastern part of the

Verkhoyansk zone in the Volgian time and spread over the whole zone in Early Cretaceous. Decollement of the sedimentary pile from the basement is typical for the Verkhoyansk zone. The general vergence is westward to the Siberian continent. Thrusts and nappes are found along the frontal part of the zone near the boundary with the Pre-Verkhoyansk foredeep. Postorogenic granites were intruded nearly 90 m.a. indicating the orogeny ended before this time.

The Chersky magmatic arc includes two units: the Ilintas volcanic belt and Kolymian batholithic belt. The Ilintas belt was formed in Upper Jurassic beginning from the Callovian. It consists of calc-alkaline volcanics of the island arc type which are interfingered with turbidites and olistostromas. Polarity of the Ilintas volcanics is under discussion now. The author prefers a Benioff zone inclination eastwards from the Verkhoyansk zone under the Kolymo-Omolon massif. Batholiths of the Kolymian belt were intruded in the Early Cretaceous time approximately along the Ilintas arc axis. They seemed to be controlled by a Benioff zone dipping westwards, i.e. in the opposite direction comparatively with the former one.

The Kolymo-Omolon massif has an ancient core more than 2500 m.a. old overlain by Late Precambrian and Paleozoic sequences. Subaerial acid volcanics of the Kedonian Suite lay unconformably over older rocks. They are an alien unit for the North-East USSR. According to paleomagnetic data (Kolesov, 1981; Khramov, 1982) the Kolymo-Omolon massif was at the 30-35°N in Late Devonian/Early Carboniferous meanwhile the East Siberia was at 60° N. Paleomagnetic poles for both units are different till Cretaceous.

The South Anui zone consists of ophiolites and island arc assemblages of the Upper Jurassic and Lower Cretaceous age. It is a suture along which the Kolymo-Omolon massif and Chukotka were collided (Seslavinsky, 1970; Natal'in,1979). An Oceanic basin existed on the place of the South Anui zone in Triassic and Jurassic. It was closed in the end of Jurassic and beginning of Cretaceous in a process of oceanic crust subduction mainly under the Kolymo-Omolon massif. The zone was deformed during a very short time interval between Neocomian and Aptian.

The Chukotka foldzone is beyond of the boundaries of the Verkhoyansk-Kolymian belt. Its Upper Paleozoic/Lower Mesozoic sequence is similar to that of the Verkhoyansk zone.

Structure of the Verkhoyansk-Kolymian belt was formed, as we have seen, during a single time episode and extremely quickly, in Early Cretaceous. In Middle Cretaceous (beginning from Aptian) the Okhotsk-Chukotka volcanic belt of the Andean type superimposed unconformably on all the former structures of different ages.

The history of the Verkhoyansk-Kolymian belt can be given schematically as follows (Fig. 6). During Late Paleozoic and Mesozoic up to the Middle/Late Jurassic boundary, the Siberian passive margin was facing north towards the Paleo-Pacific ocean. The Arctic oceanic basins were still closed at that time. Chukotka (together with Northern Alaska) was joined with the Canadian Arctic Archipelago. The Kolymo-Omolon massif was somewhere within the Paleo-Pacific but in Jurassic it was already not far from Siberia. Opening of the first Arctic oceanic basin, the Canadian basin, where magnetic anomalies from M25 to M12 (150-125 m.a.) are established, was beginning in Late Jurassic about 150 m.a. ago. The Chukotka/Alaska block moved apart from North America closing the South Anui oceanic basin and, possibly, inducing movement of the Kolymo-Omolon microcontinent towards the Siberia passive margin. The Ilintas and South Anui volcanic arcs were originated under which the oceanic crust separated the Kolymo-Omolon massif from the Chukotka/Alaska block and Siberia was subducted. The oceanic crust was completely consumed in the earliest Cretaceous and then, after very short continental collision, spreading in the Canadian basin was ceased and fold structures of the Verkhoyansk-Kolymian belt were formed. To adjust continued convergence between the Pacific (Kula) and Eurasia plates a new subduction zone appeared along the new Siberian margin. It is witnessed by the Okhotsk-

Chukotka volcanic belt.

Thus, the most peculiar feature of the Verkhoyansk-Kolymian belt is quickness of its formation. All structure-forming events including origination of island arcs, oceanic crust consumption, continental collision and folding took place during 25-30 m.a. only. As a result, three previously separated crustal blocks were gathered together and the large passive margin underwent folding. Due to short duration of deformations all the three blocks can be easily recognized in the present structure.

CONCLUSIONS

The examples considered above show that no standard pattern can be proposed for the evolution of foldbelts. Location of foldbelts along zones of lithosphere plate convergence and their origination on the place of former oceanic basins of different types due to continental collision are unvariable. It is necessary to make an individual historical analysis for each belt separately.

We have seen how different the evolution of the each of the three foldbelts is. The Uralian belt was originated on the place of island arcs and marginal basins and was being formed during the continuous continental collision, during no less than 90 m.a. long. The Central Asiatic belt, initially, also included island arcs but then active continental margins of the Nevadan and Andean types were more characteristical. No long periods of continental collision are recorded in the history of the belt. The Verkhoyansk-Kolymian belt was formed due to the one-act and very short continental collision.

Elementary geodynamic features can be recognized in all the foldbelts. These features are analogues of present spreading zones, continental rifts, island arcs, marginal basins, passive and active continental margins, however their combination and recombination in space in time can be very much different in various foldbelts.

The structural pattern of a foldbelt depends on how

long subduction activity lasted, to what type it belonged, how long continental collision was and how great was magnitude of convergence of continents.
The term "the Wilson's cycle" is used now instead of former geosynclinal cyclicity. It contains two stages, ocean opening and ocean closure, as obligatory steps in the foldbelt evolution. The existence of the second stage is doubtless. As to the first stage, ocean opening, it seems to be better interpreted as obligatory existence of an oceanic basin being independent on its age and place of opening.

REFERENCES

1. Abdulin A.A., Avdeev A.V. and Seitov N.S. Tectonics of the Sakmarian and Or-Ilek zones of the Mugodjar Mts. Alma-Ata, Nuaka, 1974, 241 p.
2. Dergunov A.B., Zaitsev N.S., Mossakovsky A.A. and Perfilev A.S. The Mongolian Hercynides and a problem of the Paleo-Tethys. - In the book: Problems of theoretical and regional tectonics. Moscow, Nauka, 1971, p. 67-103.
3. Fujita K. Tectonic evolution of the Siberian Northeast and the Arctic ocean margins of North America. - J. Geol., N 3, 1978.
4. Hamilton W. The Uralides and the motion of the European and Siberian platforms. - Geol. Soc. Amer. Bull., 1970, v. 81, p. 2553-2579.
5. Ivanov S.N., Efimov A.A., Mishkin L.M., Perfilev A.S., Rujentsev S.V. and Smirnov G.A. A nature of the Uralian eugeosyncline. - Doklady Ac. Sci. USSR, 1972, v. 206, p. 1177-1180.
6. Khramov A.N. (Ed.). Paleomagnetology. Leningrad, Nodra, 1982, 312 p.
7. Kolesov E.V. Paleomagnetism of Middle Paleozoic deposits of the Omolon massif. In: Vashchikov Ju.Ja. (Ed.). Rock magnetism and paleomagnetic stratigraphy of the east and north-east Asia. Magadan, North-East Research Institute, Ac. Sci., USSR, 1981, p. 68-74.
8. Korinevsky V.G. Geological description of the

South Mugojar Mts. In: Zonenshain L.P. and Matveenkov V.V. (Eds.). History of the Uralian paleoocean. Moscow, Institute of Oceanology, 1982, p. 85-110.
9. Krasny L.I. (Ed.). Geology of the north-east Asia. Leningrad, Nedra, 1973, 396 p.
10. Milovsky A.V., Getling R.V. Zverev A.T., Roshkovan G.R. and Svalnova V.I. PreCambrian and Lower Paleozoic of the West Kazakhstan. Moscow, University Press, 1977, 268 p.
11. Mossakovsky A.A. The Upper Paleozoic volcanic belt of Europe and Asia. - Geotectonics, 1970, N 4, p. 3-22.
12. Natal'in B.A. Eugeosynclinal zones of Chukotka and Alaska. In: Theses of the XIV. Pacific Congress, Comm. B, Section BII Continental margins, island arcs and structures of the Pacific ocean, Khabarovsk, 1979, p. 34-35.
13. Natapov L.M., Zonenshain L.P., Shulgina V.S., Surmilova E.P., Degtyarev V.S., Savosina A.K., Artemov A.V., Katz A.G. and Stavsky A.P. Geological evolution of the Kolymo-Indigirka region and the problem of the Kolymian massif. - Geotectonics, 1977, N 4, p. 18-31.
14. Parfenov L.M., Natal'in B.A., Voinova I.P. and Popeko L.I. Tectonic evolution of active continental margins in the Pacific North-West. - Geotectonics, 1981, N 1, p. 85-104.
15. Peive A.V., Ivanov S.N., Necheukhin V.M., Perfil'ev A.S., Puchkov V.N. Tectonics of the Urals (An explanatory note to the Tectonic map of the Urals in the scale of 1:1000000). Moscow, Nedra, 1977, 119 p.
16. Perfil'ev A.S. Crust formation in the Uralian eugeosyncline. Moscow, Nauka, 1979, 188 p.
17. Puchkov V.N. Bathyal complexes of passive margins of geosynclinal regions. Moscow, Nauka, 1979, 260 p.
18. Rujentsev S.V. Marginal ophiolitic allochtons (Tectonic nature and structural setting). Moscow, Nauka, 1976, 171 p.
19. Seslavinsky K.B. Structure and development of the

South Anui suture trough. - Geotectonics, 1970, N 5, p. 73-85.
20. Tíl'man S.M. Comparative tectonics of Mesozoides in the north Circum-Pacific. Moscow, Nauka, 1973, 326 p.
21. Zonenshain L.P. The geosynclinal theory and its application to the Central Asiatic foldbelt. Moscow, Nedra, 1972, 240 p.
22. Zonenshain L.P. The evolution of Central Asiatic geosynclines through sea floor spreading - Tectonophysics, 1973, v. 19, p. 213-232.
23. Zonenshain L.P., Korinevsky V.G. and Kaz'min V.G. Structure and plate tectonic evolution of the Southern Urals. In: Zonenshain L.P. and Matveenkov V.V. (Eds.). History of the Uralian paleoocean, Moscow, Institute of Oceanology, 1983, p. 5-84.

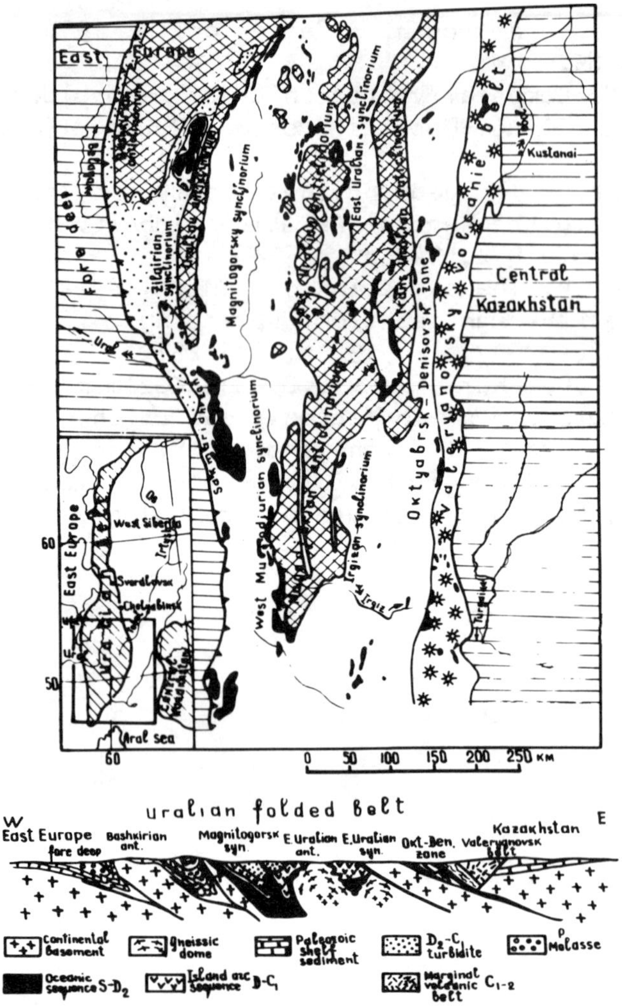

FIGURE 1. Structure of the Southern Urals.
Main structural units of the Southern Urals. Horizontally hatched are margins of the East European and Kazakhstanian continents. Cross-hatched are the Pre-Uralides, blank are the Uralides, dotted are passive margin sequences. Asterisks mark the Valeryanovsky volcanic belt. Toothed lines correspond to the main thrusts. Inset shows location of the map (squared). Hatched are outcrops of Pre-Mesozoic rocks. B - Schematical cross-section along the Kustanai latitude.

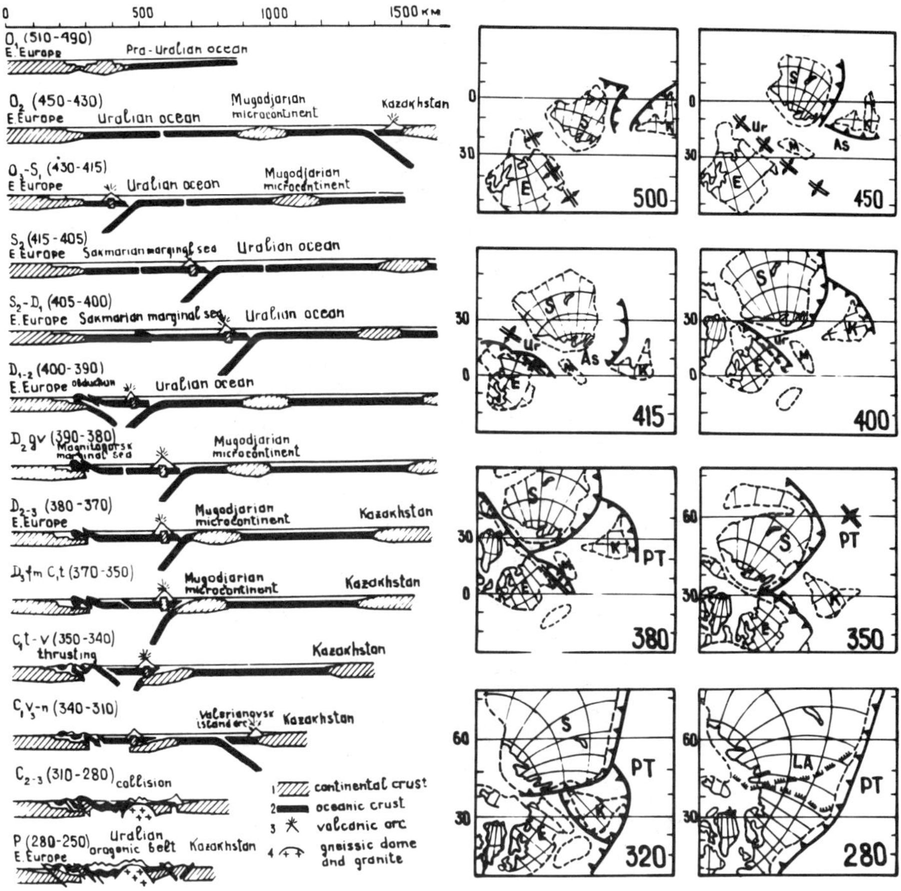

FIGURE 2. Evolution of the Southern Urals in Paleozoic A - Palinspastic cross-sections B - Palinspastic reconstructions. Divergent boundaries are shown as double lines, convergent boundaries as toothed lines. Dashed lines are continent/ocean boundaries. E - Europe, S - Siberia, K - Kazakhstan, LA - Laurasia, C - China, As - Asiatic paleoocean, Ur - Uralian paleoocean, PT - Paleo-Tethys.

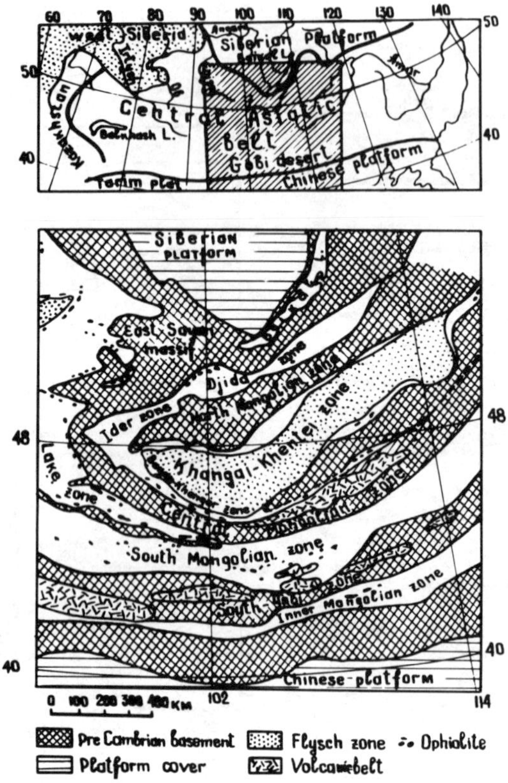

FIGURE 3. Structure of the median segment of the Central Asian belt. A - Position of the region (hatched) in the Central Asiatic belt B - main structural units.

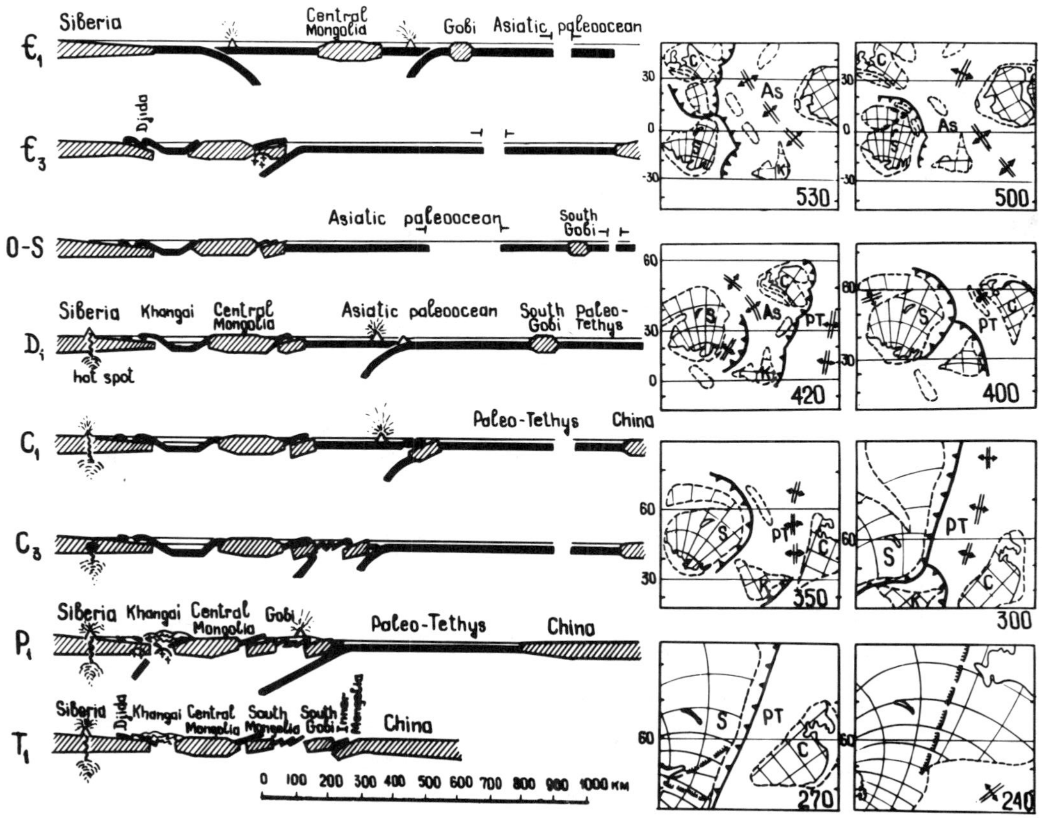

FIGURE 4. Evolution of the Central Asiatic belt in Paleozoic. A - Palinspastic cross-sections B - Palinspastic reconstructions. For legend see Fig. 2.

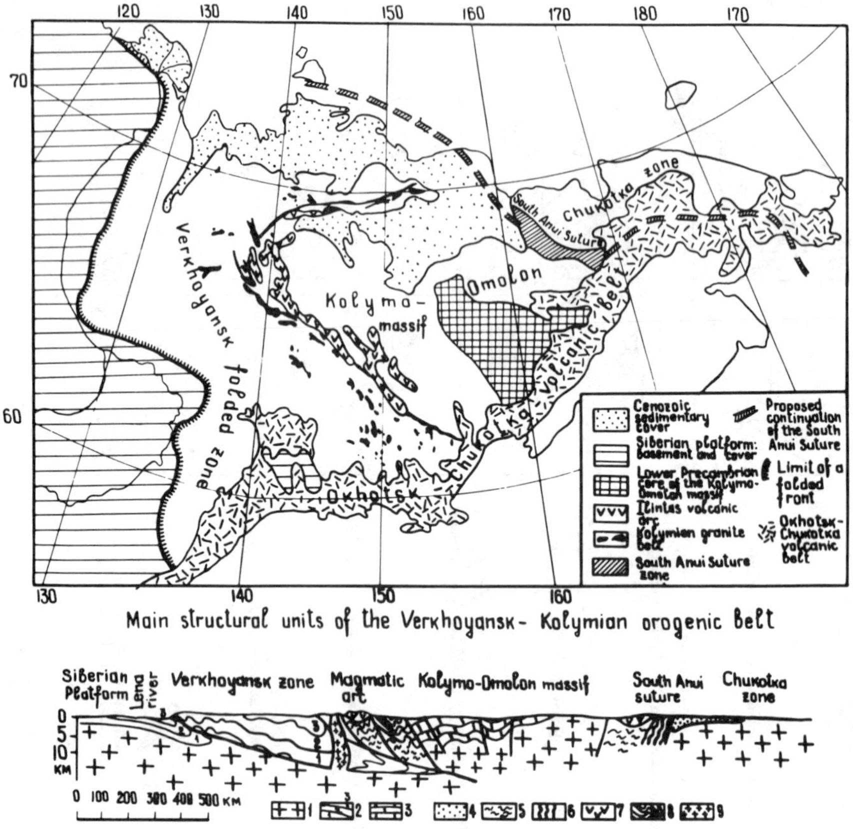

FIGURE 5. Structure of the Verkhoyansk-Kolymian belt. A - Main structural units of the Verkhoyansk-Kolymian orogenic belt B - The SW-NE cross-section.
1 - Pre Cambrian basement, 2 - sedimentary cover of the Siberian platform and Verkhoansk zone (1 - Mesozoic, 2 - Upper Paleozoic, 3 - Lower Paleozoic and Uppermost PreCambrian), 3 - Riphean and Paleozoic cover of the Kolymo-Omolon massif, 4 - Upper Paleozoic and Mesozoic sedimentary cover of the Chukotka zone, 5 - Upper Paleozoic and Lower Mesozoic fore arc accretional wedge, 6 - ophiolite, 7 - Upper Jurassic island arc volcanics, 8 - Upper Jurassic/Lower Cretaceous back arc sequence, 9 - granite.

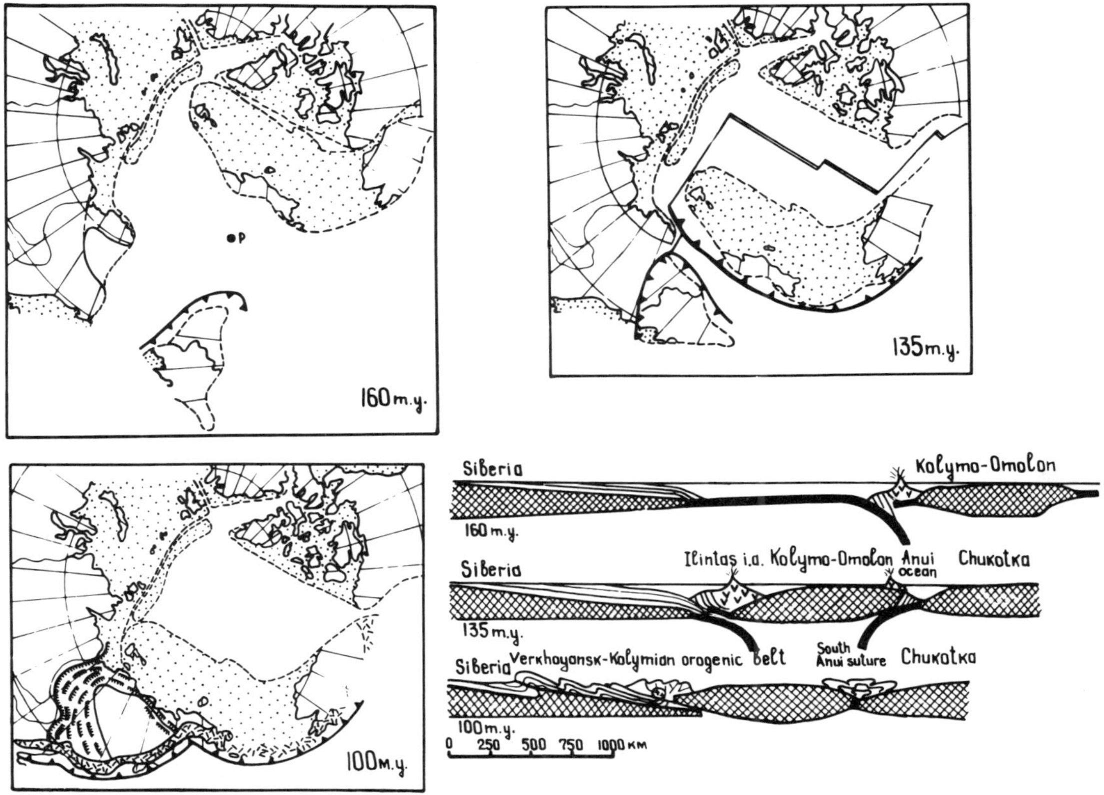

FIGURE 6. Caption overleaf.

FIGURE 6. Evolution of the Verkhoyansk-Kolymian belt in Mesozoic. A - Palinspastic cross-sections. Continental crust is cross-hatched, oceanic crust is blacked. V - mean island arcs, crests-granite. B - Palinspastic reconstructions. Dotted are present continental shelf. The Okhotsk-Chukotka volcanic belt are covered by small irregular hachures. Devergent boundaries are shown as double lines, convergent boundaries as toothed lines, collision zones as comb lines.

PERICONTINENTAL AND INTERCONTINENTAL GEOSYNCLINAL BELTS: A COMPARISON OF DEVELOPMENT FEATURES

KHAIN V.E.
Moscow State University, Moscow, USSR

The earth crust splitting to a continental and ocean hemisphere (the latter being occupied by the Pacific) outlined at least 1.4 - 1.3 billion years ago, if not earlier (2.5?). Pericontinental geosynclines developed along the Pacific periphery; the Atlantic, Mediterranean, Ural-Okhotsk and Arctic belts intercontinental geosynclines emerged and disappeared within the continental hemisphere. The formation of these geosynclines meant the disintegration of Pagaea, and their closing marked the restoration of its integrity. The conditions of the pericontinental and intercontinental geosynclines initiation and development considerably differed from each other, though the fundamental tendencies were similar. The subject of the present paper is just an attempt to sum up those differences.

1. The pericontinental geosynclines (PCG) constantly developed in the transitional continent-ocean zone, their most actively developing part on the already existing ocean crust. That is the reason why the ensimatic island arcs and fenced off type marginal seas, such as the modern Aleutian Basin of the Bering Sea, are the most characteristic of, if not exceptionally peculiar to, the PCG. But that does not exclude the manifestation of the continental crust active destruction in the PCG continental periphery, with the formation of marginal seas such as the Sea of Japan or the South China Sea, as well as microcontinental blocks split off from the continent, that is, according to the traditional geosyncline terminology, median masses, such as, for instance, Ganalsky massif in the Kamchatka peninsula, Hida in Japan, Palawan massif in the region of the Philippines. But in general microcontinents are much more characteristic of the intercontinental geosynclines (ICG), where they are widely spread, especially in the Ural-Okhotsk and Mediterranean belts. The ICG were originally initiated on

Proceedings of the 27th International
Geological Congress, Volume 7, pp. 131–137
TECTONICS

the already quite mature Early Precambrian continental crust by its riftogenous destruction. Accordingly, a crust of transitional type, that is, thinned and altered by the basic magma intrusion continental crust, is likely to be the most widespread in this type of geosynclines. On the contrary, oceanic crust has a limited development in the ICG, and authentic abyssal basins have a relatively small width. However, the overall width, especially taking into consideration its following contraction, makes it quite possible to justify the ICG belts reference to the category of oceans (Yapetus, Tethys and others).

2. The Pacific permanent existence (for all the variability of its concrete configuration) during almost 1.5 billion years predetermined as well the continuity of the geosynclinal process along its periphery. Closing of separate geosynclinal systems was immediately accompanied by initiation of the new ones in a more exterior position in relation to the continent, with the corresponding shift of the Benioff zones (Wadati-Zavaritsky-Benioff zones). The younger geosynclinal systems initiated roughly parallel to the older ones which can be very well illustrated by examples from Australia and Melanesia region. Nowadays, the active geosynclinal process continues as well on the greater part of the Pacific perimeter; some exclusions are only the Antarctica borderland and some parts of the North and South America ones which became passive or transform not long ago.

Unlike the PCG, the ICG history was an alternation of ICG expansion and closure followed by a new continental crust splitting, predominantly along sutures between the younger and the older parts. The younger geosynclinal systems did not always follow the direction of the older ones, and rather complicated plan relations could appear as a result, like, for instance, the Hercynides-Caledonides relations in the South of the British Isles, the Hercynides-Baikalides relations in the Uralo-Timan region. A marked reorganization of the mutual arrangement and relations of the geosynclinal belts of various stages was observed. Thus, the western extention of the Central Europe Hercynian belt "assimilates" in the west the former SE extention of the Caledonian Atlantic belt and farther to the south, beginning with Mexico, the extention of the North America Cordillera belt was assimilated in its turn. The Atlantic Ocean, beginning with the Early Cretaceous, cut off the Mediterranean belt extention in Central America from its main part.

The ICG interrupted development determined a distinct definition, completion of the separate stages of their development, called cycles. It was not by chance that the classic Caledonian, Hercynian and Alpine cycles were ascertained after the European ICG example. It was not accidental as well that the application of this classification while compiling tectonic maps for the Pacific Ocean segment of the Earth met great difficulties.

3. The ICG riftogenous origin conditioned a wide and prolonged development of their margins as typical passive margins, which is well defined in the modern epoch by a decisive predominance of passive margins in the frame of the Atlantic, Indian and Arctic Oceans as well as the Mediterranean Sea. Meanwhile, the Pacific Ocean margins developed uninterruptedly as active since the Late Riphean already, and it is in no way by chance that the margins of this type in other oceans are observed where the "continental barrier" between the Pacific and these oceans is either absent, like, for instance, in the region of the Scottia Sea, or is not continuous (Indonesia), or turns out to be very shallow (the Antilles-Caribbean region).

4. The folded mountain structures emerging as a result of the PCG development have as a rule an accretion origin; as for the ICG, their development results in a collision origin of these structures. The Pacific folded edifices are distinctly asymmetric: they are limited by B-type subduction zones from the side of the ocean and by A-type subduction zones from the side of the continent (according to A.W. Bally), which is most typically defined on the American side of the Pacific. In the intracontinental folded belts originated by the ICG there are the A-type

subduction zones extending from both sides, that is, on both sides the underthrust of the continental crust takes place. The continental crust often participates in the nappe structures both in shear nappes, for instance, in the Himalayas or the Scandinavian Caledonides, and the Penninic type nappes originated from recumbent folds, such as in the Alps and the Greenland and Scotland Caledonides. In two latter cases the continental crust experienced a considerable remobilization. In the PCG case the phenomenon of this type is observed perhaps only in the Canadian Cordillera (Shuswap complex).

The ICG ophiolite nappes are often transferred to large (100 km) distances overlapping in the result of obduction the former passive continental margins. The famous Middle East "ophiolite crescent" from Syria to Oman can serve as a most vivid example. The continental crust often comes to the surface from under the ophiolite nappes in tectonic windows in the form of granite-gneiss cupolas. The obduction ophiolite nappes are widely spread, as it is known, also in the Pacific periphery (Sakhalin, New Guinea, New Caledonia and others, predominantly on the western side of the ocean), but here they are emplaced either on the old island arcs or on the forearc accretion wedges.

5. A large scale transcurrent fault tectonics with displacement for not only hundreds but thousands km is likely to be most characteristic of the Circum Pacific PCG. And it is just the result of its manifestation that the Circum Pacific structure, beginning with the North American Cordillera, is depicted at present as a mosaic of blocks ("terrains") having been transformed from various remote places and being originated in various conditions, but predominantly of intraoceanic character. The displacement is considered as a function of subduction, obliquely oriented in the relation to the continental edge. It would not be right to claim that the phenomena of this type are absolutely alien to the ICG, to the Mediterranean belt for instance, where large displacement along the Vardar zone, the North Anatolian fault with its Inebolu-Varto branch and others have been already ascertained. In the

North Atlantic belt there are well known large wrench-faul-s of the Scottish Caledonides, their scale being considerably larger than it was supposed before. Displacement of the type is also known in the Ural-Okhotsk belt (Talass-Ferghana, Zhalair-Naiman, Zhungaria and others). The number of such examples is likely to grow in the future but still this number is not supposed to become comparable to the one already registered in the Circum Pacific belt. This may be evidently explained by the fact that the PCG subduction (the PCG being limited from one side by the Pacific) goes with much greater rate and, correspondingly, the displacement component parallel to the Circum Pacific strike turns out to be more substantial. In the ICG both spreading and subduction are one order lower at an average; intracontinental space is saturated by continental fragments - microcontinents, which limits the displacement scale.

6. Both in the PCG and ICG during the orogenic stage of development volcano-plutonic belts are formed over subduction zones, including large granite batholiths. In the PCG these are "liminar" subductional belts tied to active, Andean type margins; modern belts of both Americas'western peripheries, the East Asia Late Jurassic-Cretaceous belt, the East Australia Devonian belt can serve as examples. As for the ICG, intracontinental collisional belts are typical here, which are in general formed already after the closing of deep basins with oceanic crust, owing to the transitional crust absorption. The examples of such belts are: the Late Paleozoic belt extending from Europe to the Middle and Central Asia right up to Mongolia; the Cretaceous-Paleogene belt from the Iranian western borderland to the Eastern Tibet in the Zagros rear, Baludzhistan fold system and the Himalayas. Unlike in the PCG, where granitoids and calc-alkaline volcanics have in general a mantle origin, in the ICG granitoids and volcanics a component of continental origin prevails. In general, the ICG modern continental crust is predominantly a regenerated old Early Precambrian crust, while the prevailing part of the PCG modern crust is of primary oceanic origin.

7. The difference between the PCG and the ICG applies

as well to regional metamorphism manifestations. Persistent paired metamorphic belts of high pressure-low temperature and of high temperature-low/moderate pressure are most typical of the Circum Pacific PCG, where they have been discovered frist. In the ICG, the glaucophane metamorphic zones are not always accompanied by rather well defined zones of high temperature metamorphism, tied to volcanic arcs (for instance, the Alps, the Apennines, the Dinarides-Hellenides), and vice versa, zones of high temperature metamorphism are not always accompanied by manifestations of high pressure metamorphism (for instance, the Middle Europe Hercynides and the Greater Caucasus Hercynides). It evidently reflects the peculiarities of the PCG subduction which either do not favour the formation of volcanic arcs, or do not promote high pressure metamorphism or upheaval of involved formations to the day surface.

8. And finally, differences in the PCG and ICG structure and development affect as well the composition of sedimentary trough-filling strata (within these geosynclines). Relative narrowness of the ICG basins and proximity of source areas formed by crystalline basement rocks of the cratons or by redepositional products in their sedimentary cover determine the prevalence of hemipelagic deep-marine sediments over the pelagic proper ones and define a considerable amount of detrital quartz in flysch and molasse sandstones and siltstones. On the contrary, the PCG sandstones and siltstones contain predominantly greywacke, seldomer arkose material; as for the total volume of sedimentary rocks, greater part belongs to biochemogenous sediments, cherts in particular, which forms thick series in the Circum Pacific Paleozoic and Mesozoic.

9. Underlining here the main differences between the ICG and the PCG, it is necessary to make certain reservations. First, these differences may seem to be of predominantly quantitative, but not qualitative, character. This is true on the one hand, but on the other hand, quantity is transformed into quality. Second, there are no insurmountable barriers between the two types of geosynclinal belts. They are closely

connected (in the space): the eastern Tethys and the Ural-Okhotsk belt widely opened to Pacific in Late Paleozoic and Early Mesozoic; the same was with the Atlantic belt western extremity in Early Paleozoic and with the Mediterranean belt western extremity in Middle Paleozoic. In this connection the ICG may be considered as apophysis of the peripacific PCG, the peculiarities of the latter having suffered a certain reduction. Third, if we take into consideration that the Pacific Ocean is likely to have a rift origin too, the difference in the PCG and ICG formation will be reduced to the following: in case of PCG the rifting and rapid spreading considerably preceded their initiation, while in case of ICG their initiation immediately followed the rifting, and the beginning of the development approximately coincided with the beginning of the spreading.

A continuous series of destructive-constructive structures can be outlined beginning with non-inverted and inverted aulacogens (paleorifts), continuing with intracratonic (intracontinental) geosynclines (aulacogeosynclines of G.P. Leonov), ensialic and ensimatic, intercontinental geosynclines, and finishing up with pericontinental geosynclines as their most developed form, with the most active process of the continental crust accretion. It is just the existence of intercontinental and intracratonic, parallel with pericontinental, geosynclines, that can in my opinion justify the preservation of the very notion "geosyncline" at the present stage of development of our science. For only the notion of pericontinental geosyncline can be replaced by the notion transitional ocean-continent zone (a more cumbrous term, by the way), as the other two types of geosynclines (inter- and intracontinental) will not do. It should be stressed that geosynclines are not just continental crust accretion zones: accretion is by all means preceded by destruction, immediately in some cases (inter- and intracontinental geosynclines), and much earlier in other cases.

PALAEOALPINE GEODYNAMICS OF THE CARPATHIAN-PANNONIAN REGION

ZOLTAN BALLA
Hungarian Geophysical Inst., Budapest, Hungary

The Palaeoalpine stage covered the most part of the Mesozoic history of the Carpathian-Pannonian region and has been completed by the Early Cretaceous folding. During this time the main tectonic units of the region came into being. These units were regrouped and then engirdled with the Flysch Carpathian arc as a chaotic mosaic (Fig. 1). They are distributed more or less uniformly but their primary relations remain disputable. At the same time, the Palaeoalpine geodynamics in general represents a basis for understanding the subsequent development of the region ("orogeny"). In the paper a review of the facts is given mainly on the basis of an analysis of magmatic complexes, main versions of interpretation are delineated, and the most probable of them is selected by means of a comparative analysis.

MIDDLE-LATE TRIASSIC

For a long time the Middle-Late Triassic paleogeography has been serving as a basis for correlations within Alpides, and this way the East Alpine, Pannonian, Carpathian and Balkan units can be compared convincingly (Table 1).

Virtually everywhere Upper Permian and Lower Triassic are represented by terrestrial and lagoonar facies, and in rare cases of marine sequences (Southern Alps, Bükk Mts.) shallow-water sediments are only known (Pisa, 1974; Balogh, 1981). The first deep-water deposits or if only "basin facies" appear in the Middle Triassic, consequently, this time only can be regarded as the beginning of the development of the deep-water basin along the southern rim of the region discussed. Directly in the prolongation of the facial sequences ophiolitic zones are observable which could be regarded as inner, oceanic parts of the same basins.

Middle-Upper Triassic volcanites are of two principal

Proceedings of the 27th International
Geological Congress, Volume 7, pp. 139–164
TECTONICS

types: (1) volcanites of variegated composition (basalts, andesites, rhyolites, frequently alkaline up to latites and trachytes: Southern Alps - Pisa, 1974; Bükk Mts. - Balogh, 1981); taking into account the synchroneity of these volcanites with the strong subsidence, practically in all regional works rift origin of them is accepted although compositional peculiarities are not equal to present day rift volcanics; (2) basalts of ophiolitic type (Gemerides - Hovorka, 1978, 1979; Bajaník et al., 1981; Perşani Mts. - Patrulius et al., 1966; Sandulescu et al., 1981a, 1981b) which are similar to the ocean floor ones.

Most of volcanites lye within Ladinian-Carnian sediments of shallow-water (type 1) or deep-water (type 2) origin. In the Perşani Mts. Triassic basalts are cutted from beneath by a thrust plane; primarily they could be underlyed by Campilian-Anisian limestones which are now thrust over them (Patrulius et al., 1966) or by deeper horizons of oceanic crust (Sandulescu, 1981a, 1981b). In the Bükk Mts., parallel with volcanites of variegated composition there are basalts and gabbroic rocks allied with ophiolites (Balla et al., 1983). They were ranged into Triassic (Kozur-Mock, 1977; Mahel', 1978; Balla et al., 1980) but recently Early-Middle Jurassic age of them was proposed.

In the Mureş zone oldest sediments are of Late Callovian age (Bleahu et al., 1981; Cioflica et al., 1981), and the oldest datum for basalts is 180 m.y. (Herz et al., 1974). Mures zone ophiolites can be traced in the Vardar zone where the presence of the Triassic ophiolites is disputable (Hsü et al., 1977). Thus, the presence of fragments of Triassic oceanic lithosphere is nowhere proved.

East of the region discussed Triassic ophiolites appear in Northern Dobrogea. They are of the same age as other Triassic volcanites in the region and also lye within sedimentary sequences and not beneath them (Savu et al., 1980). Their prolongation in the Carpathians (Herz-Savu, 1974) is confirmed only by similarity (Savu, 1980) of titanaugites from basalts of

Northern Dobrogea and of the Black Flysch nappe but oldest sediments in the latter are Kimmeridgian in age (Sandulescu et al., 1981a).

On the other hand, compositional study of pebbles from the Cretaceous conglomerates in the Pieniny Klippen belt led to the conclusion (Misík et al.,1977) that in the Middle-Late Triassic a deep-water basin existed between the Tatrides and European plate. By means of lithological investigations, pebbles were supposed to have been derived from a land which appeared within the Pieniny basin in the Albian and disappeared in the Paleocene (Marschalko et al., 1976; Marschalko, 1979; Marschalko-Kysela, 1980). From this land being named Pieniny Cordillera and in today's structure marked only by a tectonic line, abundant and various materials of island arc origin and of Late Jurassic - Early Cretaceous age was derived (Misík, 1978). It is also exotic, therefore the problems of the outer Triassic basin will be discussed together with the Early Cretaceous situation.

JURASSIC

Based on the Early-Middle Jurassic biogeographical zonality, two areas were distinguished within the Carpathian-Pannonian region (Géczy, 1973a, 1973b; Vörös, 1977): (1) Eastern Alps, northwestern Pannonian basin and Central West Carpathians, and (2) south-eastern Pannonian basin, Apuseni Mts., Eastern and Southern Carpathians. It was supposed that the first of them was located south of the newly generated ocean while the second of them remained north of it. Paleomagnetic data from the Transdanubian Range (north-western Pannonian basin) confirm its location on the African lithospheric plate (Márton, 1981, 1983) but from all other units there are only few data which can be interpreted in various ways.

Jurassic ophiolites are most wide-spread and best studied in Romania where they are located within two zones: the Mureş and the East Carpathian ones, the latter being prolongated into Southern Carpathians (Cioflica et al., 1981; Nastaseanu et al., 1981).

With respect to the Triassic paleogeography they were of different position: the Mureş zone coincided with the deep-water zone of the Middle-Late Triassic basin ("inherited"), and the East Carpathian one came into being within its continental periphery ("superimposed"). Ophiolites from these zones are not quite synchronous: in the Mureş zone they are pre-Oxfordian in age (Savu, 1980), and in the East Carpathian zone part of them belong to the Tithonian and only part of them are of Kimmeridgian or older age (Sandulescu et al., 1981a, 1981b).

The Mureş zone ophiolitic complex is overlain by an Upper Jurassic - Lower Cretaceous island arc sequence (Savu, 1980). East and South Carpathian ophiolites are approximately synchronous with lower members of this sequence, accordingly, these two zones can be coupled to form an "island arc - marginal sea" geodynamic system which marks subduction under the European lithospheric slab. In frame of this system, Carpathian units must have been remained north of the Tethys ocean in Early-Middle Jurassic. That is the basis for the following considerations (Table 2).

The Mureş zone is located between the Carpathians and Apuseni Mts.; consequently, a notion arose on the existence of this situation also in the Jurassic (Radulescu-Sandulescu, 1973; Herz-Savu, 1974). It meant that the Apuseni unit was located south of the Jurassic ocean. But if it was the part of the European lithospheric plate (Table 1), the Jurassic ocean in no way can separate it being generated in the Triassic deep-water zone; the generation of the ocean just there is confirmed by the fact that Lower Jurassic facial sequence of the Apuseni Mts. (Bleahu et al., 1981) was inherited from the Middle-Upper Triassic one in correspondence with the "inheritance" of the Mureş zone. Therefore (version 1), if in the Jurassic the Apuseni unit remained on the European lithospheric slab, the today's mutual position of the Apuseni Mts., Mureş zone and Carpathians could arise only as a consequence of a later displacement of one of these units. In other words, in this case one of Mureş zone boundaries must be of strike-slip type. This hypothetical strike-slip displacement can be supposed along the boundary between the Mureş zone and Apuseni Mts.

But when supposing (version 2) that the Apuseni unit was already located in the south in the Triassic (Herz-Savu, 1974), the problem of correlation of Triassic facies and modern structures arises. The Vardar zone is supposed to be the prolongation of the Mureş one. With respect to it the Pelagonian massif has the same position as the Apuseni Mts. In this case it is not clear how to explain the presence of Keuper facies in the Upper Triassic of the Bihor autochton and of Gresten facies in the Liassic of the Codru system (Bleahu et al., 1981). Since the opening could take place only in the Middle-Late Triassic, this point of view involves recognition of the presence of Triassic ophiolites in the Mureş zone.

Further to the west, ophiolites and allied rocks are known in three areas: in the Bükk Mts. (Szarvaskő), in the Gemerides (The Bôrka nappe: Lesko-Varga, 1980) and in the pebbles from the Pieniny Cordillera. Ophiolites of the Bükk Mts. lye within a sedimentary sequence of Jurassic age forming nappes of southern vergence (Balla et al., 1983). In the Gemerides, Triassic age of basalts from the Meliata series does not exclude Jurassic age of ophiolites from the Bôrka nappe but there are no data to support Jurassic age of the latters. Although Early Jurassic age is confirmed for some basalts from the Pieniny Cordillera (179 m.y.: Misík-Sýkora, 1981), their ophiolitic origin is disputable (Marschalko et al., 1976) and associated serpentinites are not dated.

Since the obscurities in age determinations, it will be better to discuss the problems of Early-Middle Jurassic ophiolites in a regional framework. East of the area concerned the Mureş ophiolitic zone is located, and west of it the Penninic one does. With respect to the Triassic paleogeography first of them is inherited, and the second one is superimposed. Consequently, Western Carpathians are the area where the Early-Middle Jurassic opening transited from the Triassic deep-water basin into the European shelf. Taking into account segmentation of the structure, transform boundaries with both Mureş and Penninic zones seem to be probable.

The West Carpathian unit, thus, was located between two transforms. In such position, appearance of two opening zones instead of single ones which are observable along the strike, is hardly probable. But existence of two ophiolitic zones within the area discussed should not be excluded in general. The southern of them should correspond to the Dinaric-Subpelagonian ophiolitic belt presenting its fragment which was later torn out from the area of jointing of Dinarides with Southern Alps. Such a hypothesis involved necessity to divide the structure of the Gemer-Bükk area into "Carpathian" and "Dinaric" units, therefore this problem will be discussed together with the Early Cretaceous situation.

We have seen that in regional framework there is only a single ophiolitic zone of Early-Middle Jurassic age which represents an interstitial member between the Mureş and Penninic zones. It is usually supposed to be located north of the Tatrides, and in this case it could not coince with the Triassic deep-water basin (Bûkk-Meliata) on the southern margin of the Tatrides (Table 1). Association of Jurassic ophiolites with Triassic limestones in the pebbles from the Pieniny Cordillera can be regarded as an evidence of the Jurassic ocean opening inside of the Middle-Late Triassic deep-water basin. In the regional picture (Fig. 2, A) this belt is analogous to the Mureş one ("inherited"). But when considering the same association of Jurassic ophiolites with Triassic deep-water limestones in the pebbles a secondary feature with Jurassic ocean opening inside the Triassic shelf (Fig. 2, B), the ophiolitic belt seems to be an analogue of the Penninic one ("superimposed").

In the version 1, Bükkian ophiolites can be regarded as analogous to Dinaric ones, in the version 2 they originated in the area of jointing of the Mureş zone prolongation with the transform beyond which Pieniny ophiolites appear. When recognizing Jurassic age of the Gemeric ophiolites (Bôrka nappe), in the version 1 they can be correlated both with the Bükkian and Pieniny ones while in the version 2 they most probable correspond to the Pieniny ones.

EARLY CRETACEOUS

In the Early Cretaceous the Magura zone of the Western Carpathians was represented by a deep-water basin (Eliás, 1979), probably, with oceanic crust (Mahel', 1979). Island arc complex from the Pieniny Cordillera according to age determinations is only a little younger (mainly Neocomian: 116-139 m.y. - Misík-Sýkora, 1981) but that is hardly a discrepancy of principle. Thus, the Late Jurassic geodynamic system "island arc - marginal sea" can be recognized also in the Western Carpathians, hence both the Early-Middle Jurassic constructive plate boundary and the destructive one originated in the Late Jurassic can be traced along the whole Carpathian region.

The Early Cretaceous folding has completed the subduction process marked by the Late Jurassic island arc. Judging by position of marginal seas the subduction was directed under the European lithospheric plate, hence the deformation structures could be supposed to have vergence towards the African slab. Such a picture, however, can be observable only in few places, and the vergence is directed mostly towards the European plate.

Corresponding units of the Eastern and, probably, Southern Carpathians were located behind the volcanic arc where such reversed vergence is a normal feature (Dewey-Bird, 1970; Sorokhtin, 1979). In version 1 (Table 2) the Apuseni unit occupied similar position, in version 2 it was on the front of the island arc. The presence of a crystalline zone (Biharia nappe system) between the Mureş ophiolitic zone ("eugeosyncline") and the Codru nappe system ("miogeosyncline") corresponds to the usual scheme of the folded belt structure (Aubouin, 1965), i.e. version 2, but the absence of a fore-deep along the boundary between the Codru nappe system and the Bihor autochton contradicts this scheme. Overthrust along the boundary between the Codru and Biharia nappe systems can mask previous strike-slip displacement but no traces of it are known up to now. Thus, there is no possibility to select the versions

(Table 2) based on the structural picture, and both of them can explain the observable vergence.

In the Western Carpathians, nappes of the Tatric system manifest reversed vergence. Behind them an ophiolitic suture (root-zone of the Bôrka nappe: the Roznava line - Reichwalder, 1971) is located being accompanied by paired metamorphic belts which correspond to subduction to the north, under the Volovec anticlinorium in the North Gemerides (Hovorka, 1979). The latter is built up of Paleozoic sequences with Cretaceous granites and superimposed Alpine hT-lP metamorphism, so it can be regarded as a fragment from the basement of an Early Cretaceous island arc (Lesko-Varga, 1980). At the front of the Tatrides the Pieniny Klippen belt is located. Within it the Pieniny Cordillera supplied island arc material of Late Jurassic - Early Cretaceous age from the Albian up to the Paleocene. Corresponding subduction was directed towards the north according to the position of the Magura zone (marginal sea) with respect to the Pieniny one.

Thus, island arc of the Volovec anticlinorium and the Pieniny Cordillera were of the same age and of the same polarity. Existence of two such objects behind each other is hardly probable in general. Independently, in frame of the corresponding version 1 (Fig. 2) the southern island arc ought to have been regarded as the prolongation of the Dinarides. Today there are no data to support such a correlation.

A conclusion can be drawn: the Volovec anticlinorium and the Pieniny Cordillera are fragments of the same island arc (Lesko-Varga, 1980). In frame of both versions (Fig. 2), it was located in Jurassic north of the Tatrides. In order to occupy today's position Tatrides must have been chipped off from the subducting African plate and overthrust over the European one (Horváth et al., 1977) overjumping the island arc and tearing off its cover carrying it on their front (Lesko-Varga, 1980).

DISCUSSION

The succession of the Triassic facies in the Tatric nappes is quite definite: the higher the nappe is, the more pelagic facies compose it. South of the Roznava line a reverse succession is observable: the Meliata series is overlain by the Silica nappe (Kozur-Mock, 1973). Such a situation could arise if the Silica nappe represented the back part of the Tatric plate which was torn off from the subducting slab, and the Meliata series remained on the slab which continued subduction after the Tatric plate was torn off (Fig. 3); that is an explanation of the southern vergence of the Silica nappe (Mock, 1978).

In the frame of this model (corresponding to the version 2 in the Fig. 2), it is difficult to understand folding and charriage in the Bükk Mts. which remained far in the south on the subducting slab. When considering the southern vergence and the Dinaric facial affinity (Balogh, 1964) we can think over another possibility: folding in the Bükk Mts. could take place in the prolongation of the Dinarides (according to the version 1 in the Fig. 2). In this case, the boundary between the "Carpathian" and "Dinaric" structure could follow the so-called "Darnó line" (Zelenka, 1973).

Similar models proposed previously, either did not take into account presence of island arc rocks in the pebbles from the Pieniny Cordillera (Horváth et al., 1977) or did not explain presence of Triassic deep-water limestone there (Lesko-Varga, 1980). Association of these rocks (Misík, 1978, 1979) can be understood only in frame of an hypothesis for the existence of two deep-water basins in the Middle and Late Triassic: north of the Tatrides (with Late Jurassic - Early Cretaceous island arc; fragment of its basement being preserved in the Northern Gemerides, and its cover turned to the Pieniny Cordillera) and south of them (prolongation of the Dinaric branch of the Alpides). This hypothesis leads to the conclusion that geological formations of the Bükk Mts. were located in the direct prolongation of Dinarides in the Jurassic (version 1 in the Fig. 2) and even in the Early Cretaceous. But when considering the Jurassic ocean north of the Tatrides being "superimposed" (version 2 in the Fig. 2) Triassic deep-water limestones could be supplied on the Pieniny Cordillera only from the highest Tatric nappes (according to the facial succession). It is necessary, however, to regard as hardly possible on the Pieniny Cordillera combination of rocks derived from beneath Tatrides (Early Cretaceous island arc complex) and from their highest horizons (Triassic deep-water limestones).

The chronology of events is clearest in the Eastern Carpathians and the Mureş zone. The marginal sea growth is finished in the Tithonian (there are no younger ophiolites: Sandulescu et al., 1981a), and most of the calc-alkaline volcanites from the island arc are also of Tithonian age (only few rocks belong to the Berriasian-Hauterivian: Cioflica et al., 1981). Charriages begin in the Barremian and finish in the Albian(Bleahu et al., 1981; Sandulescu et al., 1981a). In the Southern Carpathians the structural reorganization is established approximately from the Aptian/Albian boundary (Nastaseanu et al., 1981).

In Western Carpathians, approach of the Tatrides and the Late Jurassic - Early Cretaceous island arc can be dated by the appearance of Urgonian limestones (Upper Barremian - Lower Aptian) within both of them. In deep-water Albian and Cenomanian sediments of lowermost Tatric nappes, pebbles from the Pieniny Cordillera are established with size decreasing to the south equally in different nappes (Mišík et al., 1981). Consequently, in that time at least the lowermost Tatric nappes were ready, and this process obviously was going under sea. The main charriages are usually thought to have taken place in the Turonian, however, that was most probably the time of an emergence after the charriages.

Thus, the main events of the Early Cretaceous folding are approximately synchronous both in the Eastern and Western Carpathians in spite of the somewhat younger age of the island arc volcanism of the Western Carpathians. In units of interstitial position, the Early

Cretaceous age of the folding is nowhere confirmed: in the Codru nappes there are no sediments younger than Hauterivian (Bleahu et al., 1981), similar situation is in the Mecsek Mts. (Vadász, 1969); in the Bihor autochton a weak disconformity can be seen only around the Barremian-Aptian boundary (Bleahu et al., 1981), finally, in the Villány Mts. there is no disconformity in the whole Lower Cretaceous up to the Albian (Vadász, 1960). Along the northwestern margin of these units a large basalt belt is traced which is often regarded as ophiolitic (Juhász-Vass, 1974; Szepesházy, 1977). Basalts are dated (Valanginian-Hauterivian) only in the south-western ending of the belt where they are of rift chemistry (Mecsek Mts.: Bilik, 1983). In other sites tholeiites also occur, and the age can differ a little. Basalts are accompanied by flysch and pelagic sediments (Albian - Upper Eocene), therefore, they can be regarded as fragments from the basement of this newly formed basin (Balla, 1982).

Genetic interpretation of this basin depends on the position of the Apuseni unit with respect to the Tethys (Table 2). In the version 2 (location on the African lithospheric plate) the basin opens within the subducting slab; in this case, it becomes clear why there are no evidences of Early Cretaceous folding but it is difficult to understand the opening process and the apparent relations of the Great Hungarian plain flysch with the Carpathian one. In the version 2 with location of the Apuseni unit on the European lithospheric plate, the opening takes place behind the island arc, and the basin is of marginal sea type. True, basalts can be considered too young but only with respect to the Mureş island arc and the East Carpathian marginal sea. The Mecsekian basalts are practically synchronous with island arc magmatites from the Pieniny Cordillera as well as their nearest Carpathian analogues: teschenites from the Moravo-Silezian Beskids (Mahmood, 1973) along the outer rim of the West Carpathian marginal sea.

Consequently, (1) the Apuseni and South Pannonian units must have been located on the European lithospheric plate in the Triassic and Jurassic (version 1 in the Table 2) and (2) in spite of the absence of direct data, the main folding has to be dated Early Cretaceous as well as in other areas. The site where the Apuseni and South Pannonian units were torn off from the European lithospheric slab, ought to be looked for in the area of the junction of Eastern and Western Carpathians.

The Magura and Pieniny complexes accompanying the front of the Early Cretaceous folding in the Western Carpathians take far in the rear of the East Carpathian Crystalline-Mesozoic zone which was consolidated in Early Cretaceous, and the Outer Dacidian zone accompanying the latter goes far along the front of the Magura nappe (Fig. 1). Thus, the equal main structural elements of the Eastern and Western Carpathians, i.e. the front of the Early Cretaceous folding and directly accompanying formations overlap each other like coulisses for over 200 km in the length. Thrust of the Magura and Pieniny complexes over the back of the innermost zones of the Eastern Carpathians is dated Early Miocene (Sandulescu et al., 1981a), and the strike of overthrusts is deflected from the Carpathian folded system inside with the same vergence.

Fronts of the Early Cretaceous folding would lye on the same line by means of a reconstruction of the situation before a left-lateral strike-slip displacement along the front of the Magura nappe. This oblique (with respect to the Carpathian strike) fault most probable is masked by subsequent overthrusts. The sharp turning up of this strike-slip line rights can be explained by its transition into an overthrust in the south (Fig. 1). Since the separation along this overthrust (when judging by the separation along the conjugated strike-slip fault) is at least 200 km, the latter can be regarded as the destructive microplate boundary supposed on the basis of the Middle Hungarian volcanic belt (Balla, 1982).

Junction of the Western and Eastern Carpathians along the oblique transcurrent fault requires the recognition of their approach from the large distance (some of hundred km) and of existence of an unknown area between them which was overlapped during the overthrust. This area corresponds to a section of the Early Alpine marginal sea, i.e. as a matter of fact, the fragment of the Outer Carpathians, and on its outer side the section was located from which the Apuseni and South Pannonian units were torn off in the Early Cretaceous.

CONCLUSIONS

Thus, in the Palaeoalpine geodynamics of the Carpathian-Pannonian region two stages can be delineated: ocean opening in the Triassic and Jurassic and its closing beginning from the Late Jurassic. During this process an island arc arose with a marginal sea which was opened behind it within the European lithospheric plate. The Early Cretaceous folding was conditioned by collision of continental parts of the subducting African lithospheric plate with the island arc, and the front of this folding was located approximately along the boundary between the island arc and the marginal sea.

REFERENCES

Aubouin J., 1965. Geosynclines. Development in geotectonics 1. Elsevier, Amsterdam - London - New York, p. 332.

Bajaník S., Hovorka D., Miko O., Vozár J., 1981. Predterciérny vulkanizmus Západných Karpát. In: "Paleovulkanizmus Západných Karpát", D. Stúr Geol. Inst., Bratislava, pp. 27-38 (in Slowakian with English abstract: Pre-Tertiary volcanism in the West Carpathians, pp. 38-39)(Editors: S. Bajaník, D. Hovorka).

Balla Z., 1982. Development of the Pannonian basin basement through the Cretaceous-Cenozoic collision: A new synthesis. Tectonophysics, 88 (1-2): 61-102.

Balla Z., Baksa Cs., Földessy J., Havas L., Szabó I., 1980. The tectonic setting of ophiolites in the Bükk Mountains (North Hungary). Geol. Zborn.-Geol.Carp., 31 (4): 465-493.

Balla Z., Hovorka D., Kuzmin M., Vinogradov V., 1983. Mesozoic ophiolites of the Bükk Mountains (North Hungary). Ofioliti, 8 (1): 5-46.

Balogh K., 1964. Die geologischen Bildungen des Bükk-Gebirges. Annals of Hung. Geol. Inst., 48 (2): 555 - 705.

Balogh K., 1981. Correlation of the Hungarian Triassic. Acta Geol. Acad. Sci. Hung., 24 (1): 3-48.

Bilik I., 1983. Lower Cretaceous submarine (rift) volcanism in South Transdanubia (South Hungary). In: "Proceedings of the Seventeenth Assembly of the European Seismological Commission", Akadémiai Kiadó, Budapest, pp. 569-576 (Editors: E. Bisztricsányi and Gy. Szeidovitz).

Bleahu M., Lupu M., Patrulius D., Bordea S., Ştefan A., Panin Ş., 1981. The structure of the Apuseni Mountains. Carp.-Balk. Geol. Ass. XIIth Congr., Guide to Exc. B3, Inst. Geol. Geophys., Bucharest, p. 106.

Carte tectonique internationale de l'Europe et des régions avoisinantes, 1:2 500 000, 1962, Acad. Sci. URSS, Moscou (Editor in gen.: N. Schatsky).

Carte tectonique internationale de l'Europe et des régions avoisinantes, 1:2 500 000, 2e édition, 1981. Acad. Sci. URSS, Moscou (Editors: A.A. Bogdanoff, V.E. Khain).

Cioflica G., Savu H., Nicolae I., Lupu M., Vlad Ş., 1981. Alpine ophiolitic complexes in South Carpathians and South Apuseni Mountains. Carp.-Balk. Geol. Ass. XIIth Congr., Guide to Exc. A3, Inst. Geol. Geophys., Bucharest, p. 80.

Dewey J.F., Bird J.M., 1970. Mountain belts and the new global tectonics. J. Geophys. Res., 75 (14): 2615 -2647.

Eliáš M., 1979. Notes on the paleogeographic and paleotectonic development of the Flysch Carpathians. In: "Czechoslovak geology and global tectonics", VEDA, Bratislava, 115-126 (Editors: M. Mahel', P. Reichwalder).

Ganev M., 1974. Stand der Kentnisse über die Stratigraphie der Trias Bulgariens. In: "Die Stratigraphie der alpin-mediterranen Trias", Springer, Wien - New York, pp. 93-96 (Editor: H. Zapfe).

Gochev P.M., 1979. Myastoto na Strandzha v alpiyskata struktura na Balkanskiya poluostrov. Rew. Bulg. Geol. Soc., 40 (1): 27-46 (in Bulgarian, with English summary: The place of Strandzha in the Alpine structure of the Balkan Peninsula, p. 27).

Géczy B., 1973a. Plate tectonics and paleogeography in the East-Mediterranean Mesozoic. Acta Geol. Acad. Sci. Hung., 17 (4): 421-428.

Géczy B., 1973b. The origin of the Jurassic faunal provinces and the Mediterranean plate tectonics. Ann. Univ. Sci. Budapest, sect. Geol., 16, 1973: 99-114.

Herz N., Jones L.M., Savu H., Walker R.L., 1974. Strontium isotope composition of ophiolitic and related rocks, Drocea Mountains, Romania. Bull. Volc., 38 (4): 1110-1124.

Herz N., Savu H., 1974. Plate tectonics history of Romania. Geol. Soc. Am. Bull., 85 (9): 1429-1440.

Horvâth F., Vôrôs A., Onuoha K.M., 1977. Plate-tectonics of the Western Carpatho-Pannonian region: A working hypothesis. Acta Geol.Acad. Sci. Hung., 21 (4): 207-221.

Hovorka D., 1978. The West Carpathian Meso- and Cenozoic basalts. Geol. Zborn.-Geol. Carp., 29 (1): 77-89.

Hovorka D., 1979. The West Carpathians incomplete ophiolites. In: "Czechoslovak geology and global tectonics", VEDA, Bratislava, pp. 155-166 (Editors: M. Mahel', P. Reichwalder).

Hsü K.J., Nachev I.K., Vuchev V.T., 1977. Geologic evolution of Bulgaria in light of plate tectonics. Tectonophysics, 40 (3-4): 245-256.

Juhász A., Vass G., 1974. Mesozoischen Ophiolite im Beckenuntergrund der Grossen Ungarischen Tiefebene. Acta Geol. Acad. Sci. Hung., 18 (3-4): 349-358.

Kovács S., 1982. Problems of the "Pannonian Median Massif" and the plate tectonic concept. Contributions based on the distribution of Late Paleozoic - Early Mesozoic isopic zones. Geol. Rdsch., 71 (2): 617-639.

Kozur H., Mock R., 1973. Zum Alter und zum tektonischen Stellung der Meliata-Serie. Geol. Zborn.-Geol. Carp., 24 (2): 265-374.

Kozur H., Mock R., 1977. Conodonts and holothurian sclerites from the Upper Permian and Triassic of the Bükk Mountains (North Hungary). Acta Univ. Szeged., Acta Min. Petr., 23 (1): 109-126.

Lesko B., Varga I., 1980. Alpine elements in the West Carpathian structure and their significance. Miner. Slov., 12 (2): 97-130.
Mahel' M., 1978. Geotectonic position of magmatites in the Carpathians, Balkan and Dinarides. Záp. Karp., sér. Geol. 4: 1-119.
Mahel' M., 1979. Palinspastic picture of the West Carpathians in the basic evolutionary stages. In: "Geodynamic investigations in Czechoslovakia", VEDA, Bratislava, pp. 179-186 (Editors: J. Vanek et al.).
Mahmood A., 1973. Petrology of the teschenite rock series from the type area at Czeszyn (Teschen) in Polish Carpathians. Roczn. Pol. Tow. Geol., 43 (2): 153-212.
Marschalko R., 1979. Cretaceous flysch of the Pieninian Klippen Belt and the decay of the crust in the Tatride foreland. In: "Geodynamic investigations in Czechoslovakia", VEDA, Bratislava, pp. 243-251 (Editors: J. Vanek et al.).
Marschalko R., Kysela J., 1980. Geológia a sedimentológia bradlového pásma a manínskej jednotky medzi Zilinou a Povazskou Bystricou. Záp. Karp., sér. Geol., 6: 7-71 (in Slowakian with English abstract: Geology and sedimentology of Klippen Belt and Manín Unit between Zilina and Povazská Bystrica, pp. 71-79).
Marschalko R., Misík M., Kamenicky L., 1976. Petrographie der Flysch-Konglomerate und Rekonstrukzion ihrer Ursprungszonen (Paläogen der Klippenzone und der angrenzenden tektonischen Einheiten der Ostslowakei). Záp. Karp., sér. Geol., 1: 7-120.
Márton E., 1981. Tectonic implication of paleomagnetic data for the Carpatho-Pannonian region. Earth Evol. Sci., 1981 (3-4): 257-264.
Márton E., 1983. Tectonic implication of the palaeomagnetic results in the Carpatho-Balkan and adjacent areas. Geol. Soc. London, spec. vol. (in press).
Misik M., 1979. Pieniny Klippen Belt and the global tectonics. In: "Czechoslovak geology and the global tectonics", VEDA, Bratislava, pp. 89-101 (Editors: M. Mahel', P. Reichwalder).
Misik M., Jablonský J., Mock R., Sýkora M., 1981.

Konglomerate mit exoiteschem Material in dem Alb der Zentralen Westkarpaten - paläographische und tektonische Interpretation. Acta Geol. Geogr. Univ. Comen., Geol., 37: 5-55.
Misik M., Mock R., Sýkora M., 1977. Trias der Klippenzone. Geol. Zborn.-Geol. Carp., 28 (1): 27-69.
Misik M., Sýkora M., 1981. Pieninský exotický chrbát rekonstruovaný z valúnov hornîn kriedových zlepencov bradlového pásma a manînskej jednotky. Záp. Karp., sér. Geol., 7: 7-90 (in Slowakian with German abstract: Die peininische exotische Rücken, rekonstruiert aus Gerôllen karbonatischer Gesteine kretazeischer Konglomerate der Klippenzone und der Manîn Einheit, pp. 91-111).
Mock R., 1978. Nové poznatky o juzných castiach Západných Karpát. In: "Paleogeografický vývoj Západných Karpât", D. Stúr Geol. Inst., Bratislava, pp. 321-337 (in Slowakian with English abstract: Knowledge recently gained about the southern parts of the West Carpathians, pp. 338-341) (Editors: J. Vozár et al.).
Nastaseanu S., Bercia I., Iancu V., Vlad Ş., Hârtopanu I., 1981. The structure of the South Carpathians (Mehedinţi-Banat area). Carp.-Balk. Geol. Ass. XIIth Congr., Guide to Exc. B2, Inst. Geol. Geophys., Bucharest, p. 100.
Patrulius D., Popa E., Popescu I., 1966. Seriile mezozoice autohtone şi pînza de decolare transilvana în împrejurimile Comanei (Munţii Perşani). An. Com. Geol. Rom., 35: 397-434 (in Romanian with French abstract: Les séries mésozoïques et la nappe de décollement Transylvaniae dans les environs de Comana (Monts Perşani), pp. 435-444).
Pisa G., 1974. Stratigraphische Tabelle der südalpinen Trias. In: "Die Stratigraphie der alpin-mediterranen Trias", Springer, Wien - New York, pp. 159-160 (Editor: H. Zapfe).
Radulescu D.P., Sandulescu M., 1973. The plate-tectonic concept and the geological structure of the Carpathians. Tectonophysics, 16 (3-4): 155-161.
Reichwalder P., 1971. Roznavská zlomova lînia a jej vzt'ah k sedimentácii, magmatizmu a metamorfóze.

Geol. Práce, Spr. 57: 215-221 (in Slowakian with German abstract: Die Roznava-Bruchzone und ihre Beziehung zur Sedimentation, Magmatismus und Metamorphose, pp. 221-222).

Sandulescu M., 1972. Consideraţii asupra posibilitaţilor de corelare a structurii Carpaţilor Orientali şi Occidentali. D.S. Inst. Geol., 58, 5: 125-144 (in Romanian with French abstract: Considérations sur les possibilités de corrélation de la structure des Carpates Orientales et Occidentales, pp. 145-150).

Sandulescu M., 1980a. Sur certain problèmes de la corrélation des Carpathes Orientales Roumaines avec les Carpathes Ucrainiennes. D.S. Inst. Geol. Geofiz., 65, 5: 163-180.

Sandulescu M., 1980b. Analyse géotectonique des chaînes alpines situées autour de la Mer Noire occidentale. An. Inst. Geol. Geofiz., 56: 5-54.

Sandulescu M., Kräutner H.G., Balintoni I., Russo-Sandulescu D., Micu M., 1981a. The structure of the East Carpathians (Moldavia-Maramureş area). Carp.-Balk. Geol. Ass. XIIth Congr., Guide to Exc. B1, Inst. Geol. Geophys., Bucharest, p. 92.

Sandulescu M., Ştefanescu M., Butac A., Patruţ I., Zaharescu P., 1981b. Genetical and structural relations between flysch and molasse (The East Carpathians model). Carp.-Balk. Geol. Ass. XIIth Congr., Guide to Exc. A5, Inst. Geol. Geophys., Bucharest, p. 95.

Savu H., 1980. Genesis of the alpine cycle ophiolites from Romania and their associated calc-alkaline and alkaline volcanics. An. Inst. Geol. Geofiz., 56: 55-77.

Savu H., Udrescu C., Neacşu V., 1980. Structural petrologic, geochimic and genetic study of the ophiolites in Niculiţel Zone (North Dobrogea). D.S. Inst. Geol. Geofiz., 65, 1: 41-64.

Sorokhtin O.G., 1979. Mekhanizmy obrazovaniya regionalnykh nadvigov, ofiolitovykh pokrovov i geosinklinalnoy skladchatosti. In: "Okeanologiya (Editor: A.S. Monin). Geofizika okeana (Editor: O.G. Sorokhtin), tom 2, Geodinamika", "Nauka", Moscow, pp. 294-306

(in Russian: Mechanisms of generation of regional overthrusts, ophiolitic nappes and geosynclinal folding).
Szepesházy K., 1977. Az Alföld mezozoos magmás képződményei. Földt. Közl., 197 (3-4): 384-397 (in Hungarian with English abstract: Mesozoic igneous rocks of the Great Hungarian Plain, p. 397).
Tectonic map of Czechoslovakia, 1:1,000,000, 1971. Ústr. Ústav Geol., Praha (Editors: O. Kodym et al.).
Vadász E., 1960. Magyarország földtana, II. kiadás. Akadémiai Kiadó, Budapest, p. 646 (In Hungarian: Geology of Hungary, 2nd edition).
Végh-Neubrandt E., 1972. Fauna- und Faziesverbreitung der Obertrias der Transdanubischen Mittelgebirges. Ann. Univ. Sci. Budapest, sect. Geol., 15, 1971: 111-120.
Vörös A., 1977, Provinciality of the Mediterranean Lower Jurassic Brachiopod fauna: Causes and plate-tectonic interpretation. Palaeogeogr., Palaeoclim., Palaeoecol., 21 (1): 1-16.
Zelenka T., 1973. New data on the Darnó megatectonic zone. Acta Geol. Acad. Sci. Hung., 17 (1-3): 155-172.

TABLE 1. Correlation of main tectonic units of the Alpine-Carpathian-Balkan system using the Norian isopic zones (simplified and completed after Kovács, 1982)

	NW Pannonian	Eastern Alps	Western Carpathians	SE Pannonian	Apuseni Mountains	Eastern Carpathians	Southern Carpathians	Balkan
Continental detrital deposits (mainly Keuper facies) or hiatuses		Frankenfels	Tatricum Križna	Mecsek Villány	Bihor Vălani Finiş	Infrabucovinian Subbucovinian Bucovinian	Danubian Getic Supragetic	Moesian
Main dolomite	W Bakony E Bakony Vértes Gerecse	Bajuvaricum	Choč+ +Šturec Stražov		Arieşeni			Fore-Balkan Balkan Srednogorie
Dachstein limestone	Buda Pilis	Tirolicum	Stratena		Dieva Moma			
Hallstatt limestone	Csővár	Juvavicum	Silica		Vaşcău	Rarău—Hăghimaş		SE Strandzha
Ophiolite	(SW Bükk)		(Meliata)		(Mureş)	Perşani		

Completed using the following data: Végh-Neubrandt, 1972 (NW Pannonian), Sandulescu et al., 1981b (Southern Carpathians), Ganev, 1974 and Gochev, 1979 (Balkan).

Table 2. Comparison of versions on the position of he Apuseni unit with respect to the European lithospheric slab (ELS). Assumptions: Carpathians: in T_{2-3} and J_{1-2} - on the ELS. Mures zone: J_{1-2} ophiolites - in T_{2-3} deep-water zone ("inherited"). East Carpathian ophiolite belt: marginal sea superimposed on the ELS in the back of the Mures J_3 island arc

Version	Object	Supplementary assumption	Conclusion	Correlatable object	Problems
1	Apuseni Mts.	in T_{2-3} and J_{1-2} - on the ELS	-	Eastern Carpathians	-
	Mureş zone	-	a) T_{2-3} ophiolites cannot be present b) boundary with Apuseni - left-lateral strike-slip fault	Vardar zone	strike-slip displacement along the boundary with Apuseni
2	Apuseni Mts.	in T_{2-3} and J_{1-2} - divided from the ELS	-	Pelagonian massif	presence of Keuper (T_3) and Gresten (J_1) facies
	Mureş zone	-	a) T_{2-3} ophiolites are present b) boundaries are deformed only by compression	Vardar zone	-

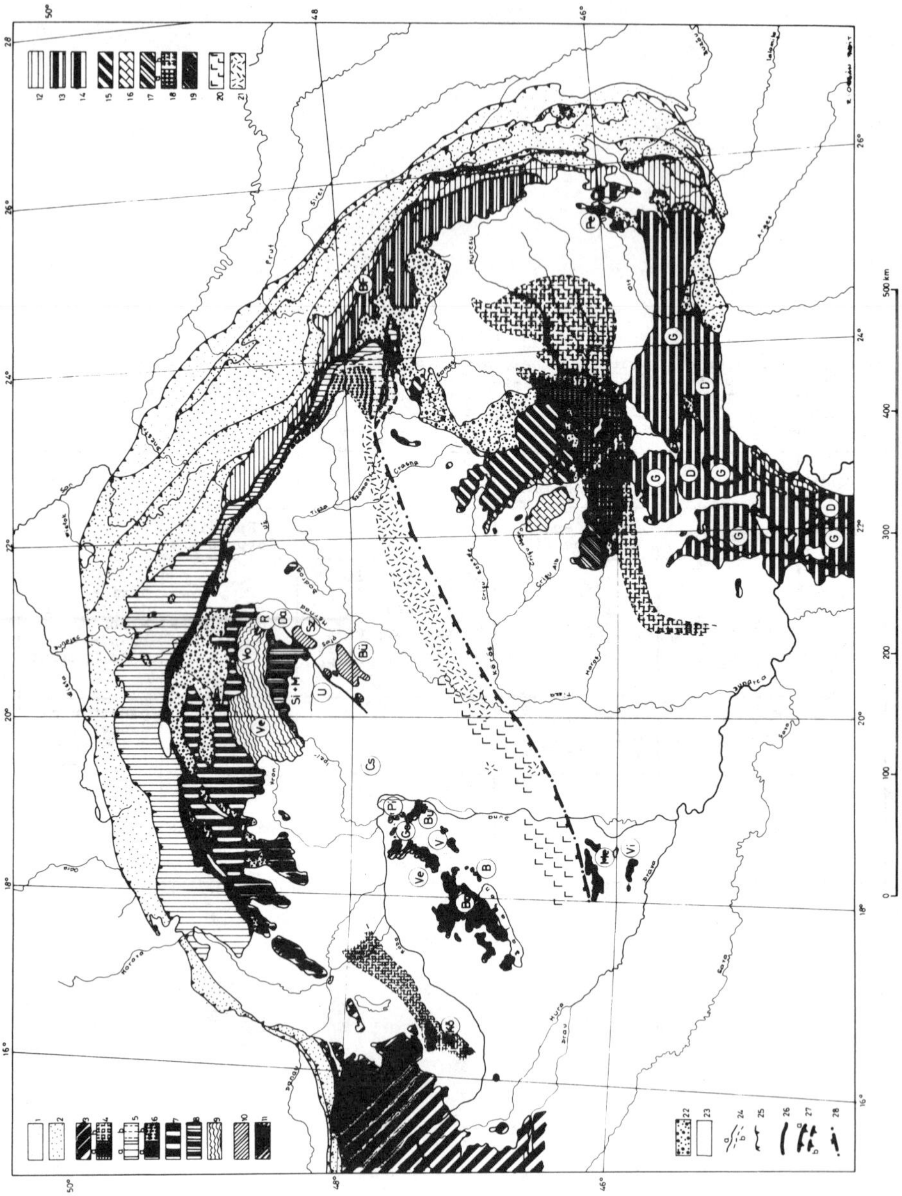

FIGURE 1. Tectonic sketch of the Carpathian-Pannonian region

Sources of data: Carte tectonique ..., 1962; Carte tectonique ..., 1981; Tectonic map ..., 1971; Balla, 1982; Bleahu et al., 1981; Nastaseanu et al., 1981; Sandulescu, 1972, 1980a, 1980b.

Legend: Outer flysch zones of the Carpathians and Alps: 1 - folded molasse, 2 - flysch (Alps - Helvetic, Western Carpathians - Silezian, Eastern Carpathians - Moldavidian); Eastern Alps: 3 - Austroalpine nappe system, 4 - Penninics, a - exposed, b - covered; Western Carpathians: 5 - Magura flysch zone (in Eastern Carpathians - Dragova-Petrova and Wildflysch nappes), a - exposed, b - covered, 6 - Pieniny Klippen belt (in Ukrainian Carpathians - Krichevo zone, in Eastern Carpathians - Botiza nappe), a - exposed, b - covered, 7 - Tatrides, 8 - Southern Gemerides (Silica nappe overlying the Meliata series, and related nappes in Northern Hungary), 9 - Veporides and Northern Gemerides; Northwestern Pannonian units: 10 - Bükkides (with Uppony-Szendrő Paleozoic), 11 - Transdanubian Range (with Balaton-Velence Paleozoic); Eastern and Southern Carpathians: 12 - Outer Dacidian flysch zone (in Ukrainian Carpathians - Chernaya Gora zone, in Southern Carpathians - Severin nappe), 13 - East Carpathian Crystalline-Mesozoic zone (in Ukrainian Carpathians - Marmarosh block), 14 - South Carpathian nappe system; Apuseni and Southeastern Pannonian units: 15 - Bihor autochton, 16 - Codru nappe system, 17 - Biharia nappe system, 18 - Mureş zone, a - exposed, b - covered, 19 - South Transdanubian hills; Mid-Hungarian mobile belt (covered): 20 - basalts (Early Cretaceous), 21 - flysch and pelagic sediments (Middle-Upper Cretaceous - Paleogene); Inner Carpathian and Pannonian region: 22 - Cretaceous-Paleogene post-tectonic cover, 23 - Neogene molasse and volcanites; Linear symbols: 24 - contacts, a - exposed, b - covered, 25 - overthrusts, 26 - transcurrent boundary between the Western and Eastern Carpathians (supposed), 27 - overthrust boundary (Early Miocene) between the West and East Carpathian complexes, a - exposed, b - covered, 28 - underthrust boundary (Miocene) between

the North-west and Southeast Pannonian units (covered).
Letters: Eastern Alps: Kő - Kőszeg unit; Western Carpathians: Ve - Veporides, Vo - Volovec anticlinorium, Si - Silica nappe, M - Meliata series, R - Roznava line, Da - Darnó line; Northwestern Pannonian units: Bü - Bükk Mts., U - Uppony Hills, Sz - Szendrő Hills, Cs - Csővár unit, Pi - Pilis Mts., Bu - Buda Mts., Ge - Gerecse Mts., Vé - Vértes Mts., Ba - Bakony Mts., V - Velence Hills, B - Balatonfő Hills; Eastern Carpathians: BF - Black Flysch nappe, Pe - Perşani Mts.; Southern Carpathians: D - Danubian autochton, G - Getic and Supragetic nappes; Southeastern Pannonian units: Me - Mecsek Mts., Vi - Villány Mts.
Notes: the symbol 13 includes Transylvanian nappes (derived from the Mureş zone) and symbols 14-18 do Laramian magmatites.

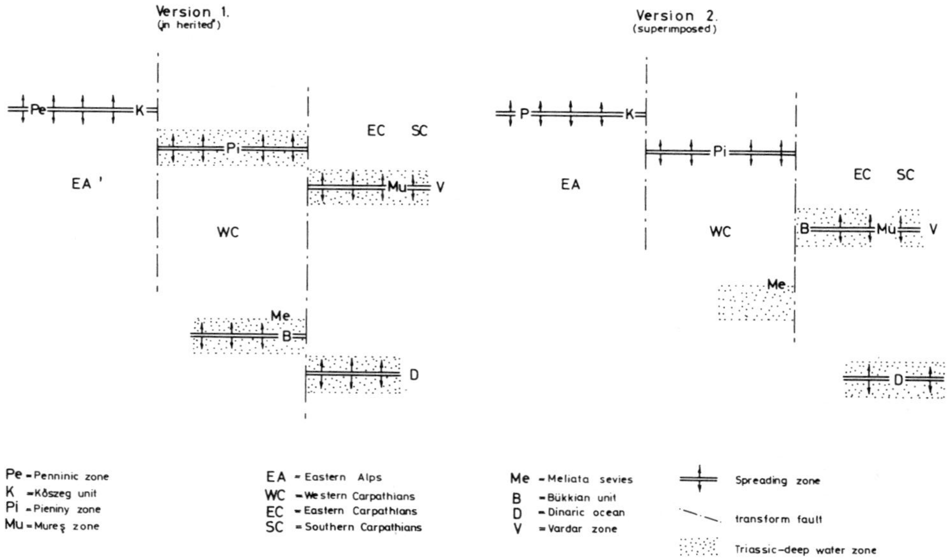

FIGURE 2. Sketch of the Position of the Pieniny Ocean in the Early-Middle Jurassic Geodynamics of the Carpathian Region

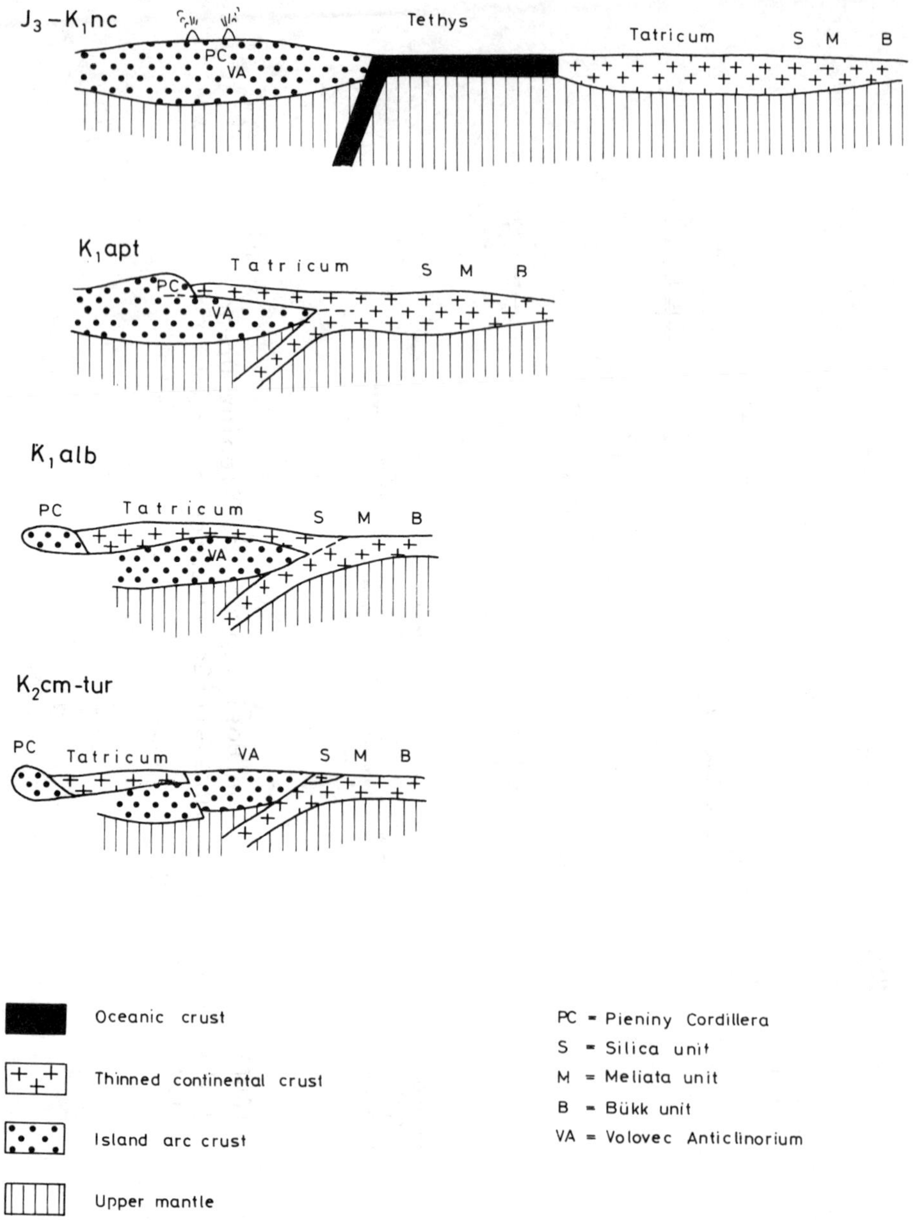

FIGURE 3. Schematic Sections Through the Western Carpathians in the Late Jurassic and Early-Middle Cretaceous.

PHYSICAL CHARACTERISTICS AND EVOLUTIONARY TRENDS OF CONTINENTAL RIFTS

IVAR B. RAMBERG[1] AND PAUL MORGAN[2]
[1] Institutt for Geologi, University of Oslo, Box 1047, Blindern, Oslo 3, Norway.
[2] Lunar and Planetary Institute, 3303 NASA Road One, Houston, Texas 77058, USA.

ABSTRACT

Rifts may be defined as zones beneath which the entire lithosphere has ruptured in extension. They are widespread and occur in a variety of tectonic settings, and range up to 2,600 m.y. in age. The object of this review is to highlight characteristic features of modern and ancient rifts, to emphasize differences and similarities in order to help characterize evolutionary trends, to identify physical conditions favorable for initiation as well as termination of rifting, and to provide constraints for future modelling studies of rifting. We characterize rifts on the basis of their structural, geomorphic, magmatic and geophysical features and discuss the diverse character of these features and their evolutionary trends through time. Mechanisms of rifting are critically examined in terms of the physical characteristics and evolutionary trends of rifts, and it is concluded that while simple models can give valuable insight into specific processes of rifting, individual rifts can rarely, if ever, be characterized by well defined trends predicted by these models. More data are required to clearly define evolutionary trends, and the models require development to incorporate the effects of lithospheric heterogeneities and complex geologic histories.

INTRODUCTION

Rifts have been defined as zones beneath which the entire lithosphere has ruptured in extension (Burke, 1977). Although the exact nature of the postulated lithospheric "rupture" may be questioned, it is generally accepted that extensional tectonic regimes have formed and form in response to plate movements and interactions and lithosphere–asthenosphere interactions, ultimately driven by energy from primordial heat and/or past or present radiogenic heat produced throughout the mantle. Other authors prefer to use the terms rift and rift valley as strictly morphological terms (e.g., Mohr, 1982). However, the common association of rifts with alkaline volcanism, high heat flow, anomalous crust and mantle, and seismicity, provide compelling evidence that rifts are not confined only to crustal levels, but that they are closely linked to dynamic processes within

Proceedings of the 27th International
Geological Congress, Volume 7, pp. 165–216
TECTONICS

the earth's upper mantle. The American Geological Institute Glossary of Geology includes both morphological and tectonic definitions of the term rift (Bates and Jackson, 1980), and for this review we have adopted the broader tectonic definition of rifts. Terms such as graben, rift, rift valley and rift system are well defined and discussed in current literature. Other terms, particularly those which deal with mode and style of rift deformation may need clarification, and are defined as used in this review in Table I.

TABLE I. Definition of commonly used terms in continental rifts

Term	Suggested reference	Short definition
Active rifting	Neumann and Ramberg, 1978 Sengör and Burke, 1978	Rifting as a result of thermal upwelling of the asthenosphere.
Passive rifting	Baker and Morgan, 1981 Morgan and Baker, 1983	Rifting as a response to a regional stress field.
Modern rift		Rift with recent tectono-magmatic activity.
Paleorift	Neumann and Ramberg, 1978a Ramberg and Neumann, 1978a	Dead (or dormant) rift.
Failed arm	Burke and Dewey, 1973 Burke 1977, 1980	Part of a triple junction not developed into spreading oceanic basin.
Aulacogen	Shatsky and Bogbanov, 1961 Milanovsky, 1981	Paleorift on ancient platform has been frequently reactivated by compressional events.

Rifts are widespread in both continental and oceanic crust where they constitute major tectonic features (ridges or depressions) with axial lengths of the order of thousands of kilometers. Because all oceanic floor is eventually consumed, only intracontinental rifts are likely to be preserved in the older parts of the geological record. Many rifts are buried under sedimentary blankets of Atlantic-type margins, cratonic basins and foredeeps. Others have been modified by subsequent tectonic events and erosion. Thus, continental rifts of various ages occur in a variety of tectonic settings, at various stages of evolution, and probably as a result of several modes of rifting. Recent studies of rifts and rift basins have revealed the occurrence of hundreds of modern and ancient rifts, ranging up to 2,600 m.y. in age (Burke, 1978: Easton, 1983: Milanovsky, 1983), demonstrating that continental rifts are much more common than previously conceived. Continental rifts, together with their compressional counterparts, the suture zones, form the basic tectonic framework of the continents.

Despite extensive research activity on continental rifts especially during the last two decades, there is still little concensus about many fundamental questions related to the process of rifting. For instance, why did some rifts stop extending while others continued and resulted in continental breakup? Why are some rifts highly magmatic whereas others appear to be "dry?" Do rifts follow a few predictable trends of evolution and what are the necessary and limiting factors for rifting? Answers to these and similar questions may be obtained in part through compilation and synthesis of the large quantity of data already available in the literature. The object of this review is therefore to highlight characteristic features of modern and ancient continental rifts, to emphasize differences and similarities in order to help characterize evolutionary trends, to identify physical conditions favorable for initiation as well as termination of rifting, and to provide constraints for future modelling studies of rifting.

An overwhelming mass of rift-related studies have emerged during the last two decades or so. Key references to some of these studies include recent conference proceedings edited by Illies (1981) and Morgan and Baker (1983). Review papers have also been assembled in the final report of the IUCG Working Group 4 on "Continental and Oceanic Rifts" edited by Palmason (1982). Paleorifts in particular have been addressed in two symposium volumes, edited by Neumann and Ramberg (1978a) and Ramberg and Neumann (1978a). Also highly relevant are separate volumes on individual grabens and rift systems, e.g., the Baikal rift (Logatchev and Mohr, 1978), the East African rift system (Baker et al., 1972: Girdler, 1972; Anon., 1980), the Oslo Graben (Dons and Larsen, 1978: Ramberg, 1976), the Rhine Graben (Illies and Mueller, 1970: Illies and Fuchs, 1975), the Rio Grande rift (Riecker, 1979), the Afar region (Pilger and Rösler, 1975, 1976), the intracontinental North Sea basin (Ziegler, 1982), and others.

DISTRIBUTION OF RIFTS

Many continental rifts are preserved in the geological record, representing a wide spectrum of evolutionary stages of rifting with ages of origin ranging from Proterozoic to Cenozoic. A global distribution map of continental rifts and rifted margins is shown in Fig. 1, taken slightly modified from Burke et al. (1978). Despite earlier suggestions to the contrary, the present configuration of rifts and rifted margins clearly reveals no preferred azimuthal trend. Statistically, all trends seem about

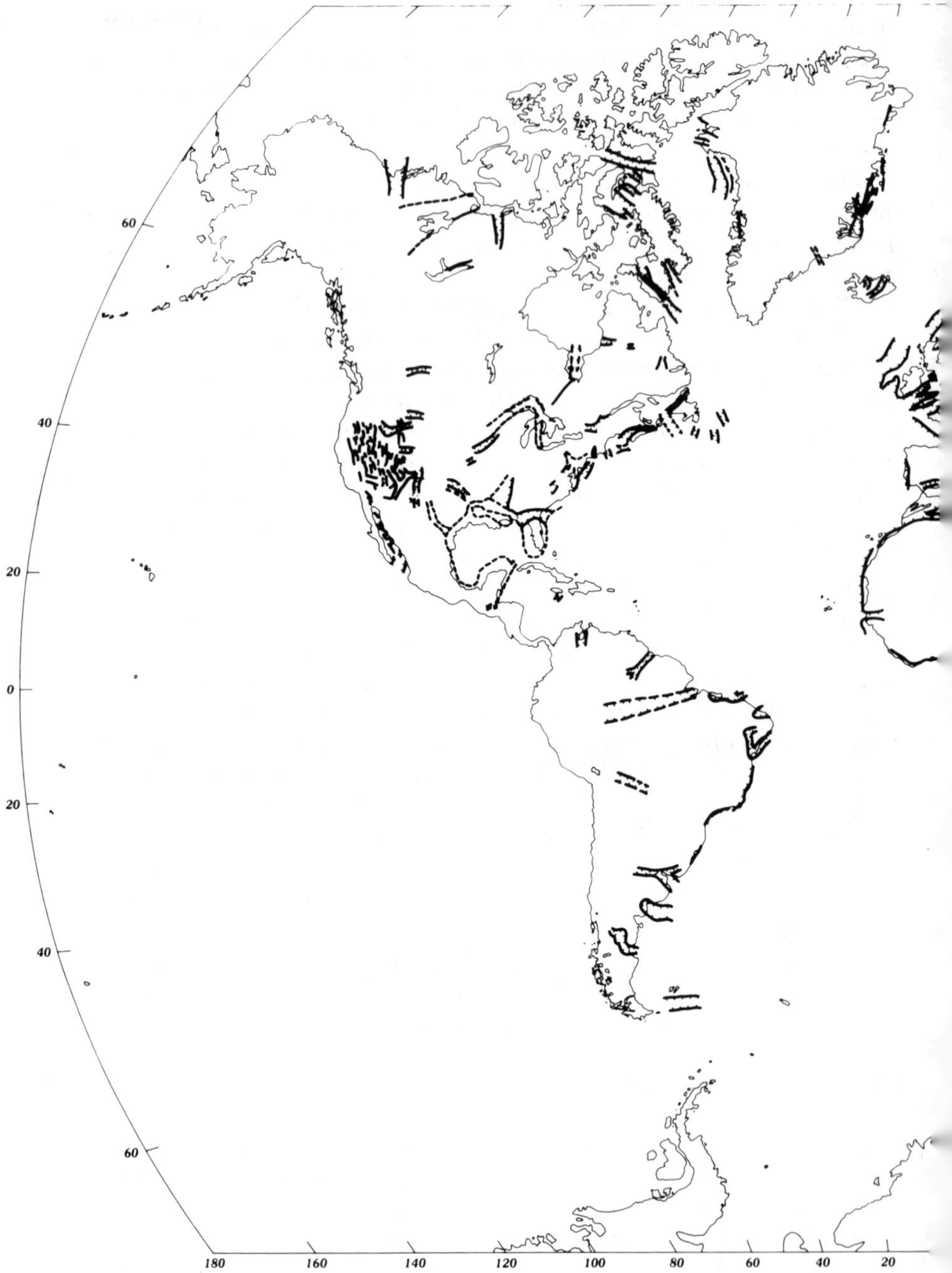

Figure 1. Global distribution of continental rifts and rifted margins (adapte

from Burke et al., 1978).

equally represented. However, Mohr (1982) in considering a selection of primarily younger rifts noted that there is a tendency for north-south axial alignment, and "In their meridional bias, continental rifts imitate major plate boundaries, many of which are related to poles of rotation located close to the poles of the Earth's rotational axis (Burke and Whiteman, 1973)." We consider the possible implications of the alignment of rift axes in more detail below in our discussion of the mechanisms of rifting.

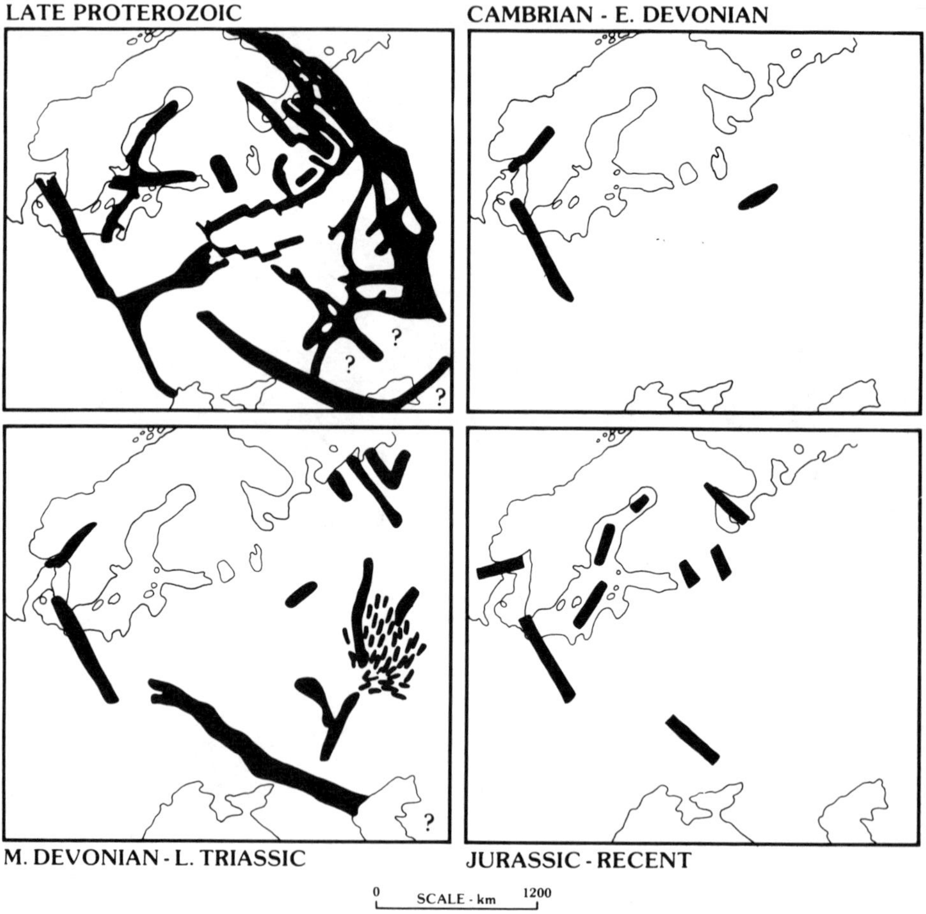

Figure 2. Development stages of rifting in the East European Platform (adapted from Milanovsky, 1981).

All continents contain rifts, frequently subdividing them into a complex system of cratonic fragments (see Figs. 2, 3 and 4). The spatial frequency of rifts apparently vary from one region to another, a phenomenon that is believed in part to reflect the varying degree of accessibility (e.g., Greenland and Antarctica), and in part difficulties in recognizing paleorifts buried below platform sediments (e.g., North America, the interior of Australia, and the Russian and Siberian platforms) or epicontinental seas (e.g., Labrador Sea, Barents Sea and Kara Sea). Certain axial trends of rifting have been observed to predominate in certain continental areas, presumably reflecting fundamental structural differences between one area and another, and perhaps also reflecting characteristic domainal differences with respect to origin and/or time of rifting. For example, an E-W axial trend, approximately perpendicular to the Andean fold belt, prevails throughout the continent of South America (Fig. 1). In contrast, a N-S axial trend predominates in the West Siberian platform (Fig. 1), and in the area of northwest Europe, as seen from the pre-drift reconstruction map of Fig. 4.

Many continents (e.g., Australia, North America and Europe) appear to provide evidence for a radial outward decrease in age of rifting and a tendency for rifts to be concentrically distributed about cratons. These observations are in accord with the idea of a spasmodic lateral migration of rift systems from areas of earlier to those of later cratonization (Khain, 1977). In Australia there is a systematic progression from the active, rifted plate margins around the circumference of the continent, inward through Cenozoic and Mesozoic rifts at the continental margin, to the older rifts of the interior (Burke et al., 1978). A similar distribution pattern exists in North America, as shown in Fig. 3 (Keller et al., 1983) as well as in Europe (Khain, 1977; Ramberg and Neumann, 1978b). In the latter area, an excellent example is provided by the sequence of rifts which start with the Riphean (Late Proterozoic) Moscow rifts in central parts of the Russian Platform (Fig. 2), jump westward to the Late Paleozoic Oslo rift, then continue westward to the Mesozoic West Norwegian trough and the main North Sea rift system, and finally to the continental margin and subsequent Norwegian Sea spreading axis between Greenland and the Eurasian continent. As a corollary of the repeated break-up and subsequent suturing of plates, the generalized distribution scheme described here will be drastically rearranged in certain regions.

Rifts appear to form at all stages of the Wilson Cycle of opening and closing of oceans (Dewey and Burke, 1974), and in a variety of tectonic settings. Many individual grabens or rifts, whether they are modern or paleorifts, appear to be just portions of extensive rift systems, like for instance the largely Mesozoic Arctic-North Atlantic rift system (Fig. 4), or the presently active East African or Salton Trough-Gulf of California rift

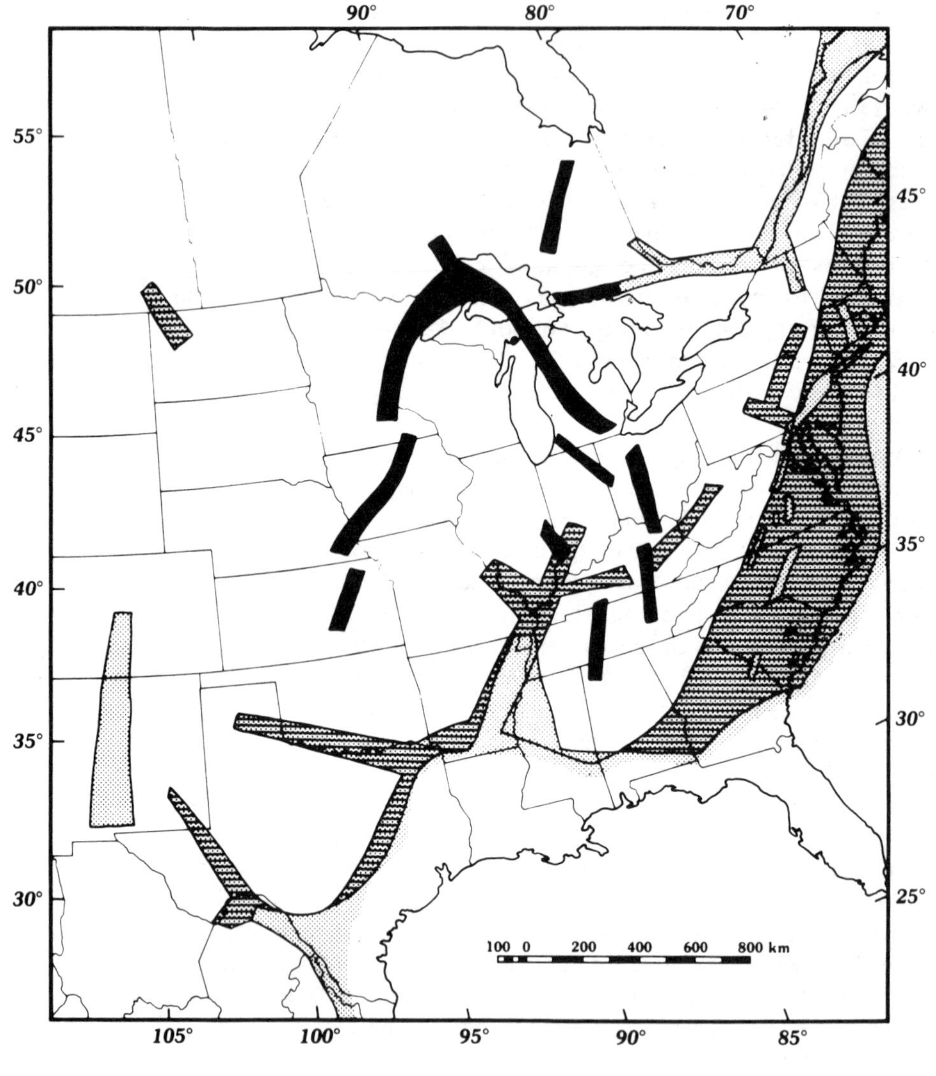

Figure 3. Development stages of rifting in the Midcontinent U.S.A. Solid pattern Precambrian rifts; Horizontal rule pattern Eocambrian rifts; Dot pattern Mesozoic and Cenozoic rifts (adapted from Keller et al., 1983).

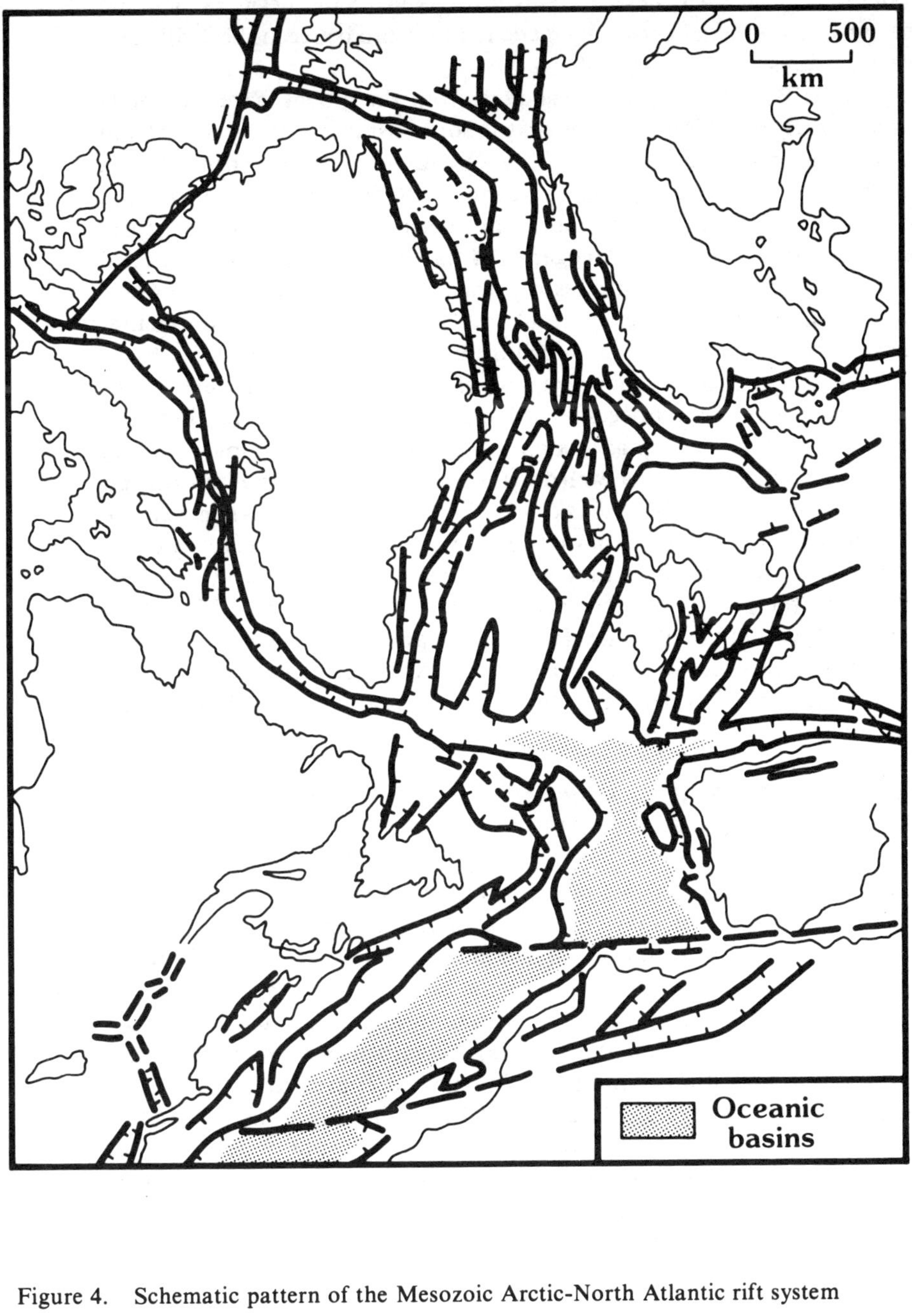

Figure 4. Schematic pattern of the Mesozoic Arctic-North Atlantic rift system (adapted from Ziegler, 1978).

systems (Fig. 1). All these systems are linked to the network of plate margins, strongly indicating a genetic relationship between these rifts and plate margins.

Other rifts seem completely isolated from the global system of plate margins (Fig. 1). This impression may in some instances merely result from inadequate mapping of the regions surrounding the mapped rifts. A number of rifts, however, may strike more or less perpendicular to a fold belt, or penetrate into continents from their continental margins, representing various classes of aulacogens and "failed rifts", respectively. Failed rifts, left over from the successful formation of an oceanic basin, are numerous, and may exceed in total length the length of the continental margin formed at rupture, as shown in Fig. 5 (Burke, 1978).

There is a marked tendency for many rifts to be reactivated. Subsequent tectonic events may not necessarily have extensional character, but may represent both strike-slip and compressional events, sometimes causing basin inversion phenomena, well documented in the Russian Platform and the North Sea (e.g., Milanovsky, 1972; Ziegler, 1982). Whenever the lithosphere is subjected to renewed deviatoric stresses, rifts are likely to act as zones of anisotropy. If they are also reasonably favorably oriented relative to the principal stress axes, the stress will be released by block movement within the rift zone. Hence, there are likely to be only a limited number of rift trends within a certain area. Also, the actual age of rifting may be increasingly difficult to determine if there are superimposed tectonic events.

CLASSIFICATION SCHEMES AND CHARACTERISTIC FEATURES

Continental rifts (and rifted margins) have very diverse characteristics, which implies that rifts may form by a variety of processes. There are numerous categories of rifts, which include paleorifts on continental margins (e.g., Atlantic margins), apparently isolated intraplate rifts (e.g., Midcontinent U.S.), rifts spatially and temporally associated with orogenic belts (e.g., Rhine and Baikal rifts), or with transform boundaries (e.g., Dead Sea, Salton Trough). Rifts may be expressed by complex structures (e.g., Western U.S. Basin and Range province), intracratonic rift-basins (e.g., North Sea basin), or possibly by hot spot traces (such as the U.S. Snake River-Yellowstone zone). This diversity causes rifts to have a variety of geological and geophysical signatures, and these signatures may be further complicated by subsequent deformational events (as for instance in aulacogens), making the identification of many rifts and rift basins a difficult and sometimes controversial task.

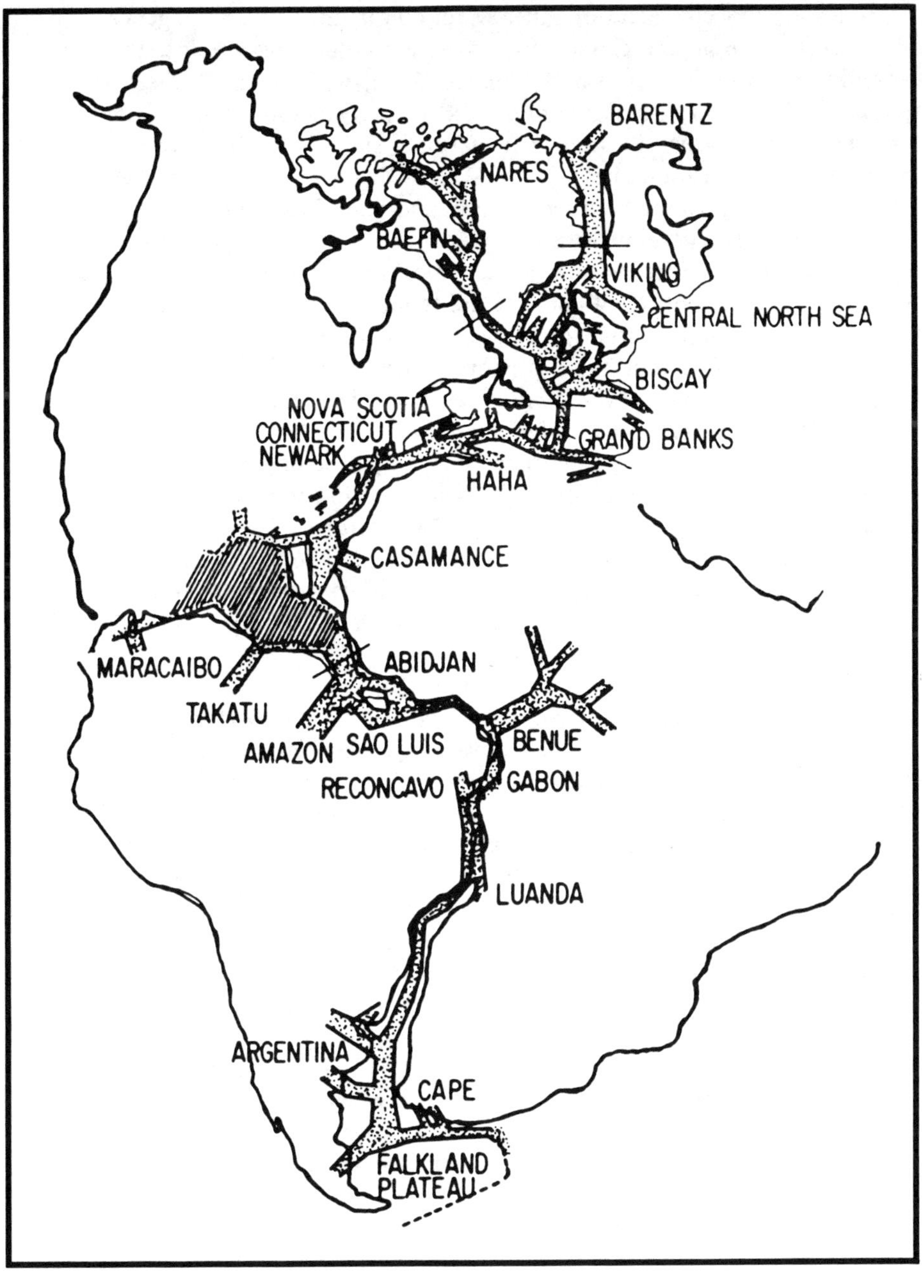

Figure 5. Failed rifts associated with the opening Atlantic (from Burke, 1978).

There are many classification schemes for continental rifts, but a generally accepted, non-genetic classification system, similar for example to that for sedimentary basins (Bally and Snelson, 1980), does not exist. Characterization of continental rifts is complicated by many factors, including the time dependent nature of some rift features, such as mean elevation, gravity and thermal anomalies (see EVOLUTIONARY TRENDS below). Therefore, distinguishing between modern and paleorifts, and preferably also between rifts with contrasting geological settings as well as between rifts in different stages of evolution is of critical importance. Burke (1977) suggested 11 classes of failed rifts extending into continents based upon varied structural and stratigraphic styles. Milanovsky (1972, 1978) has divided rift zones into four basic categories based primarily on structural and genetic criteria. Two of these categories, the continental and intercontinental rift zones, are further subdivided into 1) *epiplatform* rift zones, where rifting took place after a fairly prolonged period of platform development, and 2) *epiorogenic* rift zones, where rifting took place immediately after or during the orogenic stage. Epiplatform rifts were further subdivided between two 'end-members': a) *arch-volcanic* type rifts (major sub-aerial volcanic activity; e.g., Kenya and Ethiopia rifts), and b) *crevice* type rifts (grabens of great depth, volcanism weak or absent; e.g., Rhine and Baikal rifts).

A third approach to rift classification, recognizing the spectrum of rifting processes and settings, attempts to relate the rifts to two fundamental mechanisms of formation: the '*active* mode of rifting,' whereby rifts form as a result of asthenospheric upwelling, and the '*passive* mode of rifting,' whereby rifts form because of deviatoric stresses resulting from plate interactions at a distance (Sengör and Burke, 1978; Ramberg and Neumann, 1978b; Baker and Morgan, 1981). The 'active' mechanism implies a mantle thermal anomaly prior to rifting and which drives the rifting. In the 'passive' mechanism any mantle thermal anomaly is a response to rifting caused by upflow of the asthenosphere beneath the thinned lithosphere during rifting. Sengör and Burke (1978) suggest different sequences of events as typical for the two modes of rifting: the 'active' mechanism predicting doming–volcanism–rifting, the 'passive' mechanism predicting rifting–(uplift?)–volcanism. For active rifting, the heat flow, magnitude of uplift and volume of associated igneous rocks tend to be larger than for passive rifting. The influence of prerift structural elements is supposedly stronger in the case of passive rifting than for active rifting. Unfortunately, as emphasized by Mohr (1982), for most rifts the timing and true interrelationship of uplift, volcanism and rifting generally remain surprisingly loosely determined, and show great regional variety. However, the two postulated evolutionary patterns of rifting, representing simplified 'end-member' type mechanisms, may presently serve as useful working hypotheses.

Which, if any, of these classification systems will eventually prove the best, awaits future research. Whether rifting is caused by processes below or within the lithosphere plates, or if rifting requires a coincidence between lithospheric and sublithospheric processes, may, however, be further studied through compilation and synthesis of data already available on continental rifts. Table II is an attempt to summarize important structural, geomorphic, magmatic and geophysical features typical of modern and paleorifts. These features are further commented upon below.

TABLE II. Characteristic Features of Continental Rifts

	Modern Rifts	Paleorifts
a) *Structural & Geomorphic Features*		
Width generally 35–60 km	x	x
Length of the order of 1000 km	x	x
Development of rift valley	x	
Complex graben-like structure		x
Extensional features, normal faulting and dikes	x	x
Often occurring: asymmetric cross-sections	x	x
intrarift horsts and grabens	x	x
dog-leg patterns	x	x
polarity change along strike	x	x
Broad domal uplifts	(x)	
Thin crust, 35 km or less	x	(x)
Thickened crust		(x)
Thin lithosphere	x	
Broad early stage sedimentary basins	x	x
Broad late stage sedimentary basins		x
b) *Magmatic Features*		
Both 'wet' and 'dry' rifts	x	x
Composition diverse, predominantly alkaline (but also calc-alkaline and tholeiitic)	x	x
Bimodal igneous activity	x	x
Progression from (per-) alkaline to tholeiitic composition	(x)	(x)
Migration from peripheral to axial activity	(x)	(x)
Dike swarms, ring complexes	(x)	(x)
Subvolcanic and/or midcrustal magma chambers/ plutonic rocks	(x)	(x)
Subcrustal asthenosphere diapir	x	
Deep crustal rift 'cushion'		(x)

TABLE II. Continued

	Modern Rifts	Paleorifts
c) *Geophysical Features*		
P_n velocities less than 7.8 km/s	x	
P_n velocities greater than 8.0 km/s		x
Crustal low-velocity layers	x	
High velocity lower crustal layers		x
Teleseismic P-wave delay	x	
Long-wavelength Bouguer gravity low (low density mantle)	x	
Long-wavelength Bouguer gravity low (thickened crust)		(x)
Axial Bouguer gravity high (approx. 100 km wavelength)	(x)	x
Local Bouguer gravity highs and lows (approx. 10 km wavelength)	x	(x)
Axial Free-Air anomaly (high), decreasing with increasing isostatic compensation	(x)	(x)
Complex magnetic anomaly pattern	x	x
High surface heat flow	(x)	
Elevated lithospheric isotherms	x	
Normal heat flow and isotherms		x
Shallow Curie point depth	x	
Magneto-telluric anomalies	x	
Shallow seismicity, aligned with rift	x	
Extensional tectonics inferred from focal mechanisms	x	

x = typically occurring
(x) = somtimes occurring, lack of identification sometimes in part due to lack of exposure because of thick (post-rift) sedimentary cover.

Structural and Geomorphic Features (*Table IIa*)

As frequently stated in the literature, most continental rifts have a width of 35–60 km, apparently irrespective of age and tectonic setting. Thus, a simple relationship between rift width and crustal or lithospheric thickness, is not reflected in an age–width plot of continental rifts. Perhaps a more critical factor is the strain rate. Slow rates tend to result in wide block-faulted basins, while fast strain rates favor the development of relatively narrow, well-defined rifts, often associated with volcanism. Within an individual rift, this mechanism tends to result in a narrowing down with time, the activity becoming gradually more concentrated along high-strain axial zones.

Modern rifts, and many paleorifts *not* subjected to subsequent compressional deformation, occur almost ubiquitously in areas of apparently abnormally thin crust (and lithosphere), as indicated in Table II. Since areas of thinner continental crust are generally stronger than those of thicker crust (because of the relative strength of olivine in the mantle to quartz in the crust), unless weakened by heat, it is reasonable to assume that the thinning is not the cause, but a result of the process(es) of rifting (see discussion in MECHANISMS OF RIFTING below).

Regional uplift, if it occurs at all, is a characteristic feature only of thermally active, modern rifts. Another contrast between modern and paleorifts (if exposed) is that the rift valley morphology becomes less apparent with age due to erosion and sedimentary infilling. As rift age and depth of erosion increase one commonly observes an increasingly complex graben-like structure, with ductile fault zones, dikes and other intrusions. Well exposed Proterozoic rifts include the Athapuskow aulacogen in Canada (Hoffman, 1973, 1980; Easton, 1983) and the Gardar igneous province, Greenland (Upton and Blundell, 1978). The ratio of exposed plutonic rocks to (sedimentary and volcanic) graben fill is likely also to increase in general with depth of erosion. These changes will be accompanied by corresponding changes in the local (or short-wavelength) magnetic and gravimetric anomaly pattern (see below). Thus, despite the gradually modified morphotectonic expression observed in older rifts, the overall *similarity* of even Proterozoic rifts to modern ones suggests that similar processes of rifting may have persisted at least for the past 2,500 m.y. or so.

Rifts are commonly characteristically asymmetric, both in plan and cross-section. The North Sea Horn Graben, for example, shows many of these asymmetric features (see Figs. 6 and 7). The approximately 300 km long graben or rift zone, varies in width from about 30 km in central parts to more than 70 km toward both ends, revealing a funnel-shaped map configuration characteristic also for many oceanic ridge-rift segments, e.g., the FAMOUS area (Ramberg et al., 1978). The termination of the graben is further typified by a distinct fanning of the fault pattern and a decrease in dip-slip components along the individual faults, as seen also, for example, in many of the segments of the East African rift system (Baker et al., 1972) and at the southern end of the Rhine Graben (Illies, 1977).

The Horn Graben is outlined in its northern part by a major boundary fault to the east and by basement warping and a number of step-faults to the west (Fig. 6). To the south, the polarity is reversed, with the major boundary fault on the west side, with up to 6 km of vertical displacement at base Zechstein (Upper Permian) level. Although the graben is basically an extensional feature, the overall shape, local occurrences of compressional features and the *en echelon* distribution of minor faults suggest that a slight

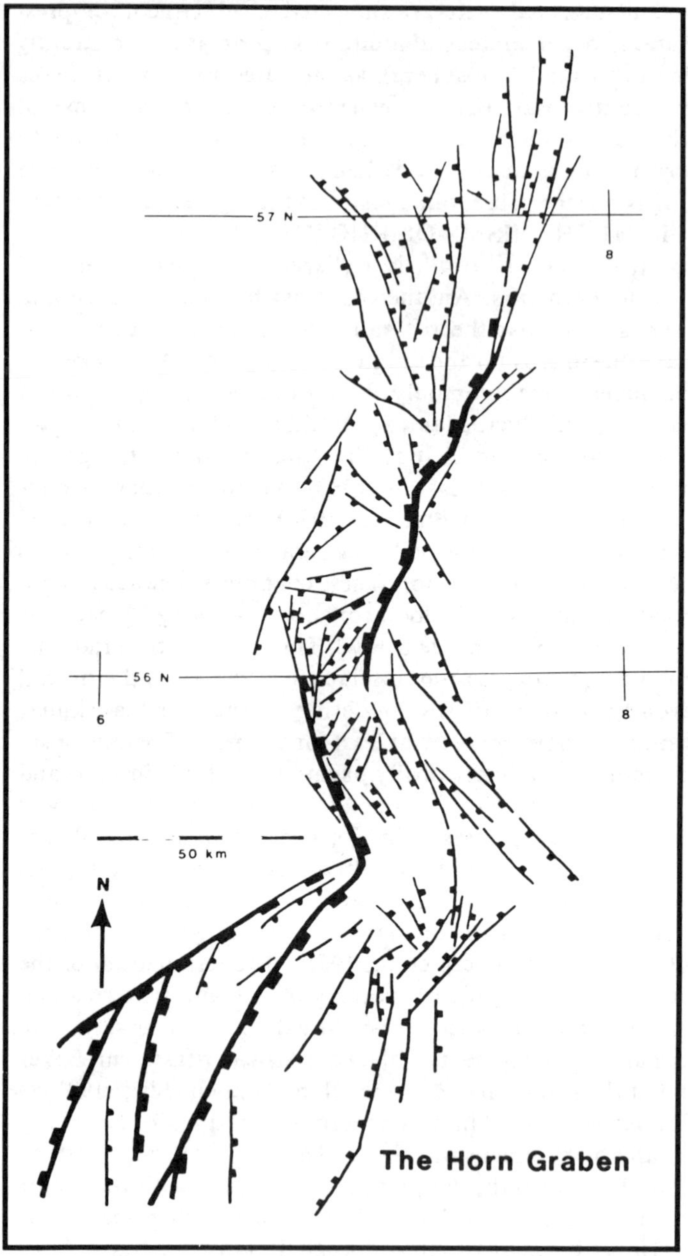

Figure 6. Fault pattern in the Horn Graben. Tick marks show down side of faults.

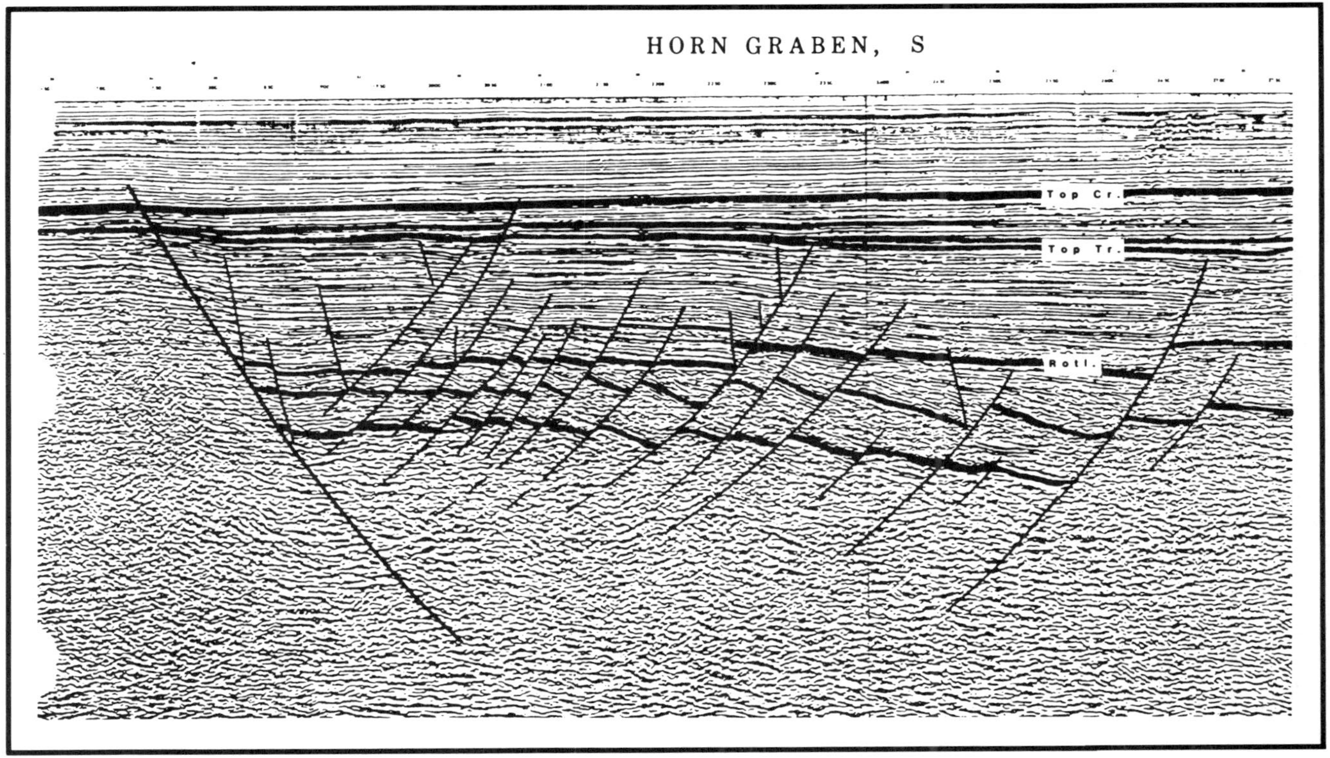

Figure 7. Interpreted seismic cross section, central part of the Horn Graben.

sinistral strike-slip movement took place during an early phase of deformation. Minor intragraben horsts are seen near the central part (south of 56° N) as well as near both ends.

Figure 7 demonstrates that the subsidence of the floor of the Horn Graben occurred in three distinguishable events (or periods) and by somewhat different mechanisms. The first two, in Permo-Carboniferous, were strongly asymmetric events, with antithetic block faulting facing the major boundary fault, and with rotation of individual blocks along listric normal faults. One or both of these early events were accompanied by the presumed sinistral strike-slip movement. The third event, Triassic in age, was apparently a simple extensional event producing uniform subsidence across the graben.

The multiple stage evolution of the Horn Graben was accompanied by Early Permian volcanism, alkaline basalts and rhyolites (Michelsen, 1982; Olsen, 1983a; Best et al., 1983). Lava flows and interbedded agglomerates, tuffs and detrital sediments presumably covered large parts of the graben floor. This activity was synchronous with similar volcanism in the Oslo Graben (Ramberg and Larsen, 1978) as well as in the Central Graben of the North Sea (Michelsen, 1982), indicating synchronous onset of rifting and volcanism in large parts of the North Sea rift system.

The structural relationships recorded in the Horn Graben, as well as the evidence of repetitious tectonic events, may be typical for many continental rifts. The observations support the idea that once a rift is established it continues to act as a major zone of weakness along which tectonic stresses transmitted through the lithosphere tend to be released. The fact that separate periods of rifting within the same region apparently develop quite differently, may also imply that the composition and structure of the lithospheric plate itself do not represent the overall controlling factors.

Magmatic Features (*Table IIb*)

The magmatic character of continental rifts varies within wide ranges, as indicated in Table IIb. Thus, in a recent comparative study of the Kenya, Rio Grande and Oslo rifts, Williams (1982, p. 219) concludes that "rifts have less in common in terms of magmatic activity than would be expected from a cursory examination of the literature." Furthermore, since igneous rocks of 'typical' rift character also occur in other geological setting, "supposedly characteristic suites of alkaline rocks do not therefore provide reliable criteria for the identification of paleorifts" (Williams, op cit). Hence, it is the total *assemblage* of physical, geological and geochemical parameters that help identify both modern and ancient continental rifts. The striking variety in, for example, the igneous expressions possibly reflects the influence of factors such as the tectonic setting, lithospheric and asthenospheric composition and thermal evolution.

It has repeatedly and correctly been pointed out that although rifts are commonly associated with extensive volcanism ('wet rifts'), other rifts are associated with very little or none ('dry rifts'). Most rifts considered to be 'active' serve as examples of the 'wet rifts', while the Western Rift of Africa, the Baikal rift, the Rhine Graben and many aulacogens are examples of the 'dry rifts' (e.g., Mohr, 1982). While the contrast in volcanic appearance between 'wet' and 'dry' rifts may reflect differing thermal conditions in the asthenosphere, it would seem that the *total magmatic* products (volcanics and intrusives) is a less arbitrary and consequently more significant measure of the magmatic activity within a rift zone. There are abundant geophysical data indicating the presence of buried volcanics, shallow and deepseated magma chambers or intrusive bodies within rift zones (e.g., Girdler et al., 1969; Fairhead, 1976; Ramberg, 1976; Sanford and Einarsson, 1981).

Thus, most rifts are characteristically associated with large volumes of igneous rocks, most of which are trapped within the crust, especially at deep crustal levels. The interesting question of why only small, or in some instances only negligible, portions of this melt reach the surface still remains. It has been suggested that 'wet rifts' are favored by high strain rates, while 'dry rifts' result in part from slow strain rates, the strain rate being linked to the dynamics of asthenospheric upwelling and magma generation (Barberi et al., 1982). Diagnostic data are not currently available to test the widespread applicability of this intriguing hypothesis, however.

Rift magmatism is generally alkaline (ranging from transitional to strongly alkaline), but tholeiitic and calc-alkaline rock types also occur. It seems, however, critically important to observe the relative timing of rifting and volcanism; for instance to discriminate between pre-rift and rift volcanism, the former of which generally reveals less alkaline affinities. In rift volcanism proper, a shift from alkali basalt to tholeiitic composition sometimes occurs with time (Gass, 1970; Ramberg and Larsen, 1978). In other instances, geographical position appears to be a dominating factor for magma chemistry, with extrusion of tholeiitic basalt near the axis of the rift, while alkali basalts are extruded on the flanks (Lipman, 1969; Baker et al., 1972; Barberi et al., 1974, 1982).

Mafic alkali rocks (alkali basalt, basanite, nephelinite, etc.) are clearly enriched in alkalies, incompatible elements (including rare earth elements (REE)), and volatiles (particularly CO_2 as compared to tholeiitic basalts, and are enriched in light relative to heavy REE (e.g., Neumann and Ramberg, 1978a). Tholeiitic continental rift and plateau basalts are enriched in alkalies and incompatible elements relative to abyssal tholeiites, and are enriched (or at least undepleted) in light relative to heavy REE. The recorded major differences between continental rift basalts may have resulted from a number of processes in operation during and / or after melting. The most important factors controlling these differences include mantle heterogeneity, depth of melting and high level fractionation for the weakly alkaline basalts (see reviews in e.g., Neumann and Ramberg, 1978a; Baker et al., 1978; O'Hara, 1977).

The often described bimodal distribution of continental rift volcanics in terms of their SiO_2 content has been discussed by Mohr (1982), suggesting that the bimodality (or 'Daly Gap') is in part an artifice caused by the use of SiO_2 as an index of fractionation, and in part a result of crustal density filtering (Gill, 1973). Mohr (op cit) states that the apparent bimodality and volumes of volcanic types need not reflect the proportion of fractionated magmas evolving in the *deep* magma chambers. In our opinion, this is supported by the fact that in the deeply eroded Oslo Graben, there is a large proportion of both intrusive and extrusive rocks of intermediate composition, effectively narrowing the 'Daly Gap'.

Igneous activity within rift zones is manifested by a number of characteristic features such as dikes and sills, dike swarms, explosion vents, volcanoes and ring complexes, and in more ancient (more deeply eroded) rifts also by composite batholiths. These volcano-tectonic features commonly display a regular spacing as well as orientation relative to rift axis. The spacing of volcanic centers has tentatively been related to crustal or lithospheric thickness (Mohr and Wood, 1976; Easton, 1983), while the orientation of the various features which may serve as paleostress indicators (Nakamura, 1977; Zoback and Zoback, 1980) have been utilized to resolve stress variations with time (Baker et al., 1978; Ramberg et al., 1984).

Geophysical features (*Table IIc*)

Geophysical signatures over modern rifts in particular strongly suggest that rifts are associated with crustal structure that departs significantly from that of standard continental crust (Table IIc). Some of the same characteristic signatures are also found in paleorifts, although they are typically less pronounced, especially for Proterozoic rifts.

The best constrained evidence for anomalous crustal and upper mantle structure beneath rifts comes from *seismic* profiling, and reveals that modern rifts are commonly underlain by crustal (and lithospheric) thinning, anomalous upper mantle, as well as shallow and mid-crustal velocity anomalies (e.g., Ansorge et al., 1970; Griffiths et al., 1971; Pilger and Rösler, 1975, 1976; Prodehl et al., 1976; Prodehl, 1981; Puzyrev et al., 1978; Mueller, 1978; Brown et al., 1980; Brocher, 1981a,b; Olsen, 1983b). Paleorifts are commonly underlain by a high velocity basal crust that is interpreted to result from crustal modification by injection of mantle derived material (e.g., Mooney et al., 1983; see also Snake River Plain, e.g., Braile et al., 1982).

The uppermost mantle below the thinned crust of modern rift zones is characterized by anomalous P-wave velocities, commonly in the range 7.3–7.8 km/s (e.g., Fuchs, 1977; Olsen, 1983b). This low velocity, low density mantle material is often interpreted as the partly melted uppermost portion of an asthenospheric diapir. Teleseismic P-delay data from the Baikal area (Zorin and Rogozhina, 1978) suggest this anomalous mantle may extend over a considerably larger area than the visible surface expression of the rift zone. Indications of a complex, lamellar crust–mantle transition zone, for instance in the Rhine Graben (Mueller et al., 1973; Werner et al., 1982), may be due to numerous basic intrusions into the lower crust. Within the larger Rhine Graben area, there is also evidence of a pronounced seismic velocity anisotropy, with maximum P_n velocities of 8.43 km/s and 7.77 km/s roughly parallel and perpendicular to the graben axis, respectively (Bamford and Crampin, 1977; Werner et al., 1982).

Velocity–depth curves from four continental rifts, shown in Fig. 8 (Olsen, 1983b), demonstrate that in addition to anomalous P_n velocities, midcrustal velocity inversions have been observed in the Baikal, Rhine and Rio Grande rifts, but not in the Kenya rift, presumably due to insufficient data. In the Rio Grande rift, evidence for the presence of a mid-crustal magma body (at about 18–20 km depth) is strong (Sanford et al., 1979; Sanford and Einarsson, 1982), although its actual thickness and regional extent is a matter of debate (and continued research). In Kenya (Fig. 8), the apparently high velocity of 6.4 km/s for crustal material above depths of 18 km has been suggested to reflect the presence of shallow high velocity basaltic material, which is consistent with interpretations of gravity (Girdler et al., 1969, Baker and Wohlenberg, 1971) and teleseismic data (e.g., Maguire and Long, 1976).

In paleorifts we do not observe anomalous upper mantle velocities. Instead, high velocity basal crust occurs on top of normal mantle (P_n ½ 8.1 km/s). In the Late Paleozoic Oslo Graben, a basal crustal body with assumed $P_n = 7.5$–7.6 km/s occurs at depths between approximately 20 and 32 km (the Moho) (Sellevoll and Warrick, 1971; Tryti and Sellevoll, 1977; Ramberg, 1972, 1976). Apparently the complex transition zone and part of the asthenosphere diapir present in modern rifts have been included in the paleorift crust, with the Moho interface re-established below the paleorift cushion, now resting on normal mantle. In the Oslo Graben, however, the crust is still thinned by several kilometers relative to the Fennoscandian Shield on both sides (Ramberg, 1976). The original crustal thinning was much more substantial, with the Moho in or above the dense basal crustal injection zone.

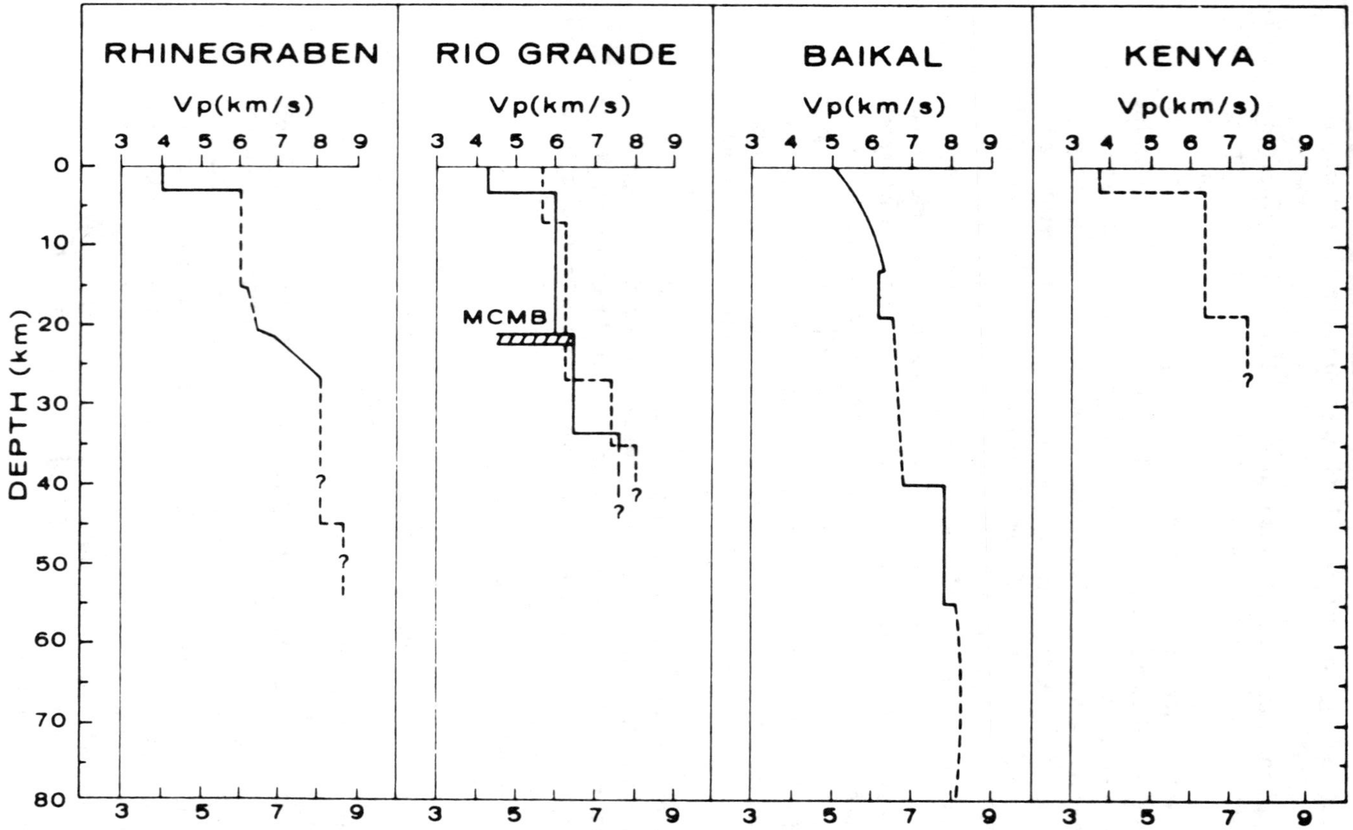

Figure 8. Comparison of compressional seismic velocity versus depth for the central axial regions of four major Cenozoic continental rifts (from Olsen, 1983b).

In an even older paleorift, the Mississippi Embayment, which is believed to be the site of a Late Precambrian continental rift that was reactivated in the Mesozoic (Burke and Dewey, 1973; Kane et al., 1981), there is also an anomalous high-velocity layer (7.3 km/s) at the base of the crust (Mooney et al., 1983). Mooney and co-workers conclude that the presence of the 7.3 km/s layer indicates that the lower crust has been altered by injection of mantle material. In contrast to the much younger Oslo Graben, the Mississippi Embayment provides evidence of a thickened crust, presumably due to subsequent deformational events and isostatic readjustment, or possibly due to underplating or intrusion into the lower crust. The spatial relationship between many old rifts and cratonic basins suggests that subsidence may be related to structural and mass changes introduced by rifting (Bally and Snelson, 1980), the positive mass anomaly causing isostatic sinking (McGinnis, 1970; DeRito et al., 1983).

Gravity studies represent another important data source for rift studies. Combined with seismic refraction and reflection data, gravity data offer regional coverage suitable for interpretation of 3-dimensional rift structures. Since gravity anomalies reflect lateral mass variations, such studies help identify the stage of evolution of modern rifts, or stage of evolution at the time of arrest of paleorifts.

The pertinent characteristic Bouguer gravity features of rifts are summarized in Fig. 9. A very long wavelength anomaly, of the order of 1,000 km (marked *1* in Fig. 9), is observed over some modern rifts, for instance the East African rifts (Girdler et al., 1969) and the Rio Grande rift (Ramberg et al., 1978). This anomaly has been interpreted as being due to a large body of slightly lower density asthenosphere replacing the overlying lithosphere. Only a few percent partial melt is required in order to cause the negative density contrast; the proportion of melt to solid asthenospheric rocks presumably increases towards the upper, gradually more decompressed portion of the rising diapir. The long wavelength gravity anomaly is not observed over ancient rifts, consistent with seismic data indicative of normal upper mantle velocities.

A broad, central gravity high, up to 100 km wide (*2* in Fig. 9) superimposed on the long wavelength low, is often found over modern rifts (e.g., Girdler et al., 1969; Khan and Mansfield, 1971, Artemjev and Artyushkov, 1971; Searle and Gouin, 1972; Ramberg et al., 1978; Werner et al., 1982), but is more commonly, and perhaps more strongly developed over ancient rifts (e.g., McLean and Qureshi, 1966; Ramberg, 1976; Halls, 1978; van Calsteren and den Tex, 1978; Keller et al., 1983). The marked Bouguer gravity high implies that the asthenosphere has penetrated the crust (creating a positive density contrast), as schematically illustrated in Fig. 9. The size of the central high also serves as a direct measure of the mass of injected mantle material, and more indirectly of the degree of crustal

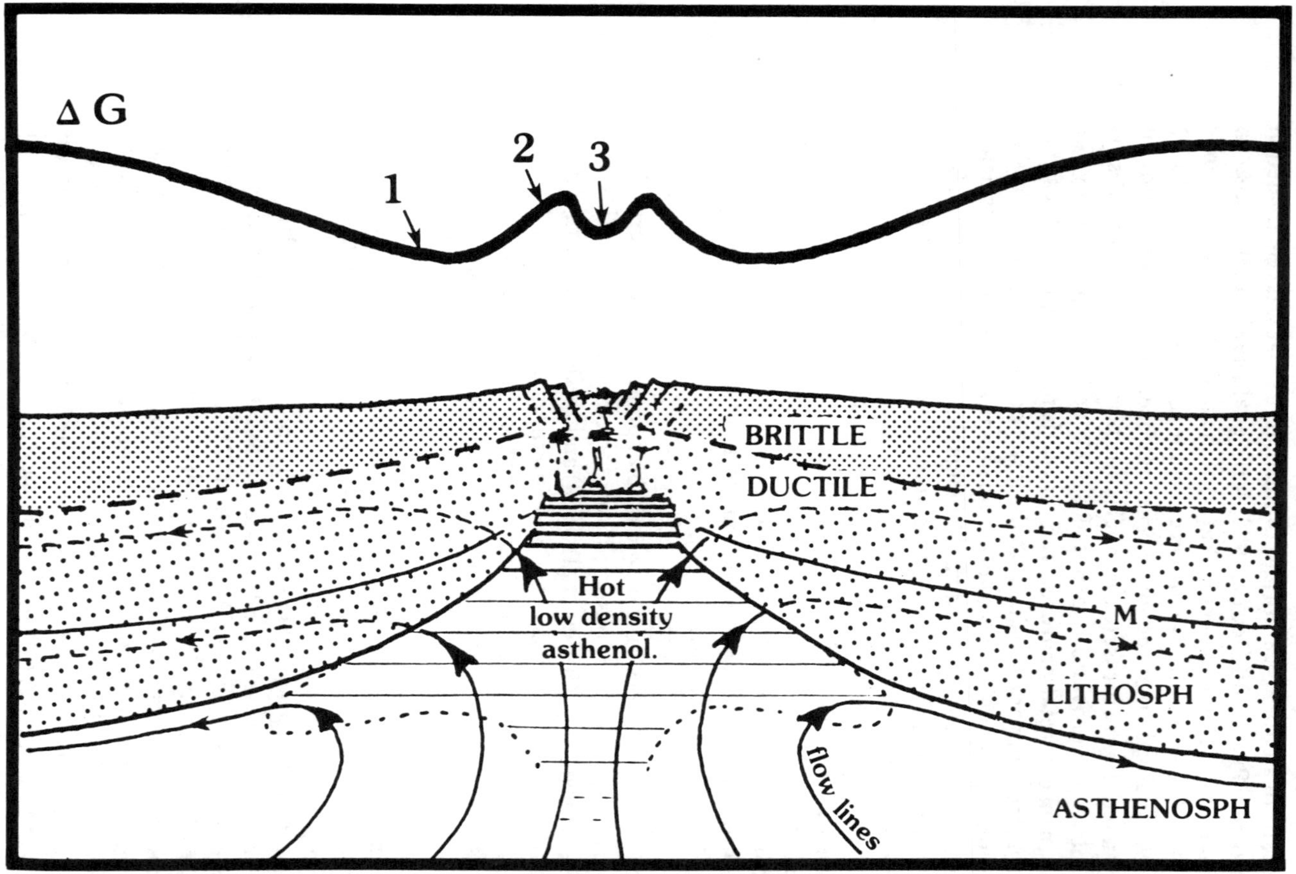

Figure 9. Schematic cross section of a continental rift with qualitative Bouguer gravity anomaly profile. (See text for explanation).

thinning. The axial intrusive complex (sometimes misleadingly termed 'axial dike') commonly occurs in the lower crust, but may extend to near the surface (Khan and Mansfield, 1971; DeRito et al., 1983). Even in areas like the Cenozoic Baikal rift where a positive gravity anomaly is not apparent in the Bouguer gravity data, a wide gravity high is apparent when a correction is made for the huge sediment accumulation approximately 6 km in thickness (Zorin, 1966, 1981; Artemjev and Artyushkov, 1971). Although a central gravity high is not found everywhere (for instance the Western Rift of Africa (Wohlenberg, 1970) or the St. Lawrence rift system (Kumarapeli, 1978)), injection of mafic rocks from subcrustal levels seems to represent a key factor in the development of many rift systems and their associated igneous rock suites, as judged from the gravimetric evidence.

The mass excess resulting from the intrusion of the relatively dense material into the crust will cause, in general, both a positive Bouguer anomaly and a positive free-air anomaly (DeRito et al., 1983). Since the former is due to the density contrast between the intrusive complex and its immediate surroundings, this anomaly will persist indefinitely once it is formed, although it may be somewhat masked by negative anomalies caused by low density sediments and intrusives (see below). The free-air anomaly will tend to diminish with time, however, reflecting the degree to which local isostatic compensation has been achieved.

A third type of gravity anomaly (*3* in Fig. 9) commonly occurs superimposed on the wider gravity high. These anomalies are sharply defined local lows, usually of the order of 10 km across. They result either from the central graben depression (usually filled with sediments, volcanics and / or water), or from shallow felsic intrusives, or the combination of all these factors. In modern rifts, the rift valley depression with low density fill is the most likely cause of local gravity lows, as observed by Bullard (1936) in the Gregory (Kenya) rift, while the effects of felsic batholiths appear predominantly in more deeply eroded paleorifts, such as the Oslo Graben (e.g., Ramberg, 1976).

Most *heat flow* data from modern continental rifts (e.g., Baikal, Basin and Range, East African, Rhine and Rio Grande rifts; Morgan, 1982, 1983) indicate that young rifts are characterized by high heat flow, with means from the grabens in the rift systems typically within the range 70–125 mW m^{-2}. Data from the essentially non-volcanic sections of the East African rift system indicate normal to low heat flow, however, indicating a strong component of volcanic heat transport in young rifts. With the exception of the Basin and Range system, and parts of the Rio Grande system, high heat flow in young rifts appears to be restricted to the grabens, and is not measured on the broad domal or plateau uplifts associated with the rift systems. This result has been interpreted as an indication that deep seated thermal processes may not be immediately manifested by conduc-

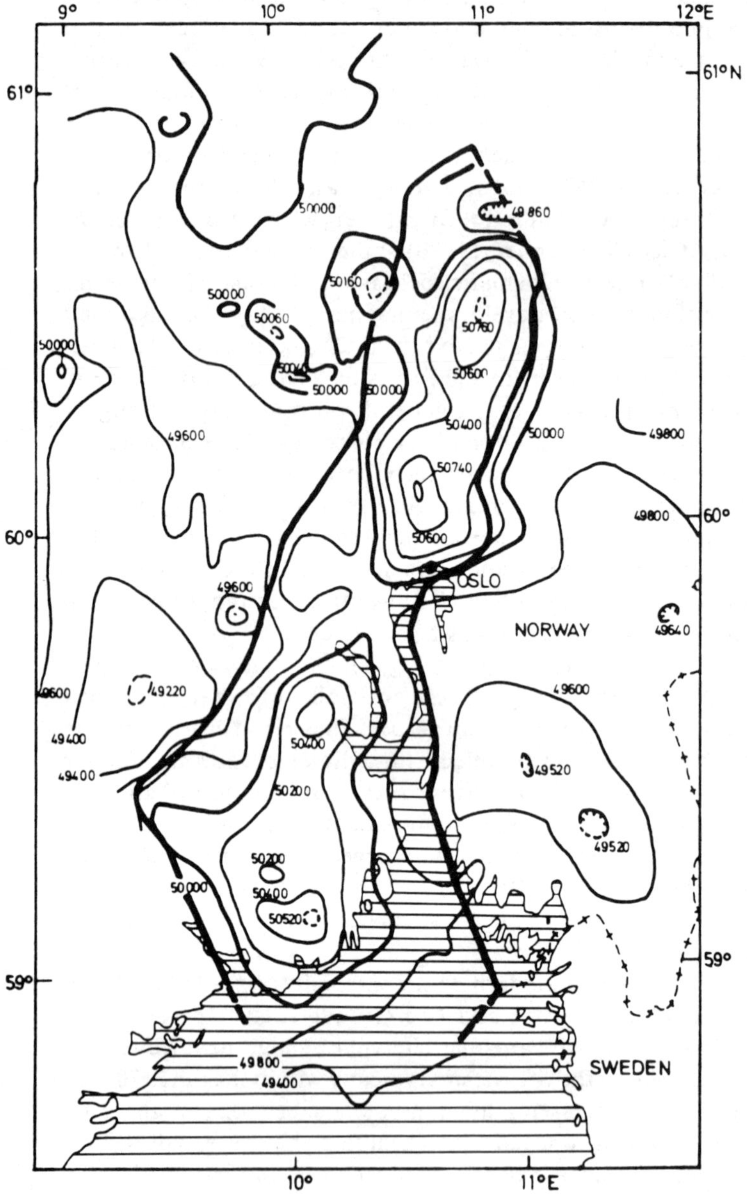

Figure 10. Total magnetic intensity map with 200 gamma contour interval over the Oslo Graben (which is contoured by thick lines) (from Husebye and Ramberg, 1978).

tive thermal anomalies at the surface in young rift zones (Morgan, op cit). High heat flow is only observed where it is convected to the surface by magmatic heat transport, and conductive surface heat flow may actually peak after the arrest of rifting due to the slow conduction of heat from depth. As the thermal time constant of the continental lithosphere is typically of the order of 150 m.y. (Blackwell and Chockalingam, 1980; Morgan, 1984), thermal anomalies associated with rifting processes are not expected in paleorifts. Small thermal anomalies are observed in paleorifts, however, but are generally attributed to thermal refraction effects due to lateral inhomogeneities in the crust (e.g., the Oslo Graben; Haenel et al., 1974, 1979; Swanberg et al., 1974), or deep groundwater circulation, possibly along reactivated rift faults (e.g., the Upper Mississippi Embayment; Swanberg et al,. 1979).

The *aeromagnetic* signatures of rifts are highly variable, depending to a large degree upon factors such as thickness of graben fill, amount of post-rift sedimentary infilling, volume and depth of igneous rocks, etc. However, the common axial intrusion complex, combined with possible dikes and volcanic products generally leads to a characteristic zone of local magnetic highs and lows following the axial trend of the rift. This is well demonstrated in the Central North American rift system (Halls, 1978). Another example is given by the Oslo Graben, shown in Fig. 10, where large magnetic anomalies follow the outline of the major paleorift batholiths (Husebye and Ramberg, 1978).

Satellite magnetic anomaly data, POGO and Magsat, have been globally correlated with rift structures (Frey, 1979; Frey and Ruder, 1981). Fig. 11 reveals the close correlation that is found between the POGO contours and the Moho topography below the Oslo paleorift. Figures 10 and 11 are from the same rift. The correspondence between geophysical signatures and tectonic structures suggests the magnetic pattern obtained from satellite data is probably controlled by lateral variations in the thickness and composition of the lower crust (Frey and Ruder, 1981).

Magnetotelluric studies have shown that conductive intracrustal layers appear to be characteristic of modern continental rifts (Garland, 1981; Jiracek et al., 1979, 1983). Although the low resistivity layers may generally be explained by free water, crustal magma, ductile zones, or concentration of certain minerals (graphite, serpentine), Jiracek et al. (1983), in a recent study of the Rio Grande rift, conclude that their data support the existence of a ductile layer at about 10 km depth, and the hypothesis of magma injection. Reviewing the deep electrical studies in five rift zones (Baikal, Rhine, Rio Grande, East Africa and Iceland), Hermance (1982) emphasizes that all these regions exhibit anomalously low values of electric resistivity, either within the crust or at high levels in the mantle. Furthermore, Hermance (op cit) maintains that the deep electrical investigations

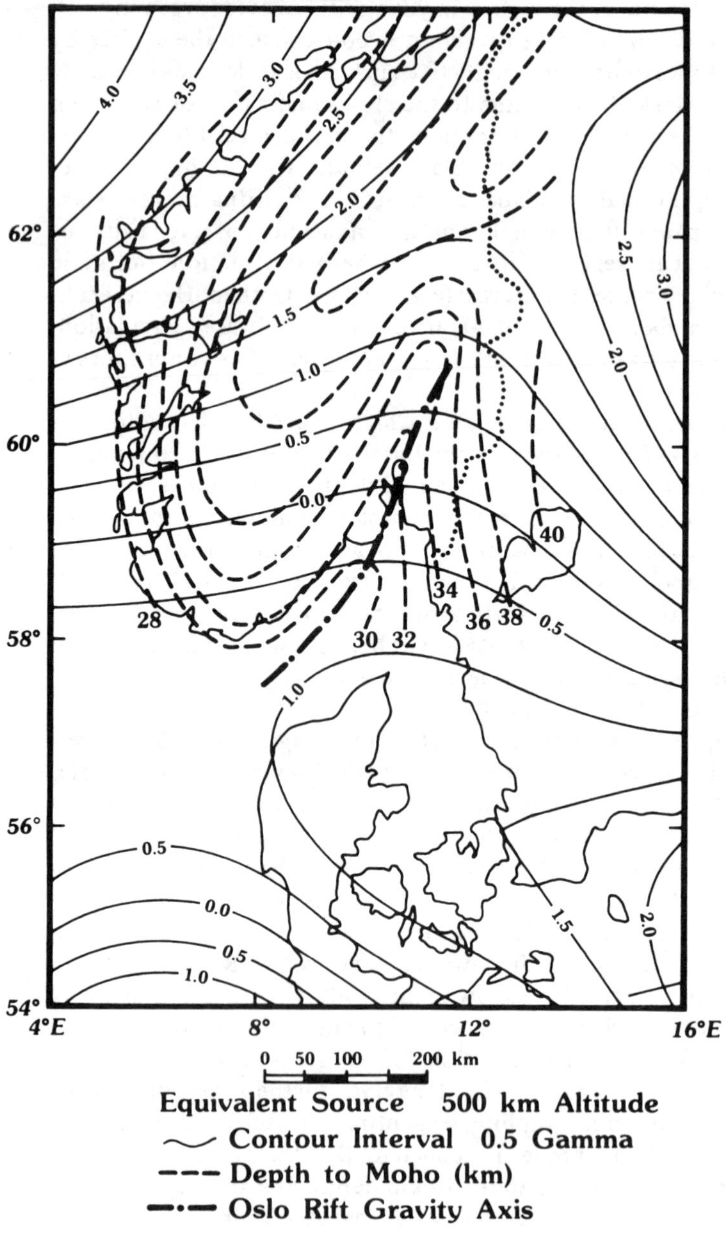

Figure 11. POGO equivalent source magnetic anomalies for southern Norway, compared with Moho topography from Ramberg (1976) (from Frey and Ruder, 1981).

support the model where ascending masses of asthenospheric material are directly coupled to the fractionation of a basalt melt and its subsequent injection into crustal levels.

The morphotectonic, magmatic and geophysical features reviewed above are predominantly taken from studies of a handful of the so-called 'major rift zones of the world'. As can be seen in Figs. 1–4, many more rifts exist, both modern and ancient, many of which are major features, comparable to the 'major rift zones'. A critical test to the validity of many of the characteristic features, or to their degree of variation, will therefore come from continued, detailed research carried out in some of the more unknown rifts of the globe.

EVOLUTIONARY TRENDS

A great variety of evolutionary trends are indicated from the data available for continental rifts, as might be expected from their diverse settings and characteristics. Ideally, a simple sequence of events might be expected for each rift or groups of rifts, with clear sequences of faulting, magmatic/volcanic activity, and regional uplift or downwarp, as suggested in genetic classifications of rifts (e.g., Sengör and Burke, 1978). Geological, geochemical and geophysical features associated with rift development might likewise be expected to follow an evolutionary sequence. In practice, however, wherever sufficient data are available to elucidate the history of a rift, complex evolutions are indicated, and although rifts may have gross evolutionary similarities, they inevitably differ markedly in detail. A comprehensive review of evolutionary trends in rifts is beyond the scope of this review article, and would almost certainly be controversial due to the data interpretation required to define these evolutionary trends. As an alternative approach we present examples of the evolutionary trends in rifts, with discussion of the problems in defining the trends, and the possible significance of the trends. For additional discussion of this subject the reader is referred in particular to Mohr (1982), Williams (1982), and other papers in Palmason (1982).

Development of Rifts–Timing of Events

A major problem in identifying the timing of events in the development of rifts is the separation of events directly related to rifting from events which are related to other tectonic activity, but which are closely related spatially and temporally to the rifting events. As noted above, rifts form in a variety of tectonic and magmatic settings, which complicate and probably modify the events of rift development. A good example of this problem is given by the Cenozoic volcano-tectonic evolution of southwestern New Mexico, USA, which has culminated in the development of the southern Rio Grande rift (Fig. 12). Although dating of magmatic and tectonic events in this region is relatively good, it is difficult to separate magmatism (if any) directly associated with the first (backarc?) extension event in this area from earlier arc(?) and non-rifting backarc(?) magmatism. Magmatism associated with the second extension event, which has formed the modern southern Rio Grande rift, is different in composition from earlier magmatism, but again does not have a clearly defined genetic temporal relationship to extension. Magmatism probably was initiated prior to the main second extension phase, which suggests an 'active' rifting mechanism, but was more abundant after the main rifting event, suggesting a passive response rather than an active source.

Similarly complex sequences of tectonism and magmatism have been elucidated for other rifts, such as the Oslo Graben (Ramberg and Larsen, 1978), the Rhine Graben (Illies, 1978), the Baikal rift (Logatchev, 1978), and the Kenya and Ethiopia sections of the East African rift system (Baker et al., 1972; Williams, 1978, 1982; Mohr 1982), although some sequence to the eruption of new magmatic products has been identified in East Africa (Wendlandt and Morgan, 1982). In many rifts (e.g., East Africa, Rhine, Baikal) broad, gentle crustal subsidence appears to preceed rifting, although this may not be a universal feature of rifting, and is impossible to qualify as a *precursor* to rifting (e.g., Mohr, 1982). In some rifts (e.g., the southern portion of the western section of the East African rift system) volcanism is sparse, and a reliable tectono-magmatic history cannot be constructed as the possibility that magmatism is represented by plutonic rather than volcanic activity cannot be dismissed. Further complications to defining evolutionary trends in rifts are created by multiple rifting events, such as in the Horn graben of the North Sea described above, or other basin forming events prior to rifting, such as Laramide compression preceeding the two extensional events in the southern Rio Grande rift (Fig. 12). In general, rifts are not characterized by any simple sequence of events in their development, perhaps due in part to the superposition of multiple rifting events and/or rifting and other tectono-magmatic events, but perhaps also due in part to the inhomogeneous nature of the lithosphere, and variations in its response to similar stimuli.

Duration of Rifting

In defining the duration of rifting, problems again arise as to the separation of features and events that are specifically associated with rifting from other events. Using a strict morphological definition of rifts

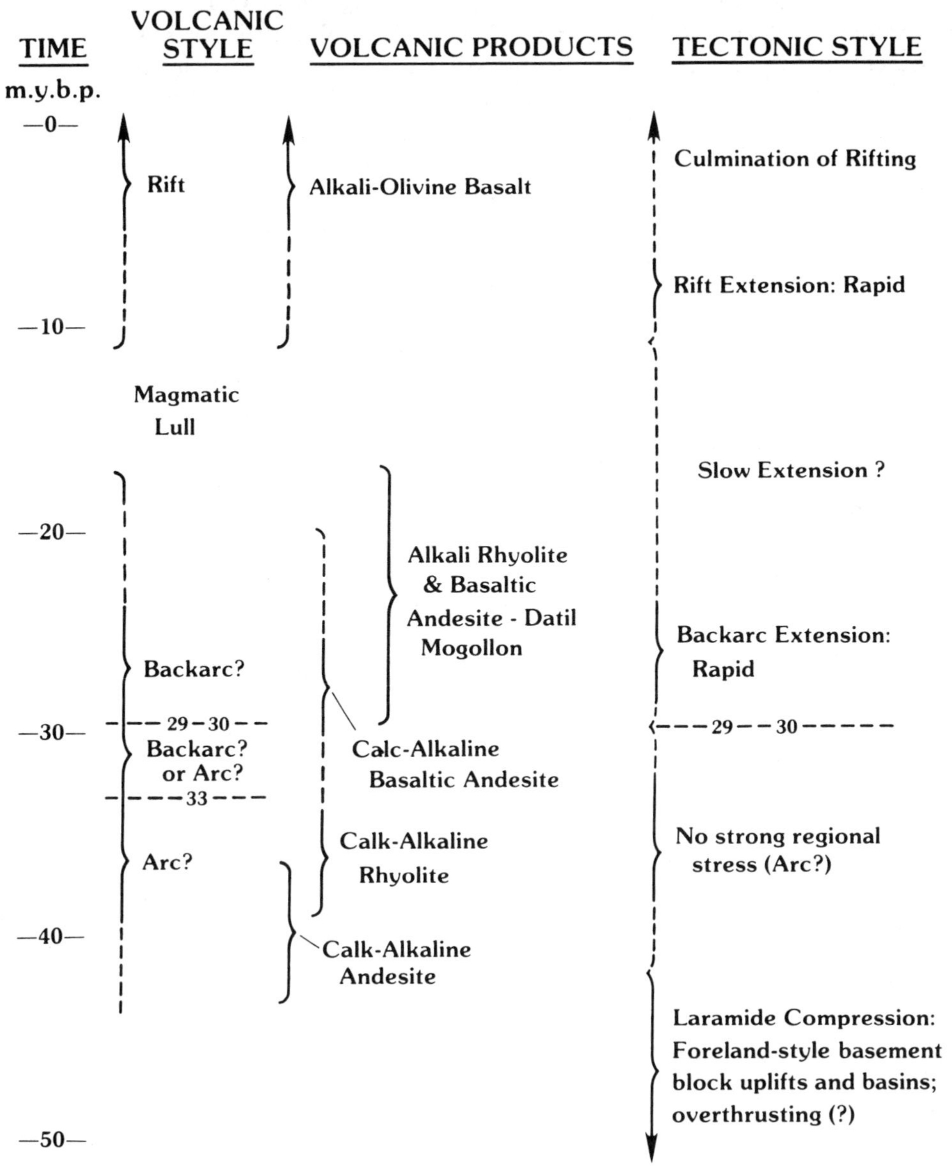

Figure 12. Volcano-tectonic evolution of the southern Rio Grande rift (W.R. Seager, personal communication, 1983).

(e.g., Mohr, 1982), wherever the dates of tectonic activity are well constrained, major graben or rift development appears to be relatively fast, on the order of ten million years in duration or less (e.g., East African rift system-Baker et al., 1972; Williams, 1978, 1982; Mohr, 1982: Baikal-Logatchev, 1978: southern Rio Grande rift - Fig. 12). Multiple rifting events, other basin forming events, or post-rifting subsidence may record a longer history of sedimentation within individual rift troughs (e.g., Horn Graben, southern Rio Grande rift), but there is no evidence that vertical movements associated with active extension are continuous for long periods of time in continental rifts. If extension continues, continental separation occurs, and the continental rift is replaced by an oceanic ridge. There is no evidence for long periods of continental extension at very slow strain rates.

In contrast, magmatic activity associated with rifting is commonly recorded as lasting much longer, sometimes tens of millions of years longer, than active extension. An igneous history lasting more than 50 m.y. is indicated for the Oslo Graben (Ramberg and Larsen, 1978), and extensive volcanic sequences have been developed in East Africa during the last 50 m.y. or so (Baker et al., 1972; Wendlandt and Morgan, 1982). Other tectonic activity, and uplift and subsidence associated with rifting may also occur over a relatively long time period, on the order of tens of millions of years. The relationships of some of these events to rifting may be tenuous, however, events preceeding rifting sometimes being unrelated to the actual rifting processes, and events subsequent to rifting being localized by the thermal and mechanical weaknesses developed in the lithosphere during rifting. Unless rejuvenated by subsequent events, thermal weakening of the lithosphere by rifting would be expected to persist for a period of the order of 100 m.y. or less. Lithospheric inhomogeneities caused by rifting may be permanent features, however, and may act to localize deformation or even igneous activity in a rift zone at any time subsequent to rifting. A rift zone thus may have a very long and complex magmatic and tectonic history, although single rifting events appear to be on the order of 50 m.y. or less in duration, and the active extension phase for rifting is generally much shorter.

Tectonism and Magmatism

The interpretation of the sequence of doming, rifting and volcanism events in rifting, as stated by Mohr (1982), is a chicken-or-egg or cause-and-effect problem in the understanding of the genesis of rifts. The basic question is whether doming and volcanism are a direct result of asthenospheric upwelling which drives rifting ('active' rifting), or merely a response of the lithosphere and asthenosphere to rifting ('passive' rifting). Doming will be considered in greater detail in the next section; we focus upon the relationship between magmatism and tectonism here.

Volumes of volcanism vary by two orders of magnitude for modern rifts, ranging from 5,000 km^3 for the Baikal rift (Logatchev and Florensov, 1978) to 500,000 km^3 for the eastern rift of the East African rift system (Baker et al., 1972). Similar variations in volumes of volcanism also occur within rift systems; for example, most of the volcanism in the East African rift system is in the northern (eastern) parts of the system, with very little volcanism to the south (west). However, care must be taken to include plutonism in analyses of the genetic relationship between tectonism and magmatism in rifting. Where erosion cuts into the basement of rifts, such as the Oslo graben, major plutonism associated with rifting is indicated. A simple calculation for the Kenya rift, assuming the dense mass in the crust beneath the rift axis indicated by gravity studies to be due to basic intrusives, suggests that the volume of intrusives in this rift exceeds the volume of volcanism by a factor of two *or more*. For rifts with less volcanism, the ratio of plutonism to volcanism may be even higher, although gravity data are ambiguous, and positive residual gravity anomalies may also be interpreted as upwarp of the Moho (e.g., Zorin, 1981). Thus the relationship between tectonism and magmatism may be obscured by lack of exposure in some rifts.

Where magmatism and tectonic activity are dated for well mapped rifts, the sequence of events generally appears to be complex (e.g., Fig. 12), and highly variable among rifts. In his review of tectono-magmatic relationships in the Oslo graben, the Rio Grande rift and the Kenya rift, Williams (1982) states: "Detailed studies have provided little support for notions that phases of salic volcanism are accompanied by pulses of domal uplift, or that the most voluminous eruptions in rifts were terminated by major faulting." Thus, a simple cause-and-effect relationship between tectonism and volcanism in rifts remains to be established.

Doming

The idea that rifts were associated with domes was widely accepted a decade or so ago (e.g., Burke and Dewey, 1973; Gass, 1976). This view has recently been challenged on two bases, however: 1) the classical Rhine-Baikal-East African rifts may represent a biased sample of continental rifts; and 2) detailed studies within these classical rifts have failed to reveal a simple pattern of uplift, magmatism and tectonism. Recent studies of sedimentary basin and passive continental margin genesis have focused attention upon the role of rifting in causing downwarp of the crust, initial subsidence resulting from crustal attenuation during rifting, with further subsidence post-rifting by thermal relaxation in the lithosphere and sediment loading (e.g., McKenzie, 1978). Rifts, or portions of rift systems exhibiting this behavior may never be exposed at the surface, and are only revealed by geophysical studies (e.g., the Horn graben, Figs. 6 and 7). Thus, there may be a sampling problem with the rift-dome association.

In addition, the association of rifts and domes in the classical modern rifts has been questioned (e.g., Williams, 1982; Mohr, 1982). There are problems associated with the interpretation of the geological record with respect to doming histories as, at best, only two reference elevation surfaces are available, a pre-rift erosion surface (e.g., the sub-Miocene erosion bevel in Kenya; Baker et al., 1972), and the modern structural elevations. Most techniques for dating and identifying 'uplift' or 'downwarp' actually record the development of local topography, and it is possible, for example, that sedimentation may be misinterpreted to indicate regional subsidence when regional uplift is actually occuring. A possible example of this misinterpretation would occur if a broad dome has a central sag, such as the Lake Victoria drainage basin in the modern East African dome. Care must be exercised in interpreting the geological record in terms of *regional* uplift or downwarp.

Where doming associated with rifting has been identified, modern investigations indicate that arching, downwarping and faulting combine to form the structural topography of the rift (e.g., Williams, 1982). Downwarping has been suggested to precede rifting in some areas, such as Kenya, the Rio Grande, Rhine and Oslo grabens (e.g., Williams, 1982; Chapin, 1979; Illies, 1978; Ramberg and Larsen, 1978), but it is difficult to constrain this downwarp as a precursor to rifting (e.g., Mohr, 1982). Some of these problems undoubtedly arise from the difficulties in separating local from regional uplift and downwarp, but it is evident that there is not a simple relationship among doming, volcanism and tectonism in most rifting events.

Geophysical Characteristics

Geophysical characteristics associated with modern rifts and paleorifts are listed in Table IIc and discussed above, and it is apparent that many of these characteristics change with time. Temporal changes in geophysical parameters are to be expected: a) during modification of the lithosphere during rifting, and b) during post-rift thermal relaxation of heating and / or thinning of the lithosphere associated with rifting. For example, the Bouguer gravity profile across a rift will change during rift development due to interplay among the three dominating factors controlling the gravity profile: 1) the asthenospheric upwelling or lithospheric thinning which causes the long wavelength gravity low, 2) dense mantle derived intrusives into the crust causing an intermediate wavelength gravity high, and 3) low density graben fill which causes a short wavelength gravity low (Fig. 9). Post-rifting lithospheric thickening will result in the disappearance of the long

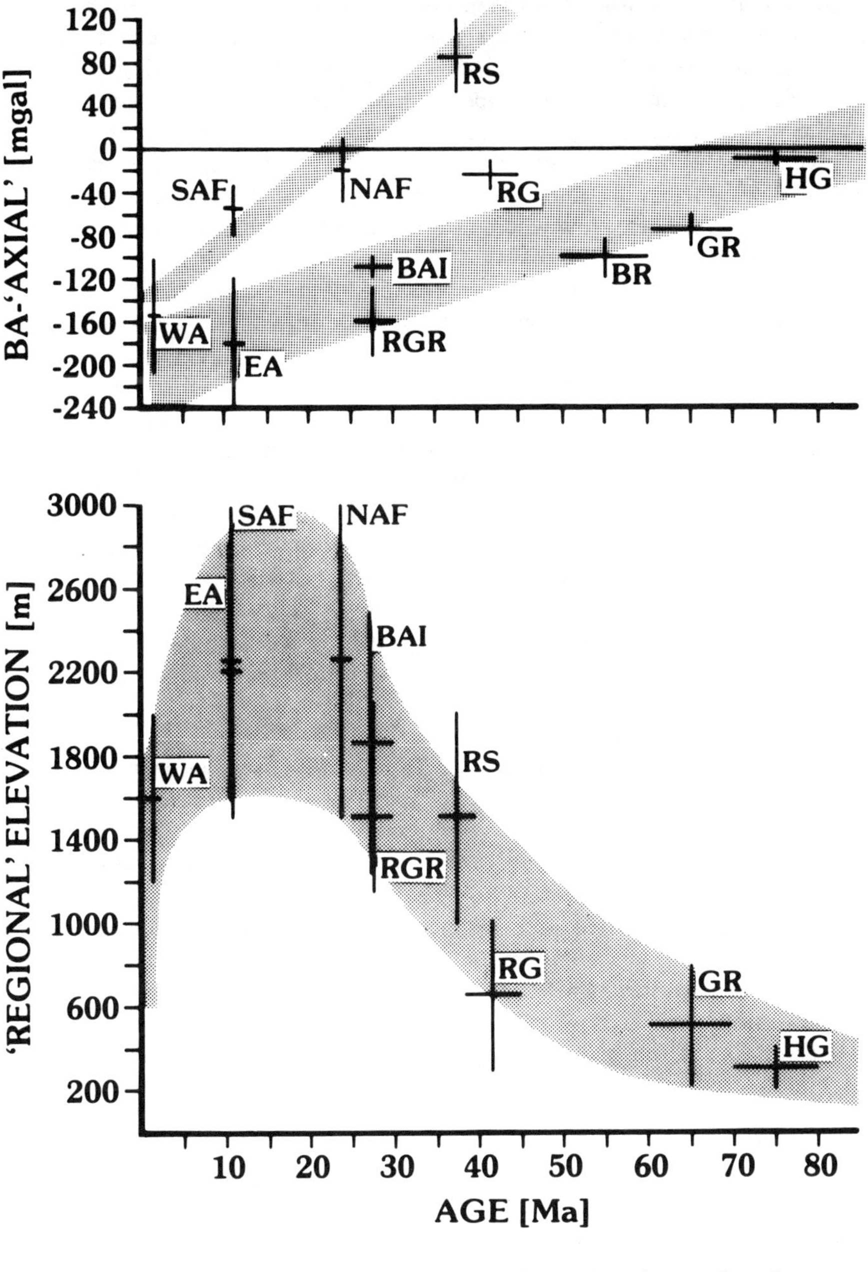

Figure 13. 'Axial' Bouguer gravity anomaly and 'regional' elevation as a function of rift age (adapted from Seidler and Jacoby, 1981).

wavelength gravity low, and regional basin subsidence and sedimentation may completely mask the intermediate wavelength gravity high due to dense crustal intrusions. Other geophysical parameters that change during and post-rifting include heat flow, upper mantle seismic velocities, uplift/subsidence and crustal and upper mantle electrical conductivity. Surface heat flow anomalies may lag behind thermal effects at depth, and other geophysical parameters dependent on temperatures at depth may change in advance of surface heat flow where heat is not convected to the surface by magmatism (Morgan, 1982, 1983).

Seidler and Jacoby (1981) recently compiled physical parameters for a number of Late Mesozoic and Tertiary rifts, and identified trends in some of these parameters with age. From this study, 'axial' Bouguer gravity anomaly and 'regional' elevation are shown as functions of the age of the onset of actual rifting in Fig. 13 (see also Seidler and Jacoby, op cit). Seidler and Jacoby draw three results from these compilations: "(1) The grouping into two families of rift zones which we call 'continental' and 'oceanic'; (2) the decay of the negative 'axial' ('continental') Bouguer anomaly with age toward zero while the 'oceanic' trend overshoots zero; (3) the rise and fall of the 'regional' altitude of the rift shoulders with age." While recognizing the limitations of their preliminary study, they conclude that their results "can be explained plausibly with an isostatic model of a hot, low-density front rising through the lower lithosphere and causing uplift, doming, and rifting of the crust and giving the characteristic gravity anomalies at the same time."

In an analysis such as that of Seidler and Jacoby (1981), it is tempting to conclude that their parameterized approach to rift development reveals the smoother fundamental processes that result in the complex surface geological record. It can also be argued, however, that the technique lacks resolution, especially with regard to problems with quantifying some of the parameters and assigning ages, and that the trends identified are fortuitous and/or do not resolve the fundamental details of rifting. As we are unable to make temporal measurements of physical parameters in a single rift over a reasonable time interval, however, the potential use of parameterized rift studies using data from many rifts of differing ages should be further investigated.

Stress Variations

Detailed information on stress field variations are not available for most rifts at present, although variations in stress field orientation have been noted for the Rhine graben (e.g., Illies, 1981) and for the western United States (e.g., Zoback et al., 1981; Eaton, 1982). These latter stress field variations have been linked to the timing, changing styles, and strike of extension in the Basin and Range province (e.g., Eaton, 1982) and the Rio

Grande rift (e.g., Golombek et al., 1983). More data are required on stress variations and the interplay of regional and local stress fields in order to interpret the significance of the timing of tectonic and magmatic events in rift development with respect to the driving mechanisms of rifting.

Although an extensional stress field is required for extension to occur in the lithosphere, there is evidence to suggest that other (compressional) stress fields may prevail during non-extensional phases of rifting. Ramberg et al. (1984) have inferred the paleostress orientations and history of the Olso graben, using modes of magmatism and faulting as paleostress indicators. They deduced the principle stress axes for the six stages of rifting in the Oslo graben defined by Ramberg and Larsen (1978). These results are summarized in Fig. 14, and indicate that both prior to and after extension in the Oslo graben, the stress field was compressional.

Strain Rate

It is commonly assumed that strain rates in continental rifts are much lower than those for oceanic spreading ridges due to the relatively small amount of extension that occurs in continental rifts (see below) in respect to the relatively long active lives of these rifts. There is also an intuitive feeling that continental spreading should be slower due to the assumed greater strength of inhomogeneous continental lithosphere compared to oceanic lithosphere (an incorrect assumption, e.g., see Lynch, 1983), and because typical oceanic plates move with higher velocities than plates with a significant amount of continental crust. Total extensions of up to a few tens of kilometers over say a lifetime of 50 m.y. for a rift implies an extension rate two orders of magnitude smaller than observed at oceanic spreading centers. Detailed studies of rift development, however, indicate that spreading may only take place during a relatively short portion of total rift development (see e.g., Figs. 12 and 14), and thus, extension rates in continental rifts may be much closer to oceanic spreading rates. If the active stage of rifting lasts 10 m.y., then extension is one order of magnitude slower than typical oceanic rates. If rifting is concentrated into a million years or less, however, extension rates in continental and oceanic environments may be similar. More detailed data on the chronology of continental rift development and the total amounts of extension in these rifts are required to resolve the difference, if any, between continental and oceanic spreading rates.

Amount of Extension

As with many aspects of rift development, the total amount of extension in most rifts is poorly defined. Mohr (1982) quotes extension estimates ranging from 5 to 70 km for the Baikal, East African and Rio Grande rifts, with sometimes a large range in the estimates for the same rift. Part of this

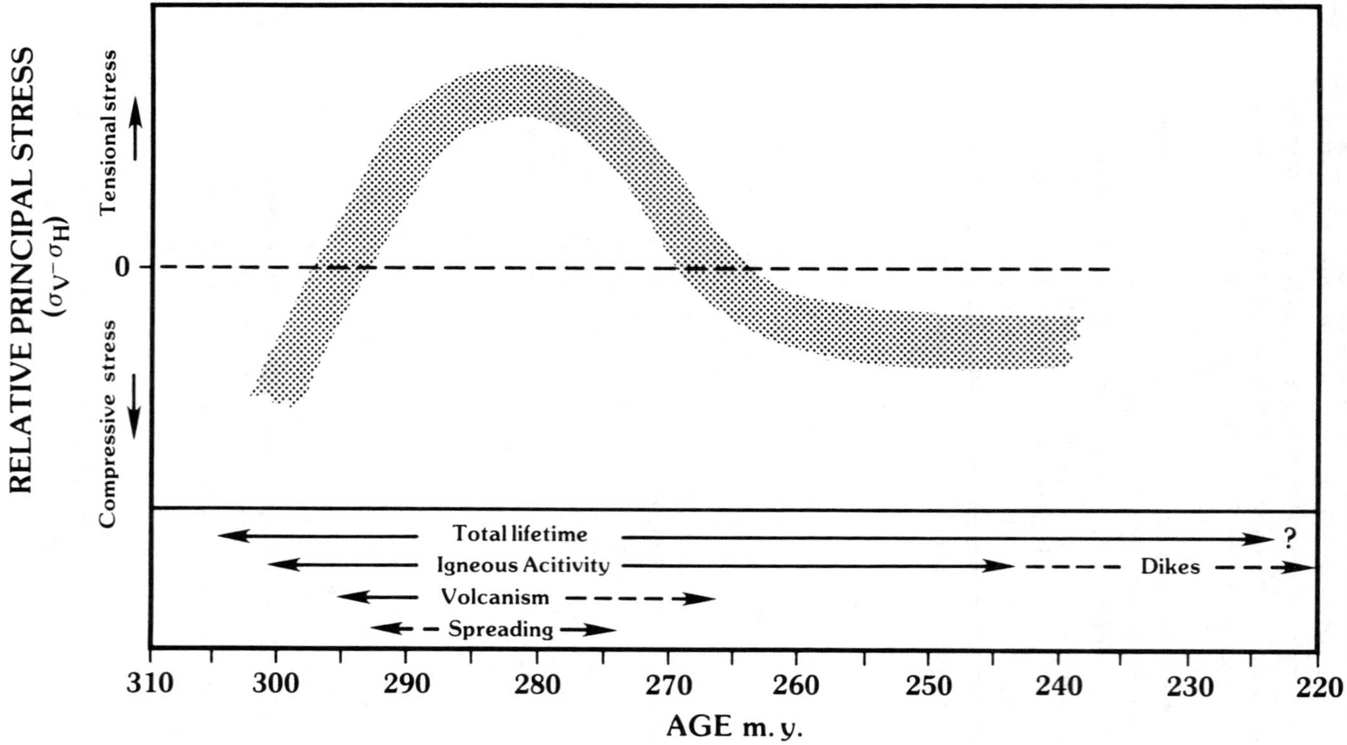

Figure 14. Variations in the relative principal stress during the development of the Oslo graben (adapted from Ramberg et al., 1984).

range is derived from different methods of estimating extension, some of which implicitly group the effects of more than one. extension or other tectonic events together.

Extension can be estimated from both geologic and geophysical data using a variety of techniques and assumptions. Geologic mapping can be used to estimate extension from the amount of movement and dip of extensional faults (e.g., Golombek, 1983) and dike widths. Poor exposure of faults and/or dikes can lead to an underestimation of extension by this technique. Gravity data can be used to estimate the amount of crustal thinning and asthenospheric intrusion during extension, from which the amount of extension can be estimated (e.g., Cordell, 1982). If crustal structure is anomalous prior to extension, however, or if all mass transfer processes, e.g., erosion, are not accurately included in the calculations, erroneous extension estimates can result from this technique. Similarly, seismic estimates of crustal thickness can yield false extension estimates if crustal structure is anomalous prior to extension. Recently McKenzie (1978) has suggested a simple lithospheric stretching model relating subsidence and heat flow to the amount of extension. Again, if crustal (or lithospheric) structure is anomalous prior to extension, an erroneous extension estimate will result from application of this model. No method of estimating extension seems to be without problems except perhaps in very local areas of application.

Other discrepancies in extension estimates are introduced by the extrapolation of local extension values to larger areas. Local strains can vary by an order of magnitude or more, as in the early extension phase on the southern Rio Grande rift (W.R. Seager, personal communication, 1983). In continental areas, total strain can locally be as much as 100% or more (e.g., Basin and Range province, Angelier and Colletta, 1983), and it has been suggested that strains are of this order or greater in continental margin rifting (e.g., the north Biscay margin, Chenet and Montadert, 1981). For most modern continental rifts, however, although commonly poorly defined, total extensions seem to be on the order of 10 km, or around 10%. Again, more data are required to reliably constrain these estimates.

MECHANISMS OF RIFTING

Recently attention has focused upon two end-member driving mechanisms for rifting, 'active rifting', in which rifting is a result of a thermal upwelling of the asthenosphere, and 'passive rifting', in which rifting is a passive response to a regional stress field (e.g., Ramberg and Neumann, 1978b; Sengör and Burke, 1978; Baker and Morgan, 1981; Morgan and Baker, 1983). Asthenospheric diapirism due to the density instability

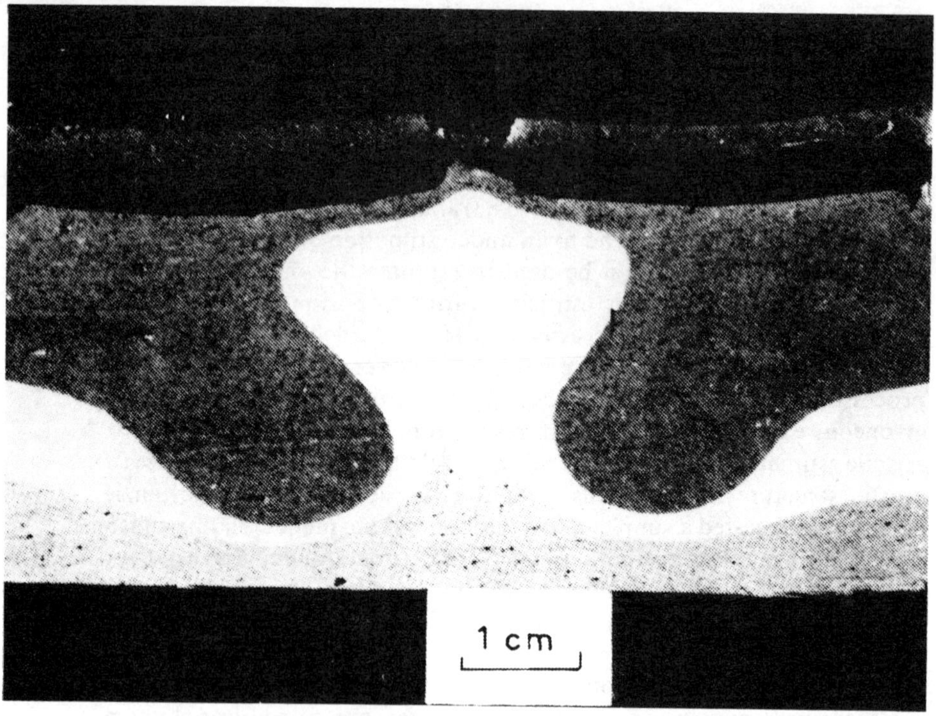

Figure 15. Cross section of centrifuge model of rift development (from Mulugeta, 1982).

which itself is caused by more dense mantle lithosphere overlying less dense asthenosphere must also be considered. Morgan and Baker (1983) suggest that "the rifting process needs to be considered in two parts: (1) What is the initiating process—active or passive? (2) Once the lithosphere has begun to thin, what are the roles of magmatism, diapirism, and of the tectonic environment?" Rifts do not develop in a homogeneous lithosphere in isolation of all other thermotectonic processes—the complexity of the development of individual rifts certainly reflects inhomogeneities in the lithosphere and perhaps also the asthenosphere prior to rifting, and the interaction of tectonic and thermal processes during rift development.

Experimental and numerical modelling of rifts (e.g., H. Ramberg, 1971, 1981; Withjack, 1979; Bridwell and Potzick, 1981; papers in Illies, 1981, and Morgan and Baker, 1983; Lynch, 1983) have provided important information bout the dynamics of rifting. In a recent experimental study, Mulugeta (1982) has carried out a series of centrifuge studies, modelling rifting caused by the ascent of viscous material induced to rise by the centrifugal body forces. Figure 15 shows a cross-section through one of his

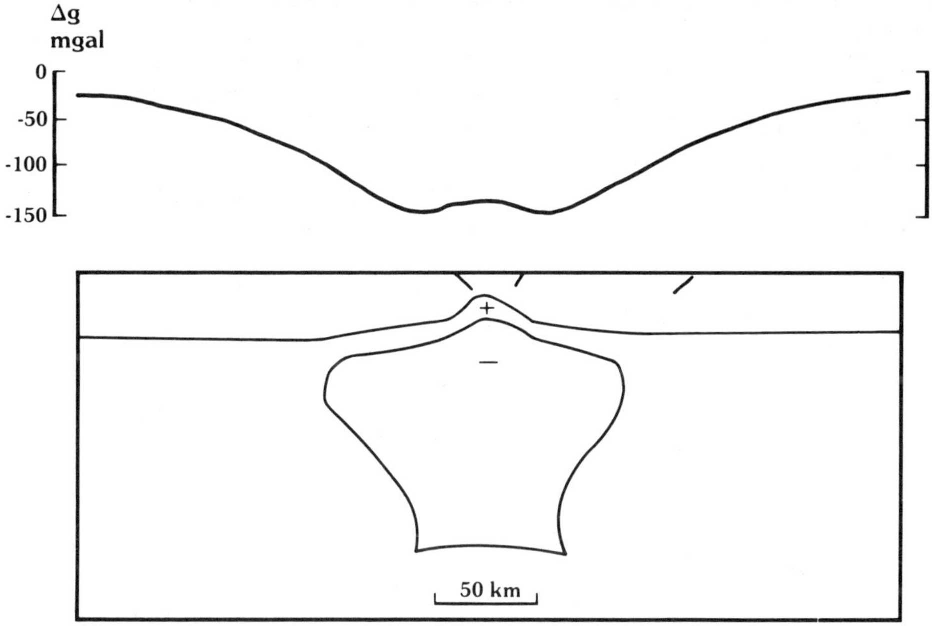

Figure 16. Computed Bouguer gravity anomaly for centrifuge model shown in Fig. 15. scaled up to natural rift dimensions (from Mulugeta, 1982).

models, illustrating the ascending and laterally spreading diapiric body, while Fig. 16 shows the calculated gravity effect of the model scaled up to natural dimensions. Mulugeta (op cit) concludes that the close resemblance in structure and gravity signature of the models to the naturally occurring rift systems suggests a close similarity of dynamics and driving forces. With the evidence that we have presented above to emphasize the diversity and complexity of continental rifts, we refer the reader to other compilations of rift model studies (e.g., Illies, 1981; Morgan and Baker, 1983), from which it will be seen that although many of the models can explain some of the features of these rifts, none of the models explain the complex geologic development of most rifts. We suggest that the apparent failure of the models to explain rift evolution is in the limitations of the models in dealing with a heterogeneous mantle and/or changing thermal and tectonic conditions. The models provide valuable insight into the processes that can potentially take place during rifting, but interactions between processes probably cause much of the complexity in rift development.

'Active' and 'passive' rifting mechanisms should not be considered in isolation of each other. It has been shown (e.g., Mulugeta, 1982), that extensional stresses simply due to thermal arching cannot explain the observed extension in rifts, and though it has been demonstrated that thermal doming creates a gravitational extensional stress field over the dome (Artyushkov, 1973, 1981; Bott, 1981), the regional stress field must also be extensional to allow extension to occur. Thus, an 'active' rift system may have a 'passive' control in the actual timing of rifting if during some time intervals extension is prevented by a regional compressive stress field. Conversely, if a plate is under a uniform regional extensional stress field, deformation will be localized where the plate is weakest, possibly by a local thermal anomaly (Lynch, 1983). Thus, a thermal 'active' system may localize 'passive' rifting. In addition, diapirism during 'passive' rifting may effectively transform the system into an 'active' rifting system. In practice, most rifts may have both 'active' and 'passive' components, the different mechanisms dominating at different stages of rifting.

Rifts may fail for one or both of two mechanical reasons; either the lithosphere strength increases to exceed the extensional stress, or the regional stress field changes. Increasing strength in the lithosphere during extension may be expected as the crust thins and 'weak' quartz-rich crustal rocks are replaced by 'strong' olivine-rich mantle rocks in the vertical section. This effect will be counteracted by a rising geotherm during rifting, however, and will only be significant at very slow extension rates or after a hiatus in extension (e.g., see Lynch, 1983). Changes in the regional stress field will cause rifting to cease if the resultant local extensional stress fall below the lithospheric yield strength, or becomes compressional (e.g., Fig. 14). Stress and strength are not independent during rifting, as strength is dependent on the geotherm, which during extension is related at least in part to the extension (e.g., Lachenbruch and Sass, 1978; McKenzie, 1978), which is in turn related to the extensional stress. For a rift to develop to a stage of continental breakup, the stress field must remain extensional for a sufficient period of time for continental separation to occur, and the extension rate must be high enough for convection associated with extension to maintain a high geotherm and weak lithosphere.

CONCLUDING REMARKS

Recent regional studies have shown continental rifts to be abundant, and detailed studies of individual rifts show them to be complex and diverse in character. Many generalizations that were made about continental rifts even as recently as a decade ago have not been supported by new data, many new rifts have been identified from which data need to be considered in analyzing rifting processes. Future studies should concentrate upon delineating the details of development of rifts, focusing not only upon classical rifts, but also upon lesser known rifts, including paleorifts. In analyzing data from rifts in terms of their development, rifts should be considered in their tectonic setting, and within the framework of other tectonic processes around them. Only with this integrated approach will it be possible to apply model studies to examine the driving mechanisms and processes of rifting and understand the full significance of continental rifting.

ACKNOWLEDGEMENTS

We thank Sharon Adlis, Donna Chady, Carl Grossman, Karen Hrametz, Marilyn Morgan and Pam Thompson for assistance with the preparation of this manuscript and figures. This review was written while one of the authors (I.B.R.) spent a sabbatical year at Rice University and Superior Oil International (both in Houston, Texas) supported in part by a scholarship from the Norwegian Council for Technical and Scientific Research (NTNF). The other author (P.M.) was at the Lunar and Planetary Institute, which is operated by the Universities Space Research Association under contract no. NASW 3389 from the National Aeronautics and Space Administration. This is LPI contribution number 517.

REFERENCES CITED

Angelier, J. and Colletta, B. (1983). Tension fractures and extensional tectonics. *Nature* 301, 49–51.

Anon. (1980). Geodynamic Evolution of the Afro-Arabian Rift System. *Academia Nazionale Dei Lincei, Atti Dei Convegni Lincei* 47, Rome, 705 pp.

Ansorge, J., Emter, D., Fuchs, K., Lauer, J.P., Mueller, St. and Peterschmitt, E. (1970). Structure of the crust and upper mantle in the rift system around the Rhine graben. In: *Graben Problems,* J.H. Illies and St. Mueller (eds). Schweizerbart, Stuttgart, pp. 190–197.

Artemjev, M.E. and Artyushkov, E.V. (1971). Structure and isostasy of the Baikal Rift and the mechanism of rifting. *J. Geophys. Res.* 76, 1197–1211.

Artyushkov, E.V. (1973). Stresses in the lithosphere caused by crustal thickness inhomogeneities. *J. Geophys. Res.* 78, 7675–7708.

Artyushkov, E.V. (1981). Mechanisms of continental riftogenesis. *Tectonophysics* 73, 9–14.

Baker, B.H. and Morgan, P. (1981). Continental rifting: progress and outlook. *Eos, Trans. Am. Geophys. Un.* 62, 585–586.

Baker, B.H. and Wohlenberg, J. (1971). Structure and evolution of the Kenya rift valley. *Nature* 229, 538–542.

Baker, B.H., Mohr, P.A. and Williams, L.A.J. (1972). Geology of the Eastern Rift System of Africa. *Geol. Soc. Am. Special Paper* 136, 67 pp.

Baker, B.H., Crossley, R. and Goles, G.G. (1978). Tectonic and magmatic evolution of the southern part of the Kenya rift valley. In: *Petrology and Geochemistry of Continental Rifts,* E. -R. Neumann and I.B. Ramberg (eds). D. Reidel Publishing Co., Dordrecht, pp. 29–50.

Bally, A.W. and Snelson, S. (1980). Realms of subsidence. In: *Facts and Principles of World Petroleum Occurrence,* A.D. Miall (ed). Can. Soc. Petrol. Geol. Memoir 6, pp. 9–94.

Bamford, D. (1976). MOSAIC time term analysis. *Geophys. J. R. Astron. Soc.* 44, 433–446.

Bamford, D. and Crampin, S. (1977). Seismic anisotropy—the state of the art. *Geophys. J. R. Astron. Soc.* 49, 1–8.

Barberi, F., Santacroce, R. and Varet, J. (1974). Silicic peralkaline volcanic rocks of the Afar depression (Ethiopia). *Bull. Volcanol.* 38, 755–790.

Barberi, F., Santacroce, R. and Varet, J. (1982). Chemical aspects of rift magmatism. In: *Continental and Oceanic Rifts*, G. Palmason (ed). Geodynamics Series Vol. 8, Geological Society of America/American Geophysical Union, Washington, D.C., pp. 223–258.

Bates, R.L. and Jackson, J.A. (eds). (1980). *Glossary of Geology.* American Geological Institute, Falls Church, Virginia, 751 pp.

Best, G., Kockel, F. and Schöneich, H. (1983). Geological history of the southern Horn Graben. *Geologie en Mijnbouw* 62, 25–33.

Bott, M.H.P. (1981). Crustal doming and the mechanism of continental rifting. *Tectonophysics* 73, 1–8.

Braile, L.W., Smith, R.B., Ansorge, J., Baker, M.R., Sparlin M.A., Prodehl, C., Schilly, M.M., Healy, J.H., Mueller, St. and Olsen, K.H. (1982). The Yellowstone–Snake River Plain seismic profiling experiment: Crustal structure of the eastern Snake River Plain. *J. Geophys. Res.* 87, 2597–2609.

Bridwell, R.J. and Potzick, C. (1981). Thermal regimes, mantle diapirs and crustal stresses of continental rifts. *Tectonophysics* 73, 15–32.

Brocher, T.M. (1981a). Shallow velocity structure of the Rio Grande rift: A reinterpretation. *J. Geophys. Res.* 86, 4960–4970.

Brocher, T.M. (1981b). Geometry and physical properties of the Socorro, New Mexico, magma body. *J. Geophys. Res.* 86, 9420–9432.

Brown, L.D., Chapin, C.E., Sanford, A.R., Kaufman, S. and Oliver, J. (1980). Deep structure of the Rio Grande rift from seismic reflection profiling. *J. Geophys. Res.* 85, 4773–4800.

Bullard, E.C. (1936). Gravity measurements in East Africa. *Phil. Trans. R. Soc. London* A 235, 445–531.
Burke, K. (1977). Aulacogens and continental breakup. *Ann. Rev. Earth Planet. Sci.* 5, 371–396.
Burke, K. (1978). Evolution of continental rift systems in the light of plate tectonics. In: *Tectonics and Geophysics of Continental Rifts,* I.B. Ramberg and E.-R. Neumann (eds). D. Reidel Publishing Co., Dordrecht, pp. 1–9.
Burke, K. and Dewey, J.F. (1973). Plume-generated triple junctions; key indicators in applying plate tectonics to old rocks. *J. Geology* 81, 406–433.
Burke K. and Whiteman, A.J. (1973). Uplift, rifting and the breakup of Africa. In: *Implications of Continental Drift to the Earth Sciences*, D.H. Tarling and S.K. Runcorn (eds). Academic Press, London, pp. 735–755.
Burke, K., Delano, L., Dewey, J.F., Edelstein, A., Kidd, W.S.F., Nelson, K.D., Sengor, A.M.C. and Stroup, J. (1978). *Rifts and Sutures of the World.* Report for Contract No. NAS5-24094 to Goddard Space Flight Center, 238 pp.
Chapin C.E. (1979). Evolution of the Rio Grande rift—a summary. In: *Rio Grande Rift: Tectonics and Magmatism*, R.E. Riecker (ed). American Geophysical Union, Washington, D.C., pp. 1–5.
Chenet, P.Y. and Montadert, L. (1981). Rifting tectonics on Galacia, Portugal and N. Biscay Margin. In: *Papers Presented to the Conference on the Processes of Planetary Rifting*, Lunar and Planetary Institute, Houston, pp. 47–50.
Cordell, L. (1982). Extension in the Rio Grande rift. *J. Geophys. Res.* 87, 8561–8569.
DeRito, R.F., Cozzarelli, F.A. and Hodge, D.S. (1983). Mechanism of subsidence of ancient cratonic rifts. *Tectonophysics* 94, 141–168.
Dewey, J.F. and Burke, K. (1974). Hotspots and continental breakup: implications for collisional orogeny. *Geology* 2, 57–60.
Dons, J.A. and Larsen, B.T. (eds) (1978). The Oslo Paleorift. *Norges Geol. Unders.* 337, 199 pp.
Easton, R.M. (1983). Crustal structure of rifted continental margins: geological constraints from the Proterozoic rocks of the Canadian Shield. *Tectonophysics* 94, 371–390.
Eaton, G.P. (1982). The Basin and Range province: origin and tectonic significance. *Ann. Rev. Earth Planet. Sci.* 10, 409–440.
Fairhead, J.D. (1976). The structure of the lithosphere beneath the eastern rift, East Africa, deduced from gravity studies. *Tectonophysics* 30, 269–298.

Frey, H. (1979). Global geophysics and geology; correlations of satellite-derived gravity and magnetic anomalies with rifts and sutures. *Eos, Trans. Am. Geophys. Un.* 60, 398.

Frey, H. and Ruder, M. (1981). Satellite geopotential anomalies and continental rifts. In: *Papers presented at the Conference on the Processes of Planetary Rifting*. Lunar and Planetary Institute, Houston, pp. 176–178.

Fuchs, K. (1977). Seismische anisotropie des obesen erdmantels und intraplatten-tektonikk. *Geol. Rundschau* 64, 700–716.

Garland, G.D. (1981). The significance of terrestrial electrical conductivity variations. *Ann. Rev. Earth Planet. Sci.* 9, 147–174.

Gass, I.G. (1970). The evolution of volcanism in the junction area of the Red Sea, Gulf of Aden and Ethiopian rifts. *Phil. Trans. R. Soc. London* A 267, 369–381.

Gill, R.C.O. (1973). Mechanism for the salic magma bias of continental alkaline provinces. *Nature Phys. Sci.* 242, 41–42.

Girdler, R.W. (ed) (1972). East African Rifts. *Tectonophysics* 15, 1–179.

Girdler, R.W., Fairhead, J.D., Searle, R.C. and Sowerbutts, W.T.C. (1969). The evolution of rifting in Africa. *Nature* 224, 1178–1182.

Golombek, M.P., McGill, G.E. and Brown, L. (1983). Tectonic and geologic evolution of the Española basin, Rio Grande rift: structure, rate of extension, and relation to the state of stress in the western United States. *Tectonophysics* 94, 483–507.

Griffiths, D.H., King, R.F., Khan, M.A. and Blundell, D.J. (1971). Seismic refraction line in the Gregory Rift. *Nature* 229, 66–71.

Haenel, R., Gronlie, G. and Heier, K.S. (1974). Terrestrial heat flow determinations from lakes in southern Norway. *Nor. Geol. Tidskr.* 54, 423–428.

Haenel, R., Gronlie, G. and Heier, K.S. (1979). Terrestrial heat flow determinations in Norway and an attempted interpretation. In: *Terrestrial Heat Flow in Europe*, V. Cermak and L. Ryback (eds). Springer-Verlag, New York, pp. 232–239.

Halls, H.C. (1978). The Late Precambrian central North American rift system—a survey of recent geological and geophysical investigations, In: *Tectonics and Geophysics of Continental Rifts*, I.B. Ramberg and E. R. Neumann (eds). D. Reidel Publishing Co., Dordrecht, pp. 111–124.

Hermance, J.F. (1982). Magnetotelluric and geomagnetic deep-sounding studies in rifts and adjacent areas: Constraints on physical processes in the crust and upper mantle. In: *Continental and Oceanic Rifts,* G. Palmason (ed). Geodynamics Series Vol. 8, Geological Society of America / American Geophysical Union, Washington, D.C., pp. 169–192.

Hoffman, P.F. (1973). Evolution of an early Proterozoic continental margin: the Coronation geosyncline and associated aulacogens of the northwestern Canadian Shield. *Phil. Trans. R. Soc. London* A 273, 457–581.

Hoffman, P.F. (1980). A Wilson Cycle of Early Proterozoic age in the northwest of the Canadian Shield. In: *The Continental Crust and its Mineral Deposits*, D.W. Strangway (ed). Geol. Assoc. Can. Spec. Pap. 20, pp. 523–549.

Husebye, E. and Ramberg, I.B. (1978). Geophysical investigations. In: *The Oslo Paleorift*, J.A. Dons and B.T. Larsen (eds). Norges Geol. Unders. 337, pp. 41–53.

Illies, J.H. (1977). Ancient and recent rifting in the Rhinegraben. *Geologie en Mijnbouw* 56, 329–350.

Illies, J.H. (1978). Two stages Rhinegraben rifting. In: *Tectonics and Geophysics of Continental Rifts,* I.B. Ramberg and E.-R. Neumann (eds). D. Reidel Publishing Co., Dordrecht, pp. 63–71.

Illies, J.H. (ed). (1981). Mechanisms of Graben Formation. *Tectonophysics* 73, 266 pp.

Illies, J.H. and Fuchs, K. (eds). (1975). *Approaches to Taphrogenesis.* Schweizerbart, Stuttgart, 460 pp.

Illies, J.H. and Mueller, St. (ed) (1970). *Graben Problems.* Schweizerbart, Stuttgart, 316 pp.

Jiracek, G.R., Ander, M.E. and Holcombe, H.T. (1979). Magnetotelluric soundings of crustal conductive zones in major continental rifts. In: *Rio Grande Rift: Tectonics and Magmatism*, R.E. Riecker (ed). American Geophysical Union, Washington, D.C., pp. 209–222.

Jiracek, G.R., Gustafson, E.P. and Mitchell, P.S. (1983). Magnetotelluric results opposing magma origin of crustal conductors in the Rio Grande rift. *Tectonophysics* 94, 299–326.

Kane, M.F., Hildenbrand, T.G. and Hendricks, J.D. (1981). Model for the tectonic evolution of the Mississippi Embayment and its contemporary seismicity. *Geology* 9, 563–568.

Keller, G.R., Lidiak, E.G., Hinze, W.J. and Braile, L.W. (1983). The role of rifting in the tectonic development of the Midcontinent, U.S.A. *Tectonophysics* 94, 391–412.

Khain, V.E. (1977). The new international tectonic map of Europe and some problems of structure and tectonic history of the continent. In: *Europe from Crust to Core*, D.V. Ager and H. Brooks (eds). John Wiley and Sons, London, pp. 19–40.

Khan, M.A. and Mansfield, J. (1971). Gravity measurements in the Gregory rift. *Nature Phys. Sci.* 229, 72–75.

Kumarapeli, P.S., (1978). The St. Lawrence paleo-rift system: A comparative study. In: *Tectonics and Geophysics of Continental Rifts*, I.B. Ramberg and E.-R. Neumann (eds). D. Reidel Publishing Co., Dordrecht, 367–384.

Lachenbruch, A.H. and Sass, J.H. (1978). Models of an extending lithosphere and heat flow in the Basin and Range Province. In: *Cenozoic Tectonics and Regional Geophysics of the Western Cordillera,* R.B. Smith and G.P. Eaton (eds). *Geol. Soc. Am. Memoir* 152, Boulder, Colorado, pp. 209–250.

Lipman, P.W. (1969). Alkalic and tholeiitic basaltic volcanism related to the Rio Grande depression, southern Colorado and northern New Mexico. *Geol. Soc. Am. Bull.* 80, 1343–1354.

Logatchev, N.A. (1978). Main features of evolution and magmatism of continental rift zones in the Cenozoic. In: *Tectonics and Geophysics of Continental Rifts*, I.B. Ramberg and E.-R. Neumann (eds). D. Reidel Publishing Co., Dordrecht, pp. 351–366.

Logatchev, N.A. and Florensov, N.A. (1978). The Baikal system of rift valleys. *Tectonophysics* 45, 1–13.

Logatchev, N.A. and Mohr, P.A. (eds). (1978). Geodynamics of the Baikal Rift Zone. *Tectonophysics* 45, 1–105.

Lynch, H.D. (1983). Numerical models of the formation of continental rifts by processes of lithospheric necking. Ph.D. Dissertation, New Mexico State University, Las Cruces, New Mexico, 290 pp.

Maguire, P.K.H. and Long, R.E. (1976). The structure of the western flank of the Gregory Rift (Kenya), part I, the crust. *Geophys. J. R. Astron. Soc.* 44, 661-675.

McGinnis, L.D. (1970). Tectonics and the gravity field in the continental interior. *J. Geophys. Res.* 77, 317–331.

McKenzie, D.P. (1978). Some remarks on the development of sedimentary basins. *Earth Planet. Sci. Lett.* 40, 25–32.

McLean, A.C. and Qureshi, I.R. (1966). Regional gravity anomalies in the western Midland Valley of Scotland. *Roy. Soc. Edinburgh* 66, 267–283.

Michelsen, O. (ed). (1982). Geology of the Danish Central Graben. *Geol. Surv. Denmark, Series B, No. 8*, 133 pp.

Milanovsky, E.E. (1972). Continental rift zones; their arrangement and development. *Tectonophysics* 15, 65–70.

Milanovsky, E.E. (1978). Some problems of rifting development in the Earth's history. In: *Tectonics and Geophysics of Continental Rifts*, I.B. Ramberg and E.-R. Neumann (eds). D. Reidel Publishing Co., Dordrecht, pp. 385–399.

Milanovsky, E.E. (1981). Aulacogens of ancient platforms: Problems of their origin and tectonic development. *Tectonophysics* 73, 213–248.

Milanovsky, E.E. (1983). Major stages of rifting evolution in the Earth's history. *Tectonophysics* 94, 599–607.

Mohr, P. (1982). Musings on continental rifts. In: *Continental and Oceanic Rifts*, G. Palmason (ed). Geodynamics Series Vol. 8, Geological

Society of America/American Geophysical Union, Washington, D.C., pp. 293–309.
Mohr, P.A. and Wood, C.A. (1976). Volcano spacings and lithospheric attenuation in the Eastern Rift of Africa. *Earth Planet. Sci. Lett.* 33, 126–144.
Mooney, W.D., Andrews, M.C., Ginsburg, A., Peters, D.A. and Hamilton, R.M. (1983). Crustal structure of the northern Mississippi Embayment and a comparison with other continental rift zones. *Tectonophysics* 94, 327–348.
Morgan, P. (1982). Heat flow in rift zones. In: *Continental and Oceanic Rifts,* G. Palmason (ed). Geodynamics Series Vol. 8, Geological Society of America/American Geophysical Union, Washington, D.C., pp. 107–122.
Morgan, P. (1983). Constraints on rift thermal processes from heat flow and uplift. *Tectonophysics* 94, 277–298.
Morgan P. (1984). The thermal structure and thermal evolution of the continental lithosphere. *Phys. Chem. Earth,* in press.
Morgan, P. and Baker, B.H. (eds) (1983a). Processes of Continental Rifting. *Tectonophysics* 94, 680 pp.
Morgan, P. and Baker, B.H. (1983b). Introduction—processes of continental rifting. *Tectonophysics* 94, 1–10.
Mueller, St. (1978). Evolution of the Earth's crust. In: *Tectonics and Geophysics of Continental Rifts*, I.B. Ramberg and E.-R. Neumann (eds). D. Reidel Publishing Co., Dordrecht, 11–28.
Mueller, St., Peterschmitt, E., Fuchs, K., Emter, D. and Ansorge, J. (1973). Crustal structure of the Rhinegraben area. *Tectonophysics* 20, 381–392.
Mulugeta, G. (1982). Experimental model study of the evolution of rift valley systems and fracture patterns in the crust. Ph.D. Thesis, University of Uppsala, Sweden, UUDMP Res. Rpt. Nos. 29–32.
Nakamura, K. (1977). Volcanoes as possible indicators of tectonic stress orientation—principle and proposal. *J. Volc. Geotherm. Res.* 2, 1–16.
Neumann, E.-R. and Ramberg, I.B. (eds). (1978a). *Petrology and Geochemistry of Continental Rifts,* D. Reidel Publishing Co., Dordrecht, 296 pp.
Neumann, E.-R. and Ramberg I.B. (1978b). Paleorifts—concluding remarks. In: *Tectonics and Geophysics of Continental Rifts*, I.B. Ramberg and E. -R. Neumann (eds). D. Reidel Publishing Co., Dordrecht, 409–424.
O'Hara, M.J. (1977). Geochemical evolution during fractional crystallisation of a periodically refined magma chamber. *Nature* 266, 503–507.
Olsen, J.Chr. (1983a). Structural outline of the Horn Graben area. *Geologie en Mijnbouw* 62, 47–50.

Olsen, K.H. (1983b). The role of seismic refraction data for studies of the origin and evolution of continental rifts. *Tectonophysics* 94, 349–370.

Palmason, G. (ed). (1982). *Continental and Oceanic Rifts.* Geodynamics Series Vol. 8, Geological Society of America/American Geophysical Union, Washington D.C., 309 pp.

Pilger A. and Rösler, A. (eds). (1975). *Afar Depression of Ethiopia.* Schweizerbart, Stuttgart, 416 pp.

Pilger A. and Rösler, A. (eds). (1976). *Afar Between Continental and Oceanic Rifting.* Schweizerbart, Stuttgart, 216 pp.

Prodehl, C. (1981). Structure of the crust and upper mantle beneath the central European rift system. *Tectonophysics* 80, 255–269.

Prodehl, C., Ansorge, J., Edel, J.B., Emter, D., Fuchs, K., Mueller, St. and Peterschmitt, E. (1976). Explosion-seismology research on the central and southern Rhinegraben: a case history. In: *Explosion Seismology in Central Europe*, P. Giese, C. Prodehl and A. Stein (eds). Springer, Heidelberg, pp. 313–328.

Puzyrev, N.N., Mandelbaum, M.M., Krylov, S.V., Mishenkin, B.P., Petrik, G.V. and Krupskaya, G.V. (1978). Deep structure from Baikal and other continental rift zones from seismic data. *Tectonophysics* 45, 15–22.

Ramberg, H. (1971). Dynamic models simulating rift valleys and continental drift. *Lithos* 4, 259–276.

Ramberg, I.B. (1972). Crustal structure across the Permian Oslo Graben based on gravity measurements. *Nature Phys. Sci.* 240, 149–153.

Ramberg, I.B. (1976). Gravity interpretation of the Oslo Graben and associated igneous rocks. *Nor. Geol. Unders. 325*, 1–194.

Ramberg, H. (1981). *Gravity Deformation and the Earth's Crust, 2nd Edition.* Academic Press, London, 452 pp.

Ramberg, I.B. and Larsen, B.T. (1978). Tectonomagmatic evolution. In: *The Oslo Paleorift*, J.A. Dons and B.T. Larsen (eds). Nor. Geol. Unders. 337, 55–73.

Ramberg, I.B. and Neumann, E. -R. (eds) (1978a). *Tectonics and Geophysics of Continental Rifts.* D. Reidel Publishing Co., Dordrecht, 444 pp.

Ramberg, I.B. and Neumann, E. -R. (1978b), Paleorift systems—introduction. In: *Petrology and Geochemistry of Continental Rifts*, E. -R. Neumann and I.B. Ramberg (eds). D. Reidel Publishing Co., Dordrecht, xix–xxvii.

Ramberg, I.B., Gray, D.F. and Raynolds, R.G.H. (1978a). Tectonic evolution of the FAMOUS area of the Mid-Atlantic ridge, lat. 35° 50′ to 37° 20′ N. *Geol. Soc. Am. Bull.* 88, 609–620.

Ramberg, I.B., Cook, F.A. and Smithson, S.B. (1978b). Structure of the Rio Grande rift in southern New Mexico and West Texas based on gravity interpretation. *Geol. Soc. Am. Bull.* 89, 107–123.

Ramberg, I.B., Larsen, B.T. and Sundvoll, B. (1984). Paleostress orientations and tectonic evolution of the Oslo Graben (in preparation).
Riecker, R.E. (ed) (1979). *Rio Grande Rift: Tectonics and Magmatism.* American Geophysical Union, Washington, D.C., 448 pp.
Sanford, A.R. and Einarsson, P. (1982). Magma chambers in rift zones, In: *Continental and Oceanic Rifts,* G. Palmason (ed). Geodynamics Series Vol. 8, Geological Society of America/American Geophysical Union, Washington, D.C., pp. 147–168.
Sanford, A.R., Mott, R.P., Shuleski, P.J., Rinehart, E.J., Caravella, F.J., Ward, R.M. and Wallace, T.C. (1977). Geophysical evidence for a magma body in the crust in the vicinity of Socorro, New Mexico. In: *The Earth's Crust: Its Nature and Physical Processes,* J.G. Heacock (ed). Geophys. Monogr. 20, American Geophysical Union, Washington, D.C., pp. 385–403.
Searle, R.C. and Gouin, P. (1972). A gravity survey of the central part of the Ethiopian rift valley. *Tectonophysics* 15, 41–52.
Seidler, E. and Jacoby, W.R. (1981). Parameterized rift development and upper mantle anomalies. *Tectonophysics* 73, 53–68.
Sellevoll, M.A. and Warrick, R.E. (1971). A refraction study of the crustal structure in southern Norway. *Bull. Seism. Soc. Am.* 61, 457–571.
Sengör, A.M.C. and Burke, K. (1978). Relative timing of rifting and volcanism on Earth and its tectonic implications. *Geophys. Res. Lett.* 5, 419–421.
Shatsky, N.S. and Bogdanov, A.A. (1961). An international tectonic map of Europe, scale 1:2,500,000. *Izv. Akad. Nauk SSSR, Ser. Geol.* (in Russian).
Swanberg, C.A., Chessman, M.D., Simmons, G., Smithson, S.B., Gronlie, G. and Heier, H. (1974). Heat flow-heat generation studies in Norway. *Tectonophysics* 23, 31–48.
Swanberg, C.A., Mitchell, B.J., Lohse, R.L. and Blackwell, D.D. (1982). Heat flow in the Upper Mississippi Embayment. *U.S. Geol. Surv. Prof. Paper 1236*, 185–189.
Tryti, J. and Sellevoll, M.A. (1977). A report on a reconnaissance seismic crustal study of the Oslo Rift. In: *The Norwegian Geotraverse Project,* K.S. Heier (ed). Unpubl. Report, Oslo 1977, pp. 151–184.
Upton, B.G.J. and Blundell, D.J. (1978). The Gardar igneous province: Evidence for Proterozoic continental rifting. In: *Petrology and Geochemistry of Continental Rifts*, E.-R. Neumann and I.B. Ramberg (eds). D. Reidel Publishing Co., Dordrecht, pp. 163–172.
van Calsteren, P.W.C. and den Tex, E. (1978). An early Paleozoic contin-xental rift system in Galacia (NW Spain). In: *Tectonics and Geophysics of Continental Rifts,* I.B. Ramberg and E.-R. Neumann (eds). D. Reidel Publishing Co., Dordrecht, pp. 125–132.

Wendlandt, R.F. and Morgan, P. (1982). Lithospheric thinning associated with rifting in East Africa. *Nature* 298, 734–736.

Werner, D., Kahle, H. -G., Ansorge, J. and Mueller, St. (1982). Mass displacements and geothermics within the upper mantle of the Rhinegraben rift system: A model to interpret data from gravity and seismology, In: *Continental and Oceanic Rifts,* G. Palmason (ed). Geodynamics Series Vol. 8, Geological Society of America / American Geophysical Union, Washington, D.C., pp. 283–292.

Williams, L.A.J. (1978). The volcanological development of the Kenya rift, In: *Petrology and Geochemistry of Continental Rifts,* E.-R. Neumann and I.B. Ramberg (eds). D. Reidel Publishing Co., Dordrecht, pp. 101–121.

Williams, L.A.J. (1982). Physical aspects of magmatism in continental rifts, In: *Continental and Oceanic Rifts,* G. Palmason (ed). Geodynamics Series Vol. 8, Geological Society of America / American Geophysical Union, Washington, D.C., pp. 193–222.

Withjack, M. (1979). An analytical model of continental rift fault patterns. *Tectonophysics* 59, 59–81.

Wohlenberg, J. (1970). On the seismicity of the East African Rift System In: *Graben Problems,* J.H. Illies and St. Mueller (eds). Schweitzerbart, Stuttgart, pp. 290–295.

Ziegler, P.A. (1979). North Sea rift and basin development. In: *Tectonics and Geophysics of Continental Rifts*, I.B. Ramberg and E.-R. Neumann (eds). D. Reidel Publishing Co., Dordrecht, pp. 249–277.

Ziegler, P.A. (1982). *Geological Atlas of Western and Central Europe.* Shell Internationale Petroleum Maatschappij. B.V., The Hague, 130 pp. + maps.

Zoback, M.L. and Zoback, M. (1980). State of stress in conterminous United States. *J. Geophys. Res.* 85, 6113–6156.

Zorin, Yu.A. (1966). On the deep structure of the Baikal depression based on geophysical data. *Izv. Akad. Nauk SSSR, Ser. Geol.* 7, 76–85 (in Russian).

Zorin, Yu.A. (1981). The Baikal Rift: An example of the intrusion of asthenospheric material into the lithosphere as the cause of disruption of the lithospheric plates. *Tectonophysics* 73, 91–104.

Zorin, Yu.A. and Rogozhina, V.A. (1978). Mechanism of rifting and some features of the deep-seated structure of the Baikal rift zone. *Tectonophysics* 45, 23–30.

RIFTING THROUGH THE GEOLOGICAL HISTORY

E.E. MILANOVSKY
Moscow State University, Moscow, USSR

Not so long ago, rift zones - global linear structures marked by horizontal extension of the Earth's crust - were regarded as tectonic formations typical only of the Mesozoic-Cenozoic or, probably, even of its latest, i.e. Neotectonic (Late Cenozoic) stage. Comparative geological studies over the last 10 years showed, however, that tectonic elements, similar to some types of recent rift zones or at least, reminding them by some features of their structure and development-ancient rift- and rift-like zones-appeared for the last 3 to 3.5 b.y., that is, over the most part of the Earth's history. But the relative role and intensity of rifting, e.d., origination and development of rift and rift-like structures, was not constant during the geological history of the Earth, and we can distinguish in it periods of considerable increase and decrease of rifting. General character of rifting and its relation to other major historical and geological processes greatly changed in time, as well as structural and evolution pattern of rift and rift-like zones. In the geological part, these zones, besides the dominant horizontal extension of the crust, also experienced episodical compression. On the whole, the more ancient paleorift structure is, the more significant is the role of compression in it. In the Late Proterozoic and Paleozoic, rifting was more closely related and similar to the syncrhonous geosynclinal process than during the Mesozoic, when kinematically they became almost "antagonistic", though complementary tectonic processes. As for the earliest period of the Earth history, tectonic processes which occurred in Archean mobile linear zones combined elements of geosynclinal and rift processes of the following geological time and they cannot be attributed either to the first or to the second type. Thus, in the early history, differences of these processes seem to be "obliterated" (Milanovsky, 1981, 1983).

Five major stages of rifting and related processes can

Proceedings of the 27th International
Geological Congress, Volume 7, pp. 217–249
TECTONICS

be distinguished in the Earth's history. These stages have typical patterns of spatial distribution, structure and tectonic evolution of inherent rift and rift-like zones. These five stages are: the Archean, Early Proterozoic, Late Proterozoic, Paleozoic and Mesozoic-Cenozoic. The first, third and fifth stages are marked by extremely intensive processes of destruction and horizontal extension of the Earth's crust which created new generations and qualitatively new types of rift and rift-like structures at each of these stages. General rejuvenation of the global tectonic structure is associated with these stages (-regeneration, according to H. Stille).Existence of these stages is marked by increase in the upper mantle heating (up to its partial or even complete melting) and by large-scale horizontal extension of the considerable areas of the crust or may be of the whole Earth surface and also by appearance and active development of numerous linear extensional zones. We believe, these facts can be most naturally explained by the hypothesis of the Earth's pulsation which occurred during its general and irregular, weak or moderate expansion.

Extensional features, appeared consequently at each stage of rifting, differed considerably in respect to their geometry, morphology, scale of extension, tectonic paragenesis, rock formation, thermal regime, accompanying magmatism and metamorphism, and evolution pattern, in particular, in respect to the relative role of the following compressional deformations. Such differences between individual rift and rift-like generations reflect the irreversible evolution of the Earth, in particular, changes in the structure and properties of the crust which was periodically subjected to destruction during some stages of its history.

ARCHEAN

Most ancient linear tectonic zones revealed in the ancient shield and platform basement are dated back to the Early (3.7 to 3.0 b.y.) and Late (3.0 to 2.6 b.y.) Archean. These zones occurred on a relatively thin and, in its deep parts, plastic (because of high thermal regime during the Archean) protocontinental curst. They combined featrues of both, rift and geosynclinal zones of the consequent geological periods, and appeared under conditions of destruction, horizontal extension, and subsidence of the mobile protocontinental crust which had high permeability that provided for the rise of the magmatic melt with dominantly basic and ultrabasic composition. At the following stages, however, horizontal extension in these zones changed to comparatively strong compression which was accompanied by considerable increase in the heating of the crust. Two types of Archean linear mobile zones can be distinguished which can be regarded as common "ancestors" of rift and geosynclinal zones of the Late Precambrian and Phanerozoic: so-called greenstone belts and granulite belts. Greenstone belts (or greenstone zones) were widely spread over the shields of North-America, East European, African, Indian, Australian and some other platforms. They are comparatively narrow (of the first tens of kilometers), though deep (5 to 15 km) troughs with the extent of several hundred km. The troughs (upward) are filled up by thick basaltic and comatiitic lavas erupted in subaqual conditions, and in the upper parts of the section by more acid low-potassic volcanite and also by clastic (conglomerates including) and siliceous deposits, with considerable concentration of jaspilites. Petrochemically, volcanic formations of Archean greenstone belts resemble volcanic series of Phanerozoic rift zones more than those of the early and middle development stage (so-called, island-arc stage) of eugeosynclinal zone (Grachev and Fedorovsky, 1981). Usually, Archean greenstone belts are presented by a group of 10 to 20 and over zones with more or less parallel, though twisting in plane axes. Greenstone belts are separated by zones of ancient, mainly lower Archean complexes of so-called "grey gneisses" of granodiorite or tonalite composition. Pebbles of these "grey gneisses" are presented somewhere in the base of the section of the greenstone troughs. These troughs seem to appear at the "grey gneiss" basement in the process of extension and desintegration of this basement by series of subparallel faults which caused numerous relatively narrow linear graben-like depressions. These troughs were filled up with the products of the erosion of the adjacent uplifts of grey-gneiss basement and volcanic material extruded through longitudinal fractures in the basement of greenstone belts. Peculiar features of greenstone belts and of their early evolutionary stages seem to relate them to Phanerozoic rift zones. A number of Soviet and foreign scientists (Windley, 1973; Sinitzyn, 1979; Grachev and Fedorovsky, 1980; and others) believe that these structures are ancient analogues to Mesozoic-Cenozoic rift zones. According to spatial diffusion of extension between numerous rift structures,regions of Archean greenstone belts resemble Cenozoic Basin and Range Province in the Cordilleran area with typical of it "keyboard" block system formed by great number of graben and horst features.Grachev and Fedorovsky believe that synchronous appearance and formation of greenstone belts over many present-day continents in the Early and Late Archean, occurred under conditions of the general slight expansion of the Earth in these epochs. At the subsequent evolutionary stages, greenstone belts experienced considerable compression. Graben-like features were transformed into significantly compressed synclines or synclinoriums, whereas their gneiss basement was increasingly heated, acquired high plasticity and was greatly transformed into ultrametamorphic granite-gneiss and granitoid complexes. The latter bulged as dome massives between greenstone belts and were intruded in their fold structures. Compressional deformations, which completed the evolution of greenstone belts, and associated granitoid plutonism are not typical of the evolution of "classical" rift zones in the subsequent time: vice versa, they characterize last evolutionary stages of geosynclinal features. Proceeding from this fact, some scientists consider greenstone belts as ancient analogues to geosynclinal structures, namely, peculiar miniature but numerous eugeosynclinal troughs. Their origination was accompanied by dissection or complete desintegration of proto-continental crustal blocks, whereas accomplishment of their evolution was marked by strong compression (collision), and, according to Dewey and Burke - by formation of numerous subduction zones during the period of collision. Their existence in the Archean, however, does not seem sufficiently grounded. Some scientists, who believe in the rift origin of Archean greenstone belts, do not associate the folding of infilling formations with the general horizontal compression of the corresponding region at the completing stage of the belts development. They explain this phenomenon by the increase of volume ("swelling") of adjacent granite-gneiss-zones due to granitization (Grachev, Fedorovsky, 1980). But, even in case this supposition is true, Archean greenstone belts considerably differed from typical rift zones as the role of large and long living faults in their structure is not so remarkable. Even if such faults occurred at various stages of greenstone belt development, they are healed with time in the result of plastic deformations of the considerably heated and rheomorphozed matter of granite-gneiss substratum.

It should be noted that, with rare exceptions, Archean greenstone belts were firmly brazed to the adjacent granite-gneiss zones after they completed their evolution and, further, as a rule, were not inherited either by rifts or any other structures.

Granulite or charnockite-granulite belts, revealed on the shields of most ancient platforms, present another type of Archean linear features which also developed under conditions of alternation of the horizontal extension and compression of the Earth's crust. Laterly, similar zones have been also found in the Archean basement of Russian plate (Kratz et al., 1979). Granulite belts sharply differ from greenstone belts as they are larger in size (width to 100 km and over, length to 1000 km and over) and are characterized by much higher metamorphism with the wide development of rocks of granulite facies. Contrast to them, metamorphism of greenstone belts, as a rule, corresponds to greenschist and, sometimes, to amphibolite facies. In case greenstone belts are linear depressions, the granulite belts in their present form are linear uplifts with outcrops of rocks initially composed very deep horizons of Archean basement.

Some researchers believe that granulite belts became already individualized in the Early Archean in the process of extension of protocontinental crust of the present-day platforms, and in the result of the crust disintegration by deep fracture zones into mega-blocks which were considerably larger than granite-gneiss blocks separated by greenstone belts. Initially, the present granulite belts seemed to be extensional zones filled up by mantle derivatives of basic and ultrabasic composition which were consequently strongly compressed and crashed. Considerably metamorphozed material of the deep parts of these belts was squeezed upward and parts overthrusted on the adjacent gneiss massives. Some other researchers, like Kröner (1977) date the deeply metamorphozed formations of granulite belts back to the Archean but believe that the uplift of this material and the overthrusting of granulite blocks on the adjacent massives occurred later, in the Early Proterozoic. This supposition is supported by the observed on some platforms non-conformed structural relations between Archean greenstone belts and granulite belts: the strike of granulite belts in many cases differ from that of greenstone belts and, in plan, cuts them. It should be noted, that, as a rule, rocks of granulite belts, besides ancient (Archean) metamorphism of granulite stage also manifest consequent retrograde metamorphic processes which occurred in the Early Proterozoic and, sometimes, repeated in the following epochs. Contrast to greenstone belts, many granulite belts had longer period of "vitality", that is, the ability to consequent regeneration and experienced later (in the Late Proterozoic and Phanerozoic) repeated tectonic - thermal activization (Khain, 1979; Dolginov et al., 1973). They predetermined the localization of numerous Late Proterozoic, Paleozoic and even Mesozoic and Cenozoic continental rift zones and also of many Mesozoic (Paleozoic, in western Australia) pericontinental rift zones, and controlled position of margins of "secondary", Mesozoic and Cenozoic Oceanic basins(Indian and Atlantic).

EARLY PROTEROZOIC

By the beginning of the Early Proterozoic (2.6 billion years ago), when Late Archean greenstone belts ceased their development, and global granitization over all present-day continents completed, large relatively consolidated protoplatform regions were formed. If compared with Archean protocontinental crust they have more thick and consolidated crust of the continental type. Later they formed the cores of ancient platforms and, in some cases, almost do not yield to them in size. On the whole, in the Early Proterozoic the greater part of protoplatform regions were uplifted. Only some of them experienced subsidence attributed to subisometric and elongated zones in which ancient complexes of protoplatform cover were formed, Usually, these complexes are rather thick and metamorphozed.

In the beginning and middle of the Early Proterozoic, between protoplatforms, several mobile protogeosynclinal regions were set and developed. Their width and length were much greater than those of Archean greenstone belts but considerably smaller than those of geosynclinal belts in Neogeicum. Protogeosynclinal zones, at least their parts within the reach for investigation, seemed to be almost entirely set at an Archean protocontinental crust. It comes from the fact that, with rare exception (Baikal-Vitim zone, for instance), no typical ophiolite associations are exposed in their cross-sections, meanwhile, they have Archean uplifted blocks within the region. Development of protogeosynclinal regions was completed by folded deformations, metamorphism of amphibolite or, more rarely, greenschist facies and by granitoid plutonism. These processes mainly occurred about 2.0 to 1.9 billion years ago (Svecofennian folding epoch at East European platform, and corresponding epochs at other platforms) and about 1.8 to 1.7 billion years ago, in some of protogeosynclinal regions.

Contemporary to the protogeosynclinal regions, mobile

linear zones of less width and length originated at the proto-platforms. Their setting and development (with exception of its completing stage) proceeded under conditions of horizontal extension. This related them to intracratonic rift structures of the consequent epochs and permitted to consider them as protorift zones. These zones are known at many ancient platforms: East European, Siberian, North American, Australian, Indian, African. Protorift zones excelled Archean greenstone belts in length, and often also in width contrast to greenstone belts, they are presented as a single zone or, rare, as several (2 to 3) closely connected zones, but nos as a compact "swarm", and have less summary areal distribution. Besides, unlike greenstone belts, protorift zones have more persistent strike and are controlled by more distinct longitudinal fractures, especially by large marginal faults. At some proto-platforms, in particular, East European, Siberian, Indian and others, we can reveal a spatial relation of protorift zones with neighbouring protogeosynclinal regions: they either branches off them, or are set at some distance from their margins and run parallel to them.; for instance, protorift zones of Kola peninsula and Karelia are parallel to the eastern margin of Svecofennian protogeosynclinal region. In some cases, it is difficult to attribute an early Proterozoic mobile linear zone to either protogeosynclinal or protorift type. Somewhere, at Australian platform, for example, we can observe a structural net of early Proterozoic protorift zones of various strike (submeridional, sublatitudinal, northwestern). It testifies that these features were set under conditions of dissection and general horizontal extension in various directions - creeping apart of - "ravelling" of the proto-continental crust.

According to the type of formations, we can distinguish two types of protorift (and also of protogeosynclinal) zones. In the first type, volcanism did not play any significant role, and the clastic sediments and also jaspilites were mainly accumulated in these zones. For example, such deposits are observed in Krivoi Rog zone, Ukrainian shield, and similar zones in the region of Kursk magnetic anomaly, at Voronezh crystalline massif. The second type is presented by sedimentary-volcanic series (basaltic rock prevailed) with associated concordant and discordant intrusive bodies of basic and ultrabasite rocks (for example, Pechenga and Imandra-Varguza trough in the north-eastern part of the Baltic shield). This type probably corresponds to the periods of greater intensity of extension and dissection of the crust (down to its bottom) by deep faults and fissures. However, it is a rare case that in the protorift zones and also in the majority of protogeosynclinal regions horizontal extension could cause complete break-up of proto-continental crust and considerable disintegration of its blocks. Sometimes, because of deep subsequent denudation, only "roots" of volcanic protorift zones are preserved as thick and, in some cases, very long dyke-like basic and ultrabasite intrusions, for instance, so-called Great Dyke of Zimbabwe.

During the early stages, protorift zones, especially of volcanic type, developed under conditions of horizontal extension and deep subsidence of their basement (5 to 10 km and sometimes to 15 km). At the later stages, the majority of protorift zones experienced considerable compression. In the zones filled with thick volcanic complexes, the compression resulted in the formation of large asymmetrical depressions cut by great marginal thrusts and complicated by secondary fractures of upthrust - and even overthrust type (for instance, Pechenega and Imandra-Warzuga depression). Contrast to this, in the troughs filled up by thin-layered, in particular, ferro-siliceous sediments, there appeared considerably compressed synclines and narrow synclinoria also cut by upthrusts - and over-thrusts (for example, Krivoi Rog zone). These completing deformations occurred in protorift zones of East European platform mainly about 2 to 1.9 b.y. ago and coincided in time with the epoch of Svecofennian folding in the protogeosynclinal region of similar name, which occupied the central part of the Baltic shield. Similar contemporaneous compressional deformations in protorift and protogeosynclinal zones was revealed at some other ancient platforms. Completing compressional deformations sharply changed the initial morphology of protorift zones. Even the Great Dyke of Zimbabwe has records of strong consequent compressional and shear deformations which caused horizontal flexures (sigmoid) complicating its rectilinear trending.

In some protorift zones, like Kuola-Vygozero, Karelia, (including its southern part, Vetreny belt), during the Early Proterozoic, cycles of extension and consequent compression of the Earth's crust occurred. These cycles are extended back to 2.6 to 2.2 b.y. and 2.2 to 1.9 b.y. and correspond to the beginning and middle of the Early Proterozoic.

Only some individual protorift zones which were set and experienced subsidence during the last third of the Early Proterozoic (1.9 to 1.65 b.y.) did not show considerable compression at its last stage of development (for instance, Bilyahchan graben, eastern part of Aldan shield). On the whole, development of narrow early Proterozoic protorift zones and more vast protogeosynclinal regions was controlled by global rhythmic alternation of extension and compression of the Earth's crust.

The scale of Early Proterozoic extensional and compressional deformations which occurred on individual protoplatforms and in protogeosynclinal regions, and also global effect of these horizontal movements seem to be less than the scale of similar processes of the Archean. First of all, this supposition concerns the summary effect of horizontal extension of the protocontinental crust in the mobile zones which was greater in the Archean as compared to the Early Proterozoic.*)

Mentioned similarity in the structure and evolution of protorift and protogeosynclinal regions and their spatial relations permits to conclude that these two major types of mobile early Proterozoic linear zones were related to each other. They originated and ceased

*) This supposition should be reviewed in case the setting of the Pacific basin will be dated back to the Early Proterozoic. Now its age is unknown.

their existence synchronously, their development stages, kinematics and the sequence of deformations were similar: Protorift zones mainly differed from protogeosynclinal regions by less size, less amplitude of tectonic deformations, in particular, by less role of horizontal compression in some most late protorift zones, and also by less intensity of metamorphism which usually reached only greenschist facies. At the completing stages of protorift zone development, acid plutonism is less intensive than in protogeosynclinal areas. Arising from their geometry, size, dominant longitudinal faults and "behaviour" at the earliest stages of evolution (which testifies dissection and horizontal extension of the crust), protorift zones are related to intracontinental zones of Neogeicum, first, to Riphean aulacogens. This dual character of early Proterozoic protorift features allow to call many of them and also similar late Proterozoic and Paleozoic intracratonic folded zones, as "aulacogen" (The term proposed by Leonov, 1980). To underline the peculiar character of early Proterozoic generations of these structures it is possible to call them protoaulacogeosynclines.

LATE PROTEROZOIC

By the end of the Early Proterozoic on the greater part of the territory of the present continents, a thick and mature continental crust was formed in the result of the transformation of early Proterozoic protogeosynclinal regions into folded systems and in the result of cratonization of adjacent protoplatforms. Soon, however, it experienced destruction, i.e. dissection by system of deep faults, horizontal extension, subsidence and reworking. These processes started in the Early Riphean, intensified in the Middle and continued into the Late Riphean. They are manifested on the surface of Riphean continental crust irregularly and embraced mainly a number of wide stripe-like zones. In the result of these processes several large-scale geosynclinal belts originated. The most elongated geosynclinal belt, - the Circum Pacific one, separated the oceanic "hemisphere" (the Pacific) from continental one. The North Atlantic, Urals-Mongolian, Mediterranean, and probably some other geosynclinal belts, branching off the Circum-Pacific one, divided the continental "hemisphere" into a number of stable blocks: ancient platforms of different size. To them belong platforms of the northern, Laurasian group (from smaller to greater): Chinese, Siberian, East-European, Greenland - North American, and also the gigantic southern superplatform of Gondwana. We cannot exclude the supposition that Gondwana was not a single monolithic platform but represented a group of three (?) large platform massives: West Gondwana (Africa-South America), East Gondwana and Antarctic divided by hypothetical geosynclinal belts - Indo-Arabian and Peri-Antarctic (or Nereis belt, according to A.N. Mazarovich, 1952).

Ancient platforms, separated from one another in the process of geosynclinal belt origination, also experienced destruction though not so intensive. They were cut by numerous comparatively narrow linear fracture zones in which horizontal extension and rather deep subsidence occurred in the Riphean. N.S. Shatsky was the first who revealed these zones in East European platform. He proposed (Shatsky, 1964) to call them aulacogens (Greek for the structural zones born by furrow). They can be regarded as ancient analogues to Cenozoic continental rift zones, in particular, to intracratonic ones. Localization of Riphean aulacogens was controlled by the existence in their basement of favourable early Proterozoic and Archean linear structure zones, and, in particularly, by ancient granulite belts predisposed to the subsequent tectonic and thermal regeneration. A comparative study of a number of Riphean and Paleozoic aulacogens carried out by Shatsky has demonstrated the great variety of their recent structures and their complicated tectonic development. Shatsky distinguished two major groups of aulacogens. The frist embraces simple aulacogens, that is, aulacogens which preserved in relatively good form initial graben-like geometry (at least, in the lower strata of sedimentary cover). The second group includes complex aulacogens which represent more large, i.e. more extended, broad and deep linear zones with complex structure. Their initial graben-like features were complicated and transformed by later folding deformations into narrow intracratonic folded zones. Riphean deposits filling these zones are usually more thick than those of simple aulacogens. In many cases, they are slightly metamorphosed, usually in greenschist facies, and cut by syntectonic and posttectonic intrusive bodies. Such properties of structure and development bring "complex aulacogens" close to intracratonic geosynclinal troughs and to epigeosynclinal folded zones originated in the result of their evolution. A series of gradual transitions occurred between typical "simple" and "complex" aulacogens. Sometimes, such transition can be traced even along the strike of one single zone. To stress the dual nature of "complex" aulacogens, their intermediate position between typical aulacogens (rift zones) and geosynclinal troughs, it seems expedient to use the term "aulacogeosyncline", introduced by Leonov, 1980, for "complex aulacogens" and preserve the term "aulacogen" only for "simple aulacogens" of Shatsky.

In Late Proterozoic, considerable differences in tectonic structure and development of Laurasian and Gondwana group of platforms, were distinctly expressed. N.A. Streiss (1964) called them "bi-polarity of the Earth's structure". Most early typical manifestations of intraplatform rifting, that is, setting and evolution of numerous Riphean aulacogens (S. Str), are mainly concentrated on East European platform and also on other three platforms of Northern, Laurasian group where they were later discovered. These ancient linear graben-like zones, usually burried, are framed by normal faults which cut the basement of each platform in various directions. Bogdanov noted that these structures are either blind branches of adjacent geosynclinal belts ("penetrating aulacogens") entering deeply into the platform's body, or narrow binding branches between two neighbouring geosynclinal belts, cutting the angular segments off the platform, or dividing them into several parts ("through-going aulacogens") or, at last, closed furrow-like depressions running parallel to geosynclinal belts which frame the platform (inner or "blind" aulacogens). These relations revealed close structural and genetic link between Riphean aulacogens on the platforms and geosynclinal belts framing them, and permit to consider their origin as tectonic manifestation of the common process of dissection, horizontal extension and deep subsidence and transformation of the Earth's crust, which manifestations on the platforms, however, are more weak than in geosynclinal belts.

Synchronous occurrence of the net of numerous graben-like zones of various strike within one platform (for instance, on East European platform)or existence of a greatly curving dendritic rift system (for example, on North American platform) shows that their onset and development occurred under conditions of some general differently oriented extension ("creeping apart") of the platform crust. Broad manifestations of basaltic volcanism in some Riphean aulacogens well agree with this supposition.

It must be mentioned, that though, on the whole, destruction of the crust and its horizontal extension during the period of geosynclinal belt generation considerably exceeds similar phenomena within ancient platforms, their areal distribution was also very irregular. It lead to their differentiation into broad geosynclinal troughs and relatively "rigid" inner massives, or even groups of neighbouring massives, separated by narrow linear zones of aulacogeosynclinal type *).

According to composition and thickness of their deposits, Riphean aulacogens usually occupied the intermediate position between the miogeosynclinal troughs and typical platform nappe which started its formation in the Riphean, mainly in pericratonic

*) After geosynclinal belts which included them died away, these regions usually were transformed into young plates.

depressions.

An active or even dominating role in the filling up of some most deep Riphean aulacogens, for instance, of Keeweenawan and Gardar aulacogens was taken up by thick volcanic formations and accompanying intrusions of basic, mainly basaltic composition close to trap formations of platforms. On the whole, these features resemble volcanic complexes of some early Proterozoic protorift zones, for example, Pechenega-Warguza, or Labrador zones but comparatively with the latter, they may also subalkaline volcanites and alkaline intrusions mainly of the central type. Late Proterozoic alkaline volcanism related to rifting was also revealed in the northern part of Siberian platform (Shpunt et al., 1982). At the end of Middle Riphean or at the end of Late Riphean - Vendian, the majority of aulacogens ceased their evolution and some of them experienced inversion of tectonic movement sign: subsidence changed to uplift, extension to compression. It led to appearance of such structures as inner swell- or horst-like uplifts, thrusts, overthrusts, and in some cases, to folding deformations of aulacogen complexes. These inversion deformations are synchronous to epochs of early Baikalian, Grenvill and late Baikalian folding and evidently are relatively weak manifestations on the platform of corresponding compressional deformations occurred in the adjacent geosynclinal belts. More broad and flat platform depressions of sineclise type were formed above some Riphean aulacogens, in case when inversion was absent or was weak, in the latest Proterozoic (Vendian) or in Paleozoic.

On most platforms of the southern, Gondwana group almost no typical aulacogens dated back to late Proterozoic are documented. Their absence can be probably explained by the fact that the most part of this large region was disposed at great distance from the recorded geosynclinal belts. Only exclusions are the northern and central parts of Australian platform which are close to the Pacific geosynclinal belt (Batten, Amadeus and other aulacogens), and Antarctic platform within which one group of late Proterozoic aulacogens was clearly connected with Rossic epigeosynclinal folded belt of

Baikalian age, and the others are located in the northern part of East Antarctica, near the shore of the Southern ocean. The last fact, as well as extensive manifestations of Baikalian isotopic rejuvenation of rocks in this peripheral zone, permit to suppose the existence of Periantarctic geosynclinal folded Baikalian belt in the Southern ocean. In the western part of Gondwana, at vast territories of present-day Afro-Arabian and South American platforms, intracratonic eusialic aulacogeosynclinal troughs are widely spread instead of geosynclinal belts and typical aulacogens. Most of these troughs are comparatively narrow (less than 100 km) and have much in common with early Proterozoic aulacogeosynclinal troughs of all Laurasian, Australian and Indian platforms. Many authors suppose that the present-day South American platform was directly attached to African platform in Late Proterozoic (as well as in Paleozoic and Jurassic). However, it seems more natural to believe that the wide zone of many hundreds kilometers (up to thousand) existed between Brazil and Equatorial Africa where the complex system of Riphean aulacogeosynclinal troughs could have been located. In the western part of Gondwana, development of these intracratonic troughs was accompanied by some periods of compression: at the end of Early Riphean (in Africa) and, at the end of Middle Riphean (in South America). It was completed by intensive folding and broad manifestation of tectonic and thermal activization in a number of zones in Africa and America at the end of Riphean and in Vendian.

On the whole, horizontal extension of the crust of platforms which occurred in aulacogens during the Riphean seemed to be not completely compensated by late Riphean and Vendian compression. So, we can suppose some general horizontal extension of the crust of ancient platforms in Late Proterozoic. Simultaneous extension of many aulacogens which had very different trends within the same platforms could give evidence for their expansion and "creeping apart" in various directions. It seems also undoubtful that the extension of geosynclinal belts and increase in their area during their onset in the Riphean was not completely compensated by compressional deformations occurred at the end of Baikalian "cycle", as development of these belts with exception of comparatively small segments was not finished in the Late Riphean-Vendian and was activized with more energy in the Paleozoic.

PALEOZOIC

During the Paleozoic, geosynclinal belts, set in the Riphean, continued their development. In the Early Paleozoic, after Baikalian folding, many of their segments experienced one more strong extension which led to the break up of thinned continental crust and formation of ophiolite troughs with the crust of oceanic type. Paleozoic intracratonic rifting mainly occurred in the northern, Laurasian platforms. It had much in common with late Proterozoic rifting but had less intensity and distribution. No new intracratonic rift zones originated in the Paleozoic but many ancient aulacogens (though not all of them) regenerated after more or less long pause in their development. In some cases, in Ien-shan paleorift zone, in the northern part of Chinese platform, aulacogen subsidence continued almost without interruption from the Proterozoic to Paleozoic. Zones, touched by Paleozoic rifting, contrast to Riphean rifting, occupied not the entire area of platforms but only some of their segments adjacent to geosynclinal belts framing them. On the East European and Siberian platforms framed by Paleozoic epigeosynclinal folded areas of various age, some aulacogens and fracture zones showed poorly manifested Caledonian "cycle". Then, in the Middle-Late Devonian, there occurred strong regeneration of a set of Riphean aulacogens which was accompanied by outbursts of alkaline-basic and alkaline-ultrabasic magmatism. These events presented manifestations of the global tectonic-magmatic activization occurred in the Devonian. Contrast to the Late Proterozoic, duration of recurrent Paleozoic cycles of intensive subsidence in regenerated aulacogens was usually considerably less (tens or, maximal hundred million years) in comparison with several hundreds million years in the Riphean. Total amplitude of their subsidence, however, was almost the same as of Late Proterozoic cycle, and the rate of subsidence in the Paleozoic was even considerably higher than that in the Riphean. But the subsidence in the longest and wide Paleozoic paleorift, located on East European platform, that is, in the Dnieper-Donetz aulacogen, continued intensively, however, from the late Devonian to early Permian and completed by inversion and folding in its deepest eastern part - Donetz aulacogeosynclinal zone - in the middel of Permian. Nevertheless, till the latest Mesozoic, eastern parts of this aulacogen experienced repeated, gradually attenuated compressional deformations. Tectonic inversion, accompanied by compressional deformations and origination of narrow inner linear uplifts, also occurred in other Paleozoic aulacogens of Siberian (Patom-Viluy Irkineev, and others) and North American platforms (Bootia-Cornwallis, Vichita). In some aulacogens inversional deformations like formation of swell over graben, upthrust - overthrust displacements, etc. turned to be very far extended in time from the stage of intensive subsidence which occurred in the Middle Paleozoic, mainly in the seconf half of the Devonian. They continued through the Mesozoic (late Triassic, Jurassic, Cretaceous) and, somewhere, even in the Cenozoic (for instance, Zhiguli upthrust). Early and Middle Paleozoic rifting was weak or almost absent on most of the southern, Gondwana platforms. Australian platform is one of the few exceptions. Aulacogens of Central Australia (Amadies and others) adjacent to East Australian folded system were activized in the Cambrian and completed their development by compressional deformations in the Carboniferous. Other exceptions are aulacogens in the north-western part of African platform (Ugarta, Antiatlas) and in the western part of South American platform (Chikitos) adjacent to Paleozoic belts of the Atlas and Andes, respectively. Some Paleozoic aulacogens, no doubt, existed in the inner part of Gondwana supercontinent. Cape zone in the south of Africa and Sierra de la Ventana in the south-east of South America are of this type. Their development was completed by folding in the Triassic. Some aulacogens of north-western and western Australia set up in the Ordovician or Silurian. Occurrence of these Paleozoic intra-Gondwanian aulacogens and aulacogeosynclinal zones allow us to suppose the existence of some larger unknown Paleozoic mobile zones in the southern part of the present Atlantic and eastern part of the Indian ocean. These aulacogens could be branches of this large mobile zone. In the Paleozoic, magmatism related to rifting on ancient platforms definitely acquired subalkaline or alkaline character which was preserved for the following Mesozoic-Cenozoic megastage. On the whole, during almost the entire Paleozoic episode of aulacogen development on ancient platforms, there seemed to be neither significant expansion of their area nor of the total area of geosynclinal belts which framed these platforms, nor of the entire surface of the Earth.

The epiorogenic rift zones and rift systems, first appeared in the Paleozoic, represented a qualitatively new type of rifting. They began their formation within those geosynclinal belts or in some of their segments which completed their geosynclinal subsidence, were subjected to strong compressional deformations and transferred into folded areas during the Paleozoic.

Most ancient epiorogenic rift zones appeared in Vendian - early Cambrian, just after Baikalian orogenesis, in the Baikal-Patom region, East Siberia. Within the Urals-Mongolian belt, just after Baikalian folding, a number of other rift-like zones also appeared but they were "camouflaged" in the result of the successive tectonic transformations as, these Vendian-early Paleozoic, rift-like deformations represented the essence of the geosynclinal process regeneration. More young, Devonian generation of epiorogenic rift zones is most widely presented in North Atlantic Caledonian folded belt. This middle Paleozoic epiorogenic rift system includes grabens of East Greenland, Spitzbergen, Norway, Scotland, Northern Appalachians filled up with thick psephytic and often volcanic formations. These linear rift depressions were formed under horizontal extension associated, in some places, with large strike-slip displacement (Scotland). During the Devonian, in Caledonian folded systems of the Urals-Mongolian belt structures characterized by not linear but more complex winding pattern, depressions of irregular shape appeared. They were associated with block uplifts, volcanic belts and regions, etc. (Caledonides of Kazakhstan, Northern Tien-Shan, Sayany-Minusinsk region). Similar block extensional structures and regions of subaerial volcanism were also formed after Hercynian folded deformations during the Late Paleozoic in some regions of the Urals-Mongolian belt, on the northern flank of the Mediterranean belt (Central Europe), etc. Volcanism which was widely spread in epiorogenic rift zones differed from alkaline volcanism of the Paleozoic as well as Cenozoic intracratonic rift zones by its dominantly calc-alkaline character.

MESOZOIC AND CENOZOIC

Mesocenozoic epoch is marked in the Earth's history by considerable reorganization of its general tectonic pattern which, according to its role in the Earth's history, can be compared with appearance of geosynclinal belts and separation of ancient platforms in the Riphean. During the Mesozoic and Cenozoic, the whole area on which geosynclinal process occurred became more and more narrow. At the same time the secondary oceanic basins appeared and gradually enlarged, and the basement of the ancient Pacific basin was rejuvenated. Large-scale volcanic processes gradually expanded over the entire basement of the oceans and over large non-geosynclinal regions of continents. In the Late Proterozoic and Paleozoic, the main tectonic telements of the Earth, besides the Pacific basin, were geosynclinal belts and ancient platforms, but in the Mesozoic-Cenozoic this main role was taken up by oceanic basins and continental massives which were formed partially in the result of coupling of one or several (Eurasia) platforms with adjacent epigeosynclinal folded systems, and partially in the result of break-up of earlier greater platforms (Gondwana, and Greenland-North American). At last, one of the most important elements of the Mesocenozoic tectonic history was the appearance and development of numerous rift zones, both within continents and oceans, which played the main role in the structure and formation of the oceanic bottom. By the latest Paleozoic we can attribute appearance (or regeneration) some segments of East Gondwana of several rift zones adjacent, contrast to more ancient epi-platform rift zones, to the periphery of the recent Indian basin and not connected with any geosynclinal belt. They seemed to be genetically associated with initial stages of Gondwana break-up. Most ancient among these rift zones are early Gondwana grabens of Indian platform and "Karroo" grabens in the south-eastern part of African platform, longitudinal West Mozambique graben systems first covered by the shallow sea in the middle Permian and some contemporaneous graben-like troughs on Antarctic platform.

In the Mesozoic, the area of rifting manifestations continued to widen and their variety and intensity sharply increased. Simultaneously, in the early Triassic, a prolonged epoch of intensive trappean magmatism on continents started. It was almost completely absent during the entire Paleozoic. In the Mesozoic and early Cenozoic, Eocene including, areals of trappean volcanism "wandered" over the Earth's surface successively embracing Siberian, African, Antarctic, South American, Indian, Greenland platforms and adjacent folded regions. At least since the middle (?) of Jurassic, even much more extensive basaltic eruptions began in all oceans (in young oceans as well as in the ancient Pacific one) which continued into the recent time and exceed trappean volcanism of platforms by 15 to 20 times.

Contrast to Riphean and Paleozoic aulacogens which were structurally and genetically related to geosynclinal belts, Mesozoic and also Cenozoic intracontinental, intracratonic rift zones are connected with the basins of the Atlantic and Indian oceans as if "flowing" into them and no doubt, are genetically related with the process of their formation. With very rare exception (Cretaceous Benue rift western Africa) they did not experienced for the present tectonic inversion and compression and developed under regime of irregular extension and subsidence.

In even more close relation with formation of young oceanic basins are numerous Mesozoic pericontinental rift zones tracing along the western and eastern shores of India and Africa, along the eastern shores of North and South America and western shore of Australia. To considerable extent they are superimposed on the marginal parts of platforms separated in the process of Gondwana supercontinent break-up, but some of them (for instance, on the eastern periphery of North America) are attributed to Paleozoic folded structures and have epiorogenic character.

Most magnificant manifestations of Late Mesozoic and Cenozoic rifting are within Atlantic, Indian and Arctic basins. Formation of these great basins cannot be explained only by intraoceanic type of rifting, i.e. by spreading in mid-oceanic ridge zones, as considerable peripheral parts of their floor were formed in the result of subsidence, destruction, dispersed rifting ("creeping apart"), and transformation of the Earth's crust. The role of these processes was especially high in the formation of the North Atlantic, eastern Arctic, and north-eastern and south-western parts of the Indian ocean floor. Large-scale destructive processes in the Earth's crust, intensive rifting, gigantic Mesocenozoic volcanism in the oceans and also within platforms, and occurrence of young folded belts (marginal volcanic belts) and considerable reduction of the total area of geosynclinal belts, if compared to the Paleozoic, all these facts suppose that, on the whole, Mesocenozoic epoch was characterized by higher thermal regime in the upper mantle, and episodically, in the crust, and by general (though complicated by a number of pulsations) expansion of the Earth (Milanovsky, 1982). These processes can be compared with analogue processes probably occurred in the Archean and Late Proterozoic and Archean.

Though North Atlantic and the Urals-Mongolian geosynclinal belts ceased their active existence in the Paleozoic or by the Early Mesozoic, two other geosynclinal belts - the Pacific and Mediterranean-Indonesian - continued their development during the Mesozoic and Cenozoic. Major epochs and phases of compressional deformations coincided in time with phases of less intensity or even pauses in rifting and volcanic activity in other regions of the Earth which seemed to reflect global phases of compression. Contrary to this, during "anorogenic" periods alternating with periods of compression in these geosynclinal, and, later orogenic belts occurred horizontal extension expressed in particular, in the opening of ophiolite zones (in the Mesozoic), in the appearance and activization of epiorogenic rift zones (in the Late Ceonozic), in the large outbursts of volcanic activity, etc. During such epochs in the Late Cenozoic, within the Cordilleran Mesozoic epigeosynclinal orogenic belt, there appeared Basin and Range Province rift system, Rio-Grande rift zone and intracontinental rift zone of the California Bay which is a continuation of East Pacific rift belt within North American continent, and also the chain of epiorogenic graben-like rift depressions in the South American Andes. After the completion of intensive folding, thrust and nappe deformations in the western part of the Mediterranean orogenic belt, in Pliocene-Anthropogen, processes of destruction, extension and subsidence of numerous depressions were strongly activated. The same epoch was marked by rifting and related volcanism intensification in Afro-Arabian, Baikal and other intracontinental Cenozoic rift belts, as well as spreading increase in all intraoceanic rift belts and intercontinental rift zones. At the same time, in the Pacific belt and over the most part of Mediterranean-Indonesian orogenic belt (except its Caucasian-Himalayan segment) compressional deformations ceased or sharply decreased. It is possible to suppose, therefore, that last 4 to 5 m.y. correspond to the last phase of some general expansion of the Earth, and that in the Mesozoic-Cenozoic epoch, on the whole, global process of extension and, in particular, rifting dominates over compression. Contrast to the Proterozoic and Paleozoic, in this epoch, rifting was not directly connected with the development of geosynclinal regions and belts and was not subordinated to it as an accompanying process on the platforms which framed them. On the contrary, it was the process of extension and destruction of the crust, genetically and in time connected with formation of "secondary" oceanic basins. As compared with the Late Proterozoic and Paleozoic, Mesocenozoic rifting was a tectonic process independent of the geosynclinal belt development, and, to some extent, kinematically opposite to it, so-called complementary or "antagonistic" process. These geosynclinal and rifting processes interact in space and alternate in time. Phases of rifting activization alternate with phases of compressional paroxisms in geosynclinal regions, and on the whole, in Mesozoic-Cenozoic, the total effect of the Earth's expansion seemed to exceed the total effect of its compression.

Thus, rifting as a geological process experienced complex evolution through the Earth's history.

In the Archean, the development of linear tectonic zones, that is greenstone and granulite belt, combined the features both of rift and geosynclinal processes of the following epochs. Early Proterozoic mobile zones - protogeosynclinal regions and protorift zones, developed under the similar tectonic regime but differed by their size, scale, and intensity of deformations and thermal processes. In the late Proterozoic and Paleozoic, continental intracratonic rifting, which acquired the form of aulacogen and aulacogeosynclinal zone development, was a process related and subordinated to a geosynclinal process. During the Mesocenozoic epoch of the Earth's history, rifting for the first time increased considerably and became independent of geosynclinal process and is believed to be one of the main forms of continental crust destruction and of the formation of the oceanic basins which probably occurred during some general expansion of the Earth.

REFERENCES

Burke K.C., Dewey I.F., Kidd W.S.F. Dominance of horizontal movements, arc and microcontinental collisions in the later permobile regime. In: B.F. Windley (ed.). The early history of the Earth.

Bogdanov A.A. Trait fondamentaux de la structure et de l'évolution de la plateforme de l'Europe orientale. C.R. Soc. Geol. France. 1962, fas. 10, p. 229

Dolginov E.A., Moralev V.M., Ponikarov V.L. On the types of early Precambrian structures. Geotectonics, 1973, N 2, p. 8-19 (in Russian).

Grachev A.F., Fedorovsky V.S. On the nature of greenstone belts in the Precambrian. Tectonophysics 73 (1981), p. 195-212.

Khain V.E. Metamorphic regeneration of the regions of repeated regeneration and the problem of new formation of oceanic basins. Geotectonics, 1969, N 3, p. 31-38 (in Russian).

Kratz K.O., Berkovsky A.N., Bondarenko A.P. e.a. Main problems of the geological structure of the Russian plate. Leningrad. Publ. House "Nauka", 1978, 120 p. (in Russian).

Kroner A. Precambrian crustal evolution in the light of plate tectonics and the undation theory. Geologie en mijnbouw. Vol. 58 (2), 1979, p. 231-240.

Leonov G.P. Historical geology. Principles and methods. Precambrian. Moscow Univ. Publ. House, 1980, 344 p. (in Russian).

Masarovitsch A.N. Grundlagen der regionalen Geologie der Erdteile. Deutsch. Verlag d. Wiss. Berlin, 1958, 530 s.

Milanovsky E.E. Rifting development in the Earth's history. In: Ramberg and Neumann (eds.). Tectonics and Geophysics of Continental rifts. 1978. Reidel Publ. Co. Dordrecht, p. 385-399.

Milanovsky E.E. Major stages of rifting evolution in the Earth's history. Tectonophysics. 1983. Vol. 94 (1-4), p. 599-608.

Milanovsky E.E. Problems of the tectonic development of the Earth in the light of concept on its pulcations and expansion. Rev., Geol. dunam. geogr. phys. Vol. 22,

fasc. I. Paris, 1980, p. 15-27
Milanovsky E.E. Aulacogens of ancient platforms: problems of their origin and tectonic development. Tectonophysics, 73 (1981), p. 213-248.
Milanovsky E.E. Rifting in the Earth history (rifting on the ancient platforms). Moscow, "Nedra", 1983, 280 p. (in Russian).
Salop L.I. Geological development of the Earth in precambrian. Leningrad, Publ. House "Nedra", 1982, 344 p. (in Russian).
Shatsky N.S. On the depressions of Donetz type. In: Shatsky. eleated works, v. 2., Moscow, Publ.House "Nauka", 1964, p. 544-553.
Sinitsyn A.V. Problem of origin of precambrian greenstone belts. Geotectonics, 1979, N 6, p. 3-19 (in Russian).
Shpunt B.R., Shapovalova L.G., Shamshina E.A. Late Precambrian of the northern part of Siberian platform. Novosibirsk, Publ. House "Nauka", 1982, 226 p. (in Russian).
Stille H. Geotectonische Gliederung der Erdgeschichte. Berlin, 1944.
Streiss N.A. On the origin of Gondwana. In: 22 Int. Geol. Congr. Report of Soviet Geologists, problem 9. Gondwana. Moscow, Publ. House "Nauka", 1964, p. 7-47 (in Russian).
Windley P.F. Crustal development in precambrian. Phil. Trans. Roy. Soc. Lond., Vol. 273, N 1235, 1973, p. 321-341.

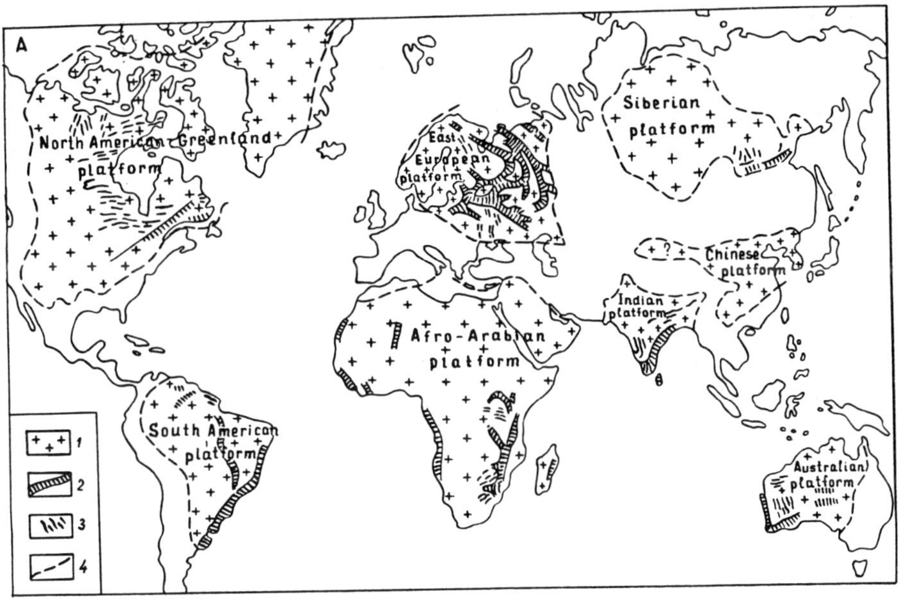

FIGURE 1. Early Archean (?) charnokite-granulite belts and early and late Archean greenstone belts on ancient platforms. (Note: Platforms and mobile belts on Figures 1-6 are shown in their modern positions, without probable horizontal displacements). 1 - Archean gneissic basement of ancient platforms partly reworked in early and late Proterozoic, without subdivisions; 2- Charnokite-granulite belts; 3 - Greenstone belts; 4 - Margins of ancient platforms delineated in late Proterozoic.

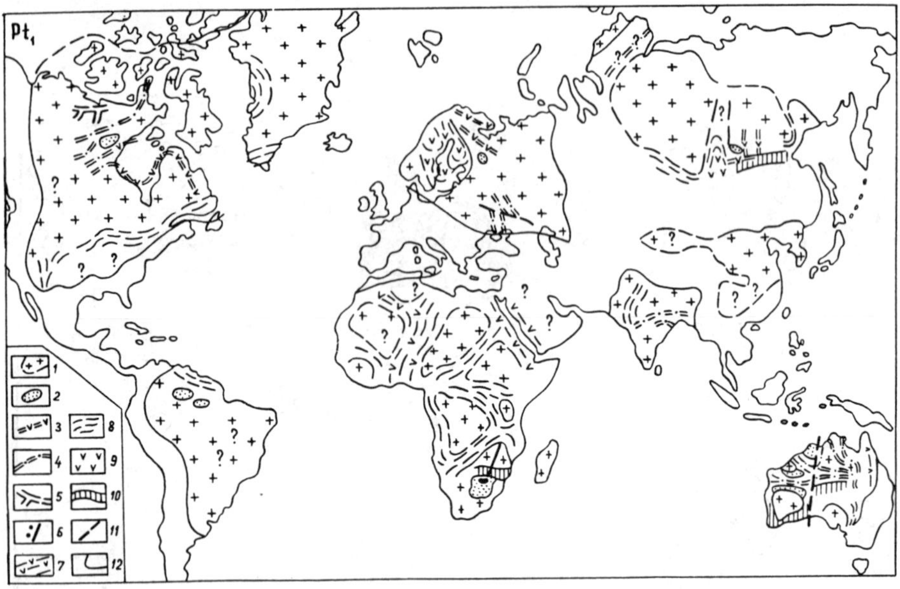

FIGURE 2. Early Proterozoic rift-like zones in the structure of ancient platforms. 1 - Protoplatform areas; 2 - Their depressions (protosyneclises) filled with volcanic-sedimentary deposits; 3 - rift-like zones (aulacogeosynclines) with volcanic-sedimentary infilling, turned into intracratonic folded zones at the end of the early Proterozoic; 4 - The same, with sedimentary infilling; 5 - aulacogens of the upper part of the early Proterozoic; 6 - great basic-ultrabasic intrusives; 7 - Epicratonic protogeosynclinal areas with volcanic-sedimentary infilling; 8 - The same, turned into intracratonic folded zones; 9 - Protoeugeosynclinal zones developed on the crust of oceanic (?) type; 10 - Zones of tectono-magmatic activation during the early Proterozoic; 11 - Some great fracture zones, possibly of strike-slip type; 12 - Margins of ancient platforms delineated in the late Proterozoic.

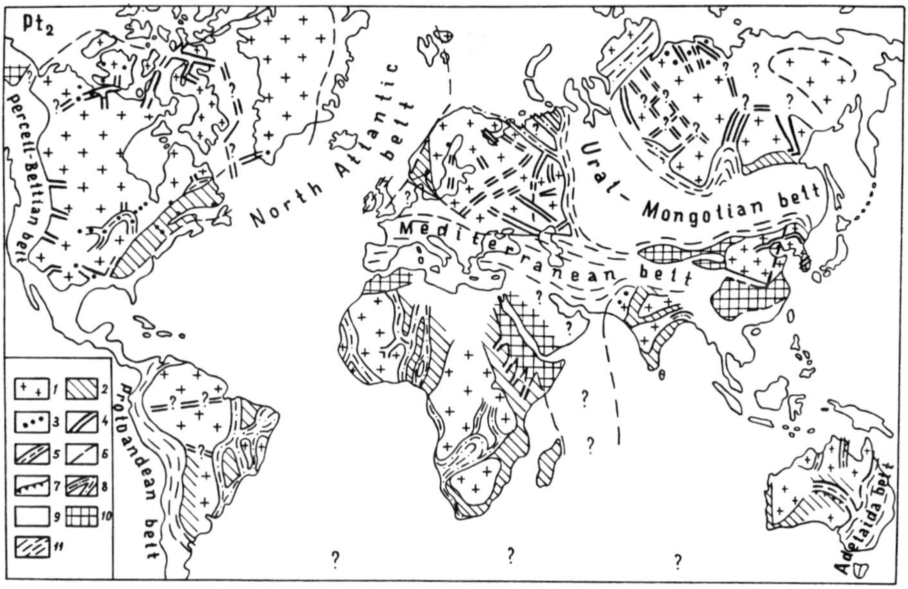

FIGURE 3. Rift and rift-like zones on the ancient platforms in the late Proterozoic. 1 - Ancient (pre-Riphean) platforms; 2 - Zones of tectono-magmatic activation on the ancient platforms during the late Proterozoic (about 1,000 and 500-600 m.y. ago); 3 - Riphean and Vendian magmatism on the platforms; 4 - Aulacogens; 5 - The same, subjected to inversion about 1,000 or 500-600 m.y. ago; 6 - Great steep fault zones; 7 - Thrusts and nappes; 8 - Aulacogeosynclines subjected to early or late Baikalian folding; 9 - Late Proterozoic geosynclinal belts; 10 - Their areas with manifestations of early Baikalian folding; 11 - The same, of late Baikalian and Salairian (Sardian) folding.

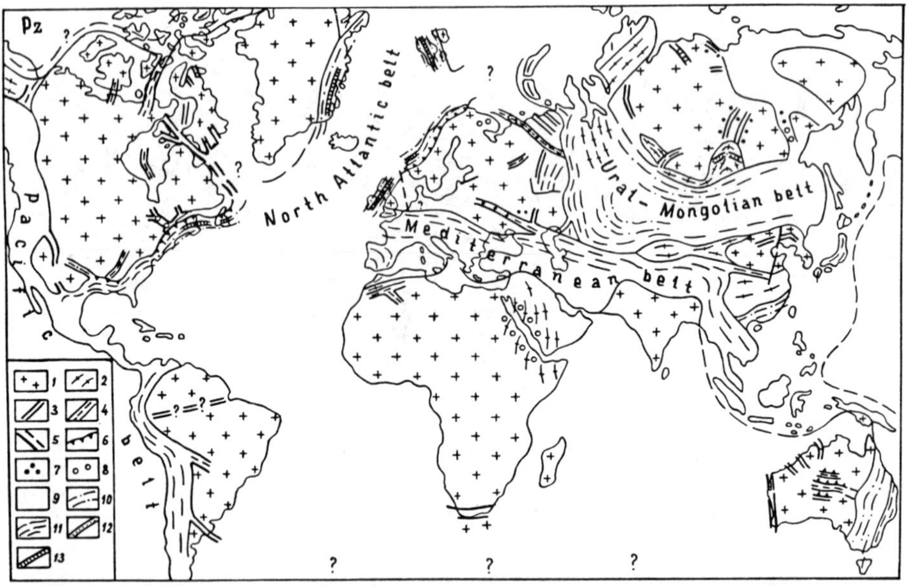

FIGURE 4. Rift zones on the ancient platforms and folded areas in the Paleozoic (up to the beginning of the Permian). 1 - Ancient platforms; 2 - Areas of Baikalian folding; 3 - early and middle Paleozoic aulacogens (mainly regenerated); 4 - The same, subjected to inversion in middle or late Paleozoic; 5 - Great steep fault zones; 6 - Thrusts and nappes; 7 - Paleozoic volcanism on the platforms; 8 - Paleozoic intrusions on the platforms; 9 - Paleozoic geosynclinal belts; 10 - Their areas which completed the geosynclinal development with Caledonian folding; 11 - Their areas which completed the geosynclinal development mainly with Hercynian folding; 12 - Cambrian epiorogenic rift-like zones on the Baikalides; 13 - Devonian (and partly Carboniferous) epiorogenic rift zones on the Caledonides.

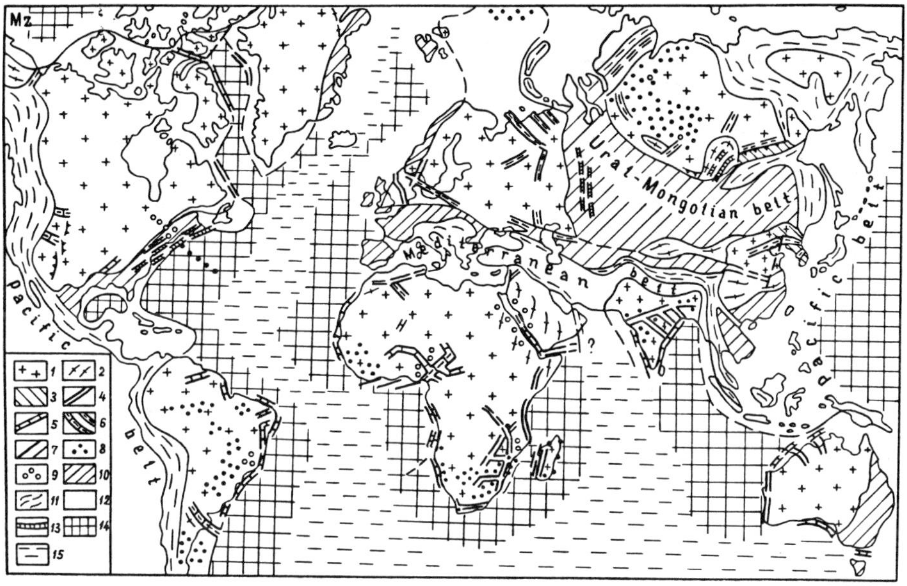

FIGURE 5. Rift zones on the ancient platforms and folded areas in Permian and Mesozoic. 1 - Ancient platforms; 2 - Areas of Baikalian folding; 3 - Zones of Mesozoic tectono-magmatic activation on the platforms; 4 - Late Paleozoic-early Mesozoic rift zones on the platforms; 5 - The same, late Mesozoic; 6 - Manifestations of Mesozoic tectonic inversion in the Paleozoic aulacogens and Mesozoic rift zones on the platforms; 7 - Great steep fault zones; 8 - Permian and mainly Mesozoic volcanism on the platforms; 9 - Mesozoic intrusions on the platforms; 10 - Paleozoic epigeosynclinal folded belts; 11 - Geosynclinal belts which completed their development with Mesozoic (Cimmerian or Laramian) folding; 12 - The same, which continued their development in the Cenozoic; 13 - Mesozoic epiorogenic rift zones in the Paleozoic folded belts; 14 - Peripheral zones of oceanic basins formed or rejuvinated (Pacific) in the late Mesozoic; 15 - Zones of probable most intensive manifestation of the Earth's expansion in the Mesocenozoic.

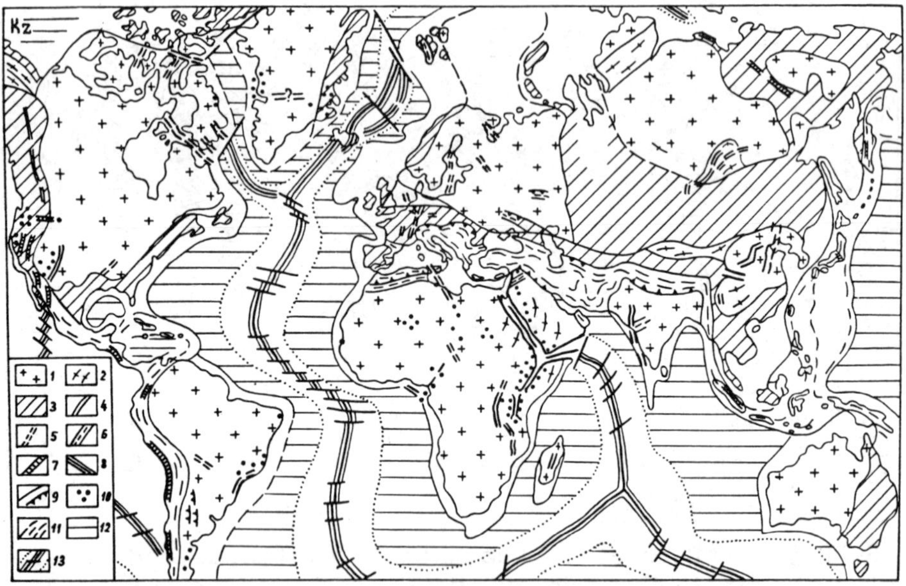

FIGURE 6. Rift zones on the continents and oceans in the Cenozoic. 1 - Ancient platforms; 2 - Areas of Baikalian folding; 3 - Paleozoic and Mesozoic epigeosynclinal folded belts; 4 - Late Cenozoic rift zones on the platforms; 5 - Ancient rift zones of platforms rejuvinated in the Cenozoic; 6 - Rift zones of platforms subjected to tectonic inversion in the Cenozoic; 7 - Late Cenozoic epiorogenic rift zones in the Mesozoic folded belts; 8 - Late Cenozoic intercontinental rift zones; 9 - Great steep faults and thrusts; 10 - Cenozoic volcanism on the platforms; 11 - Cenozoic (Alpine) epigeosynclinal orogenic areas and contemporary geosynclinal areas; 12 - Peripheral zones of oceanic basins formed of rejuvinated (Pacific) areas in the Mesozoic; 13 - Cenozoic intraoceanic rift belts.

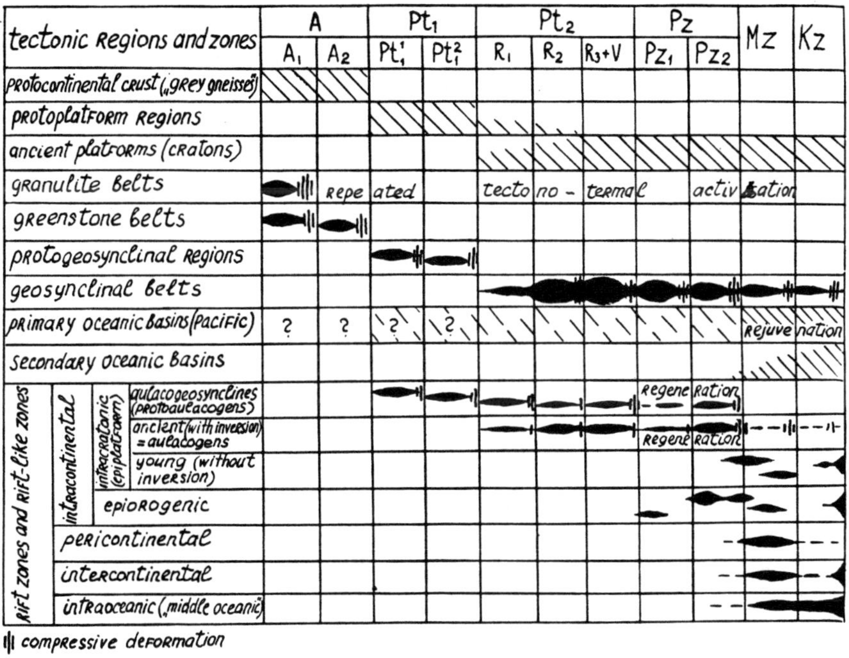

FIGURE 7. Diagram showing the development of different types of rift-like and rift zones and other types of tectonic regions in the Earth's history.

STRUCTURE AND EVOLUTIONARY STAGES OF THE BAIKAL RIFT

N.A. LOGATCHEV and YU.A. ZORIN
Institute of the Earth's Crust U.S.S.R. Academy of Sciences Irkutsk (U.S.S.R.).

ABSTRACT

The structure of the crust indicates that the depressions of the Baikal rift zone have been formed due to irregular extension. The general pattern of the zone is controlled by heterogeneities of the crust. Mechanical replacing of the lithosphere by the asthenosphere seems to be the cause of the extension.

INTRODUCTION

During the recent 20 years the Baikal rift zone has been intensively studied by geological and geophysical methods. These investigations, as the authors believe, made it possible to understand not only the mechanism of the East Siberian rift formation, but also the nature of continental rifting as a whole. The paper briefly presents the ideas obtained from these studies.

THE SAYAN-BAIKAL DOMAL UPLIFT

The Sayan-Baikal domal uplift, where rift valleys are located, extends in the northeastern direction for 1500 km. It is rather complex in morphology widening to as much as 300 km in some places and narrowing to 150 km in others. Wide segments of the uplift associate with maximum heights of the relief (up to 2500 to 3000 m). There are three culminations along its strike, namely (from SW to NE), the Sayan-Khamardaban, the Barguzin-Ikat and the Kodar-Udokan elevations. They are separated from each other by the Selenga and Vitim saddles (Fig. 1).

The central segment of the Sayan-Baikal domal uplift is situated in the Late Proterozoic fold belt and runs in parallel to the structure of the same age, as well as to the margin of the ancient Siberian platform. Its southwestern segment propagates into

Proceedings of the 27th International
Geological Congress, Volume 7, pp. 251–267
TECTONICS

the Caledonian fold belt, and its northeastern segment intersects the Late Proterozoic folded structures at a right angle and penetrates into the Archean Aldan shield of the Siberian platform.

STRUCTURE OF THE RIFT BASINS

There is a complex branching system of rift basins separated from each other by mountain links. The majority of the largest depressions (South Baikal, Barguzin, Muya, Chara and Tokka basins) is situated near the axes of the domal uplift. Some depressions (North Baikal and Upper Angara basins) are displaced to the northwestern slope of the uplift.

The Baikal depression is the largest one. It consists of two echelon-like basins (South and North) separated from each other by the diagonal uplift of the submarine Akademichesky range. Each basin is 350 km long and 45 to 50 km wide. The other depressions are 100 to 200 km long and 35 to 40 km wide (Fig. 2). According to geophysical data, the thickness of sediments amounts to 6,000 m in the South Baikal basin, to 4,000 m in the North Baikal basin and to 1,500 to 2,500 m in the other rift depressions (Zorin, 1971). These values are in line with the data obtained from drilling. A borehole drilled in the Selenga delta (central part of South Baikal basin) has not reached the basement at a depth of 3,000 m, though the borehole has dissected sediments dating from the Pleistocene to the Early Oligocene (Eocene?). Drilling in the Tunka depression was stopped in Early Miocene sediments at a depth of about 2,000 m.

The basins are separated from the domal uplift shoulders by a system of normal and strike-slip faults forming steps. The faults are branching and crossing each other. The amount of vertical displacement is usually larger on the northwestern slopes of the basin, therefore the depressions are uniformly asymmetric. Ruptures in the basin crystalline basement are combined with gentle bends (Florensov, 1969; Logatchev and Mohr, 1978). The existence of normal faults indicates the formation of the basins under extension conditions.

SEDIMENTARY INFILL OF THE BASINS AND EVOLUTIONARY STAGES OF THE RIFT ZONE

Continental sediments of the rift depressions are subdivided into two series. The lower one dating back to the Oligocene - Early Pliocene (from 35 to 3 or 5 m.y. ago) is composed of sandstones, aleurolites and clays. Beds of limestones, marls, diatomites and brown coals are rather rare. The sediments are of lacustrine, palustrine and fluvial origin. An absence of marginal coarse-grained facies in exposed segments of the lower series in the southern framing of Lake Baikal indicates that in the time under review the basins were wider, but not very much, as the sediments are not found on the tops of the surrounding mountains where pre-rift Upper Cretaceous - Paleogene erosion surfaces have been preserved in some places. It should be noted that the clastic material composing the series is badly sorted, that testifies to short distances of its transportation. The material was supplied from the adjacent ridges. Consequently, the Sayan-Baikal domal uplift existed at that time, but its heights were comparatively small.

The upper series dating from the Middle Pliocene to the Quarternary is characterized by general coarsening of sediments. This is most clearly in evidence in peripheral sections of the basins where pebbles, conglomerates, gravel and large-block sediments appear. Therefore, in the Middle Pliocene the boundaries of the basins were similar to those of the Present, and the contrasts in altitudes of the rift valleys and the adjacent mountains became greater. In the central segment of the depressions lacustrine and fluvial sediments, such as sands, aleurolites and clays, continued to play the main role. The rate of sedimentation in the upper series was by an order higher than that in the lower series (Logatchev, 1968; Logatchev and Mohr, 1978).

Thus, the composition of sedimentary infill of the depressions indicates that two stages on the rift zone evolution took place. At the first stage (Oligocene - Early Pliocene) the rates of downwarping of the depression bottoms and the growth of the domal uplift were rather low. At the second stage (Middle Pleistocene - Recent times) the rates of the depression downwarping and the growth of the domal uplift increase.

The development of the rift zone is going on at present, as evidenced from the data both on the Recent movements and on the high-level seismicity. In earthquake foci extension stresses are acting. They are oriented across the strike of the depressions (Misharina, 1967; Logatchev and Mohr, 1978).

VOLCANICS

Alkaline olivine basalts mainly occur in the Baikal rift zone. Tholeiitic basalts, trachybasalts, trachytes and basanitoids are of subordinate meaning. As a whole, Cenozoic volcanics relate to midly alkaline series. Basaltic sheets and volcanic cones form three separate fields (Sayan-Khamardaban, Vitim and Udokan). The alkaline volcanicity began simultaneously with the development of the rift depressions, but its main bulk manifested itself in the Miocene (Sayan-Khamardaban and Udokan fields). In the Pliocene and Holocene the intensity of volcanicity in the two regions became less, but it increased in the Udokan field.

The volcanic fields are confined to the above three widest and highest segments of the Sayan-Baikal domal uplift. Nevertheless, these fields are rather considerably displaced to the southeast relative to the axis of the domal uplift (Fig. 1). No other connections between the Cenozoic volcanicity and morphological features are established. Lava sheets do not usually extend into the rift basins, excepting the Tunka depression, in the northern segment of which, as judged by drilling data, volcanics compose about 20% of the infill. At the same time in the southern segment of the depression a borehole did not meet any volcanics. Basalts within the limits of this depression are characterized by a slight tholeiitic tendency, as compared with mean composition of the volcanics typical of the Sayan-Khamardaban field. This tholeiitic tendency may evidence a decrease in depth of a melting source beneath the depression, as compared with the depths beneath the surrounding ridges.

INTRUSIONS AND DEEP FAULTS BENEATH RIFT BASINS

There are no direct indications of the presence of large Cenozoic intrusions in the Baikal rift zone. Only swarms of thin basaltic dykes are found in the southern framing of the Baikal depression and in the vicinity of the Tunka and Chara depressions (Sherman et al., 1976; Kisselev et al., 1979). A neck-like picritic body and dykes found on the Bolshoy Ushkany Island are tentatively dated from the Eocene (Yeskin et al., 1978).

Within the well studied Baikal depression heat flow anomalies, no more than 30 km wide, are revealed (Golubev and Osokina, 1980). Such narrow anomalies, if they are not connected with hydrothermal systems, which unlikely exist within the South Baikal with its rather thick layer of unconsolidated sediments, may be caused by heat sources located in the crust. Intrusions which have injected rather recently and have not had time enough to cool may be such sources. For the Selenga delta interpretation of geothermal and magmatic data was made which proved the probability of such an assumption. The width of the great dyke is estimated as 7 to 10 km, the depth of its uppermost part amounts to 6 km, and the time passed from its injection is about 3 m.y. (Zorin et al., 1979).

As high values of the heat flow are peculiar to all the rift depressions which were studied by geothermal methods (Logatchev and Mohr, 1978), one may consider great dykes to be located below each of such depressions. These intrusions are evidently confined to renewed or to newly formed deep faults. The assumption of correspondence of each large depression to a deep fault or to a few fault seems to be rather verisimilar as the general pattern of rift valleys looks like a system of branching faults (Fig. 1).

STRUCTURE AND MAGNITUDES OF EXTENSION OF THE CRUST

In the Baikal zone and the adjacent areas a great number of deep seismic soundings was carried out using reflected and refracted waves from explosions (Logatchev and Mohr, 1978; Puzyrev et al., 1974). Beneath the South Baikal basin the Moho was identified at a depth of 34 km, while under the rift shoulders the crustal thickness is 42-46 km. It should be noted that the initial information on the crust thinning below the major rift depressions was obtained from gravimetric data (Zorin, 1971). This suggests ductile extension of the crust (at least in its lower part) beneath the rift valleys.

Based on conditions of conservation of the crustal volume one may roughly estimate the magnitude of the ductile extension. The thickness of the crust (without a lens of sediments) below the South Baikal is 28 km. Its width amounts to 50 km becoming 40 km without the great dyke. The area of a cross section of the crystalline crust below the basin considered amount to $28 \times 40 = 1120$ km^2. If the crustal thickness beneath the depressions before the extension was similar to that of the present day below the rift shoulders (44 km), the initial width of the block corresponding to the basin had to be equal to 25 km. Consequently, the magnitude of the ductile extension is 15 km. If we add the width of the great dyke, we shall obtain the general extension magnitude equal to 25 km. The latter figure characterizes only the South Baikal segment of the rift zone. In the Tunka, Barguzin, Muya and Chara depressions, as judged by similar estimations, the extension does not exceed 10 to 15 km, amounting to a few kilometres in the area of interdepressional mountain links (Zorin, 1981).

Such variations in the extension along the rift zone strike can scarcely be caused by irregularities of the stress field. More probable that they are produced by mechanical inhomogeneities of the crust.

One of the types of the crustal inhomogeneities, namely, deep faults enclosing great dykes, is mentioned above. The substance of the intrusions crystallizes rather quickly regenerating the continuity of the crust and heating it at narrow segments which obtain the possibility to deform plastically due to decrease of effective viscosity. After the cooling of the crust a renewal of ruptures and secondary magma intrusions became possible.

Such faults are evidently not the single type of inhomogeneities which contribute to extension and rifting. Small depressions of the Baikal rift zone (Tsipa and Tsipikan basins) have formed at places where the uppermost crust is composed of granites. The last studies indicate that the basement of the North Baikal depression is also composed of granites. Such a regularity may be explained by the following reasons. Granites have lower density than metamorphosed rocks and form rather significant mass deficiency which should be compensated by uprise of the Moho discontinuity due to isostasy. Consequently, before the rifting the crust below large granite bodies should be thinner than below the adjacent regions. It is natural that under extension conditions segments of the thinner crust begin primarily to deform and to break up. There are other lithological and structural inhomogeneities which may decrease the strength of the crust in some parts.

UPWELLING OF THE ASTHENOSPHERE

According to deep seismic sounding data (Puzyrev, 1981), the velocities of P-waves at the Moho below the Siberian platform and the Transbaikal region of moderate tectonic activity range from 8.1 to 8.2 km s^{-1}. Beneath the Baikal rift zone the velocity decreases as low as 7.7 km s^{-1}. The region with lower velocities at the Moho roughly coincides with the area of the domal uplift. At the depth of about 60 km beneath the rift zone the boundary with P-wave velocity 8.1 km s^{-1} is detected which, in opinion of S.V. Krylov et al. (Puzyrev, 1981), corresponds to the normal mantle. However, superdeep seismic soundings (Yegorkin, 1980) indicate that at the same depth below the Siberian platform the P-wave velocities range from 8.4 to 8.6 km s^{-1}. Thus, if one takes the platform as a unique reference region, it should be considered that seismic velocities in the mantle below the rift zone are anomalously low at the depth of 60 km and more.

This conclusion is supported by the fact that at seismic stations located in the rift zone the P-wave delays from distant earthquakes and explosions are constantly observed. As compared with the data from the stations situated on the platform, the delays for subvertical seismic rays are about 1 s (Rogozhina, Kozhevnikov, 1979). If one assumes that about 30% of the anomalies in the travel time are due to the crust, the thickness of the layer where such anomalies originate should be no less than 150 to 200 km at the above mentioned velocities in the upper mantle.

On this ground we may consider that under the Baikal rift zone, as well as under the other continental rifts, the asthenospheric upwelling exists and its roof reaches the base of the crust (Logatchev et al., 1978). The width of its top evidently conforms to the width of the Sayan-Baikal domal uplift. The layer in the upper part of the upwelling with P-wave velocity of 7.7 km s^{-1} may be associated with differentiation of the asthenospheric material. It should be noted that this upwelling participated in isostatic compensation of the domal uplift (Zorin, 1971).

DEVELOPMENT OF THE ASTHENOSPHERIC UPWELLING AND THE NATURE OF THE EXTENSION

So, the width of the top of the asthenospheric upwelling exceeds by an order the values of the extension which is necessary to form the largest rift depressions (25 km). Therefore, the extension, if it was caused by outside phenomena (e.g. global plate interaction), could not make room large enough for enclosing such a wide asthenospheric upwelling. In addition, assuming such a "passive" extension, it is impossible to explain the disappearance of the mantle portion of the lithosphere with the preservation of the continental crust.

Based on these simple geometric relations, the Baikal rift should be considered as active. The growth of the asthenospheric upwelling results in the formation of the Sayan-Baikal domal uplift due to the isostasy, whereas the spreading of the upwelling, due to striving of the mechanical system to the minimum of energy, caused the extension of the crust and formation of rift depressions. Evidently, the mechanism of replacing of the lithosphere by the asthenosphere is bound to exist.

On the basis of construction of the numerical models the possibility of formation of the upwelling through partial melting of the lithosphere was checked. If the heat flow into the base of the lithosphere increases to 100 mW m^{-2}, the lithosphere thickness decreases from 120 km to 40 km (thickness of the crust) within 100 m.y. This considerably exceeds the duration of the rift zone evolution. At the end of the process under review the value of the regional surface heat flow above the centre of the upwelling should reach 90 mW m^{-2}, which is close to the theoretical values of the steady-state heat flow with the present position of the asthenospheric roof in the rift zone (Zorin and Osokina, 1981). However, similar and sometimes higher values of the heat flow are typical of the local geothermal anomalies connected with great dykes. In the major part of the Baikal domal uplift out of the limits of such geothermal anomalies the heat flow usually varies from 60 to 65 mW m^{-2}. Consequently, the partial melting of the lithosphere due to increase of the heat inflow resulted from appearance of hot spots in the mantle cannot be the main formation mechanism of the asthenospheric upwelling. But the process under consideration can result in some thinning of the lithosphere. This creates initial disturbance which was necessary for realization of mechanical intrusion of the anomalous material, supplied by the hot spot, into the lithosphere under the influence of gravitational instability.

Mechanism of pure diapirism, which suggests only ductile flow of lithospheric material, is not acceptable enough for explanation of asthenospheric upwelling, because in this case the total magnitude of crustal extension must be of the same order as the width of the top of the upwelling. As mentioned above, such a ratio is not typical of the Baikal rift zone.

Evidently, making room for such a wide asthenospheric intrusion occurred on account of collapse of lithospheric blocks which sank in the asthenosphere and were finally dissolved. A ductile flow of the lithospheric material was of a secondary meaning in making room for the upwelling, but it played a significant role in the formation of the rift depressions. One may believe that the growth of the asthenospheric upwelling took place at the first stage of the rifting when the height of the Sayan-Baikal domal uplift was progressively increasing and the depressions were slowly developing. When the asthenospheric material reached the Moho its upward movement stopped over the vast region, as the asthenospheric density exceeds the crustal density. At the same time, the spreading of the asthenospheric upwelling material began which resulted in more intense extension of the crust and in quicker formation of rift depressions, i.e. the transition to the second rifting stage took place.

Such an outline of development of a deep-seated structure, which correlates well with the rifting stages, suggests rather recent (3 to 5 m.y. ago) appearance of the asthenosphere beneath the Moho. It is reasonable to check this outline numerically via constructing a two-dimensional model of transient temperature field of the lithosphere.

Bearing in mind the mentioned above mechanism of making room and insignificant extension we may neglect deformations of the lithosphere and model the asthenospheric upwelling evolution through compusolry upward displacement of isotherm of the mantle solidus (1200°C). Initial distribution of the temperature with the depth in the mantle was deduced from the data on the stable Siberian platform adjoining the rift zone. The lithosphere thickness below the platform is assumed to be

120 km, the heat flow is 46 mW m^{-2} and the thermal regime is close to a steady-state one. Under these conditions thermophysical properties of the crust and mantle were chosen (Zorin and Osokina, 1981). Upwelling of the asthenosphere was simulated by a growth in theight of the rectangular jut of the solidus isotherm. The width of the upwelling was 250 km, the time of its roof uprise from a depth of 120 km to the base of the crust (40 km) with constant rate was assumed to be 30 m.y., which corresponds to the duration of the first rifting stage. Temperatures in the lithosphere above the solidus isotherm were calculated via numerical solidus of the differential equation of conservation of energy. When the upwelling top reached the Moho its upward movement ceased, however the calculations of the thermal field proceeded till the new quilibrium.

Correspondence of the thermal field to the present conditions was determined from coincidence of the theoretical and the observed regional heat flow (Fig. 2). Simultaneously at the Curie depths below the rift zone and the Siberian platform the theoretical values of the temperature became approximately the same. The time passed from the moment of the asthenospheric approach to the crustal base is estimated as 3.5 m.y., which is in line with the duration of the second stage of the Baikal rift development. Therefore, the above outline of the deep-seated structure evolution as a whole is supported by results of modelling of the transient temperature field of the lithosphere. The crust below the Sayan-Baikal domal uplift is continuing to be heated on a regional scale (Fig. 2).

The present theoretical temperatures in the lithosphere below the Baikal rift zone and the Siberian platform, as well as the surface heat flow are shown in Fig. 3. The eastern boundary of the asthenospheric upwelling is not shown, as the behaviour of the lithosphere-asthenosphere boundary is not studied enough to the southeast from the domal uplift.

In this version of the model the existence of the great dyke in the crust beneath the South Baikal depression is taken into account. Its intrusion has caused an anomalously high heating of the adjacent local part of the crust, which is cooling at present (the dyke temperature in the model was not preserved stable, as the long-existing convection is impossible in such a narrow chamber). The presence of small marginal picks on the theoretical heat flow curve is connected with the heat flowing round the lens of low-conductivity sediments in the Baikal basin. The general correspondence of the theoretical heat flow to its observed values is quite satisfactory (Fig. 3).

CONCLUSION

Thus, the available geological and geophysical data indicate that the Baikal rift development results from mechanical replacement of the lithosphere by the asthenosphere material. This replacement occurs along the wide weak zone which was formed near the boundary between two different segments of the lithosphere, one of which corresponds to the Siberian platform and the other corresponds to its folded surrounding. The evolution of the wide asthenospheric upwelling has produced an isostatic uprise, i.e. the formation of the Sayan-Baikal domal uplift. The growth of the upwelling (first stage of evolution of the zone) and its further spreading (second stage) resulted in the extension of the crust across the uplift strike. In connection with inhomogeneities of the crust (faults, intrusions, etc.) the extension was non-uniform along the strike of the zone. This led to formation of complicated branching system of depressions separated from each other by mountain links. Similarity of the majority of continental rifts to the Baikal one (Milanovsky, 1976; Logatchev, 1978) indicates that the mechanism considered above fits in general features the explanation of continental rifting. It should be mentioned that the idea on the active nature of continental rifts does not contradict the main postulates of the plate tectonics. The rates of horizontal movements of lithospheric plates which include continents usually do not exceed 1 to

2 cm/y. The evolution of Cenozoic rifts lasts 15 to 40 m.y. During this period the asthenospheric plate with the developing rift can move by 150 to 800 km. If the width of the hot spot, which supplies material of the anomalous mantle to the asthenosphere amounts to 200 to 300 km, the rift zone may maintain its positional connection with these ascending flow. Besides, the anomalous mantle material migrating along the asthenosphere must come to a nearest "trap", i.e. to the zone of the thinned lithosphere, where the rift is developing (Artyushkov, 1979).

ACKNOWLEDGEMENTS

The authors are indebted to Mrs. S.V. Lepina for development of computer programs and calculations. Thanks are also due to Mrs. G.A. Vladimirova for preparation of the English version of this paper.

REFERENCES

Artyushkov, E.V., 1973. Geodynamics. M., Nauka, 327.

Florensov, N.A., 1969. Rifts of the Baikal mountain region. Tectonophysics, v. 8, 443-456.

Golubev, V.A. and Osokina, S.V., 1980. Distribution of the heat flow and the nature of its local anomalies in the Lake Baikal region. Izv. Akad. Nauk SSSR, Fizika Zemli, No. 4, 63-75.

Kisselev, A.I., Medvedev, M.E., Golovko, G.A., 1979. Volcanicity of the Baikal rift zone and the problems of magma formation. Novosibirsk, Nauka, 198.

Logatchev, N.A., 1968. Sedimentary and volcanogenic formations of the Baikal rift zone. In: Baikal rift. M., Nauka, 72-101.

Logatchev, N.A., 1978. Main features of evolution and magmatism of central rift zones. In: Tectonics and Geophysics of Continental Rifts. Reidel Publishing Company, 351-366.

Logatchev, N.A. and Mohr, P. (Editors), 1978. Geodynamics of the Baikal rift zone. Tectonophysics, v. 45, 1-105.

Logatchev, N.A., Rogozhina, V.A., Solonenko, V.P. and

Zorin, Yu.A., 1978. Deep structure and evolution of the Baikal rift zone. In: Tectonics and Geodynamics of Continental Rifts. Reidel Publishing Company, 49-61.
Milanovsky, E.E., 1976. Rift zones of the continents. M., Nedra, 279.
Misharina, L.A., 1967. Stresses in the crust in rift zones. M., Nauka, 134.
Puzyrev, N.N. (Editor), 1981. Bowels of Lake Baikal from seismic data. Novosibirsk, Nauka, 105.
Rogozhina, V.A. and Kozhevnikova, V.M., 1979. Region of the anomalous mantle beneath the Baikal rift. Novosibirsk, Nauka, 104.
Sherman, S.I. Medvedev, M.E., Ruzhich, V.V., Kisselev, A.I. and Shmotov, A.P., 1973. Tectonics and volcanicity of the southwestern part of the Baikal rift zone. Novosibirsk, Nauka, 136.
Yogorkin, A.V., 1980.Structure and properties of the upper mantle. In: Seismic models of the lithosphere of main geological structures in the USSR area. M., Nauka, 161-171.
Yeskin, A.S., Bukharov, A.A., Zorin, Yu.A., 1978. Cenozoic magmatism on Lake Baikal. Dokl. Akad. Nauk SSSR, v. 239, 926-929.
Zorin, Yu.A., 1971. Recent structure and isostasy of the Baikal rift zone and the adjacent areas. M., Nauka, 167.
Zorin, Yu.A., 1981. The Baikal rift: an example of the intrusion of asthenospheric material into the lithosphere as the cause of disruption of lithospheric plates. Tectonophysics, v. 73, 91-104.
Zorin, Yu.A., Golubev, V.A. and Novoselova, M.R., 1979. Geophysical data of Late Cenozoic intrusions below Lake Baikal. Dokl. Akad. Nauk SSSR, v. 249, 81-85.
Zorin, Yu.A. and Osokina, S.V., 1981. Model of transient temperature field of the crust in the Baikal rift zone. Izv. Akad. Nauk SSSR, Fizika Zemli, No. 7, 3-14.

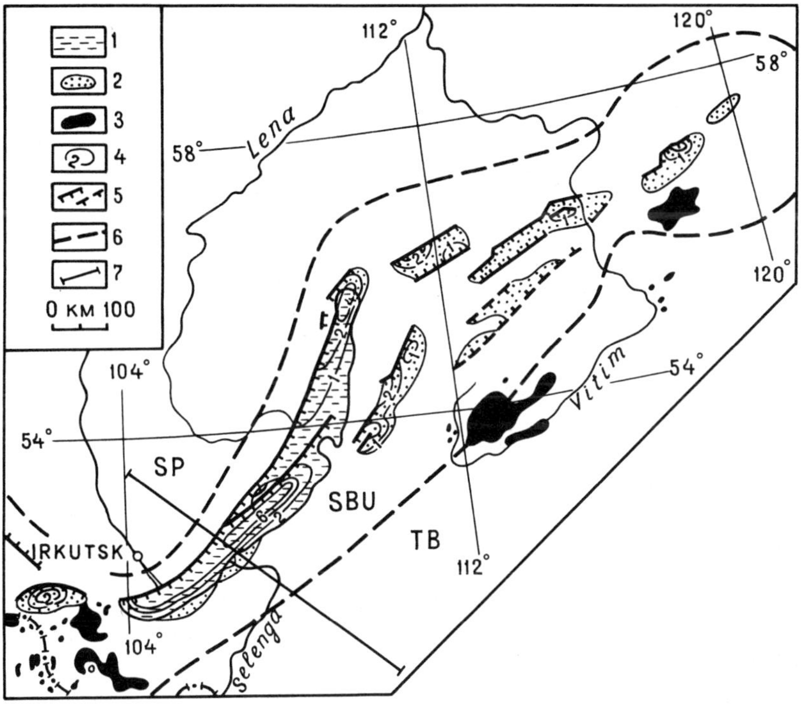

FIGURE 1. Schematic map of the Baikal rift zone and adjacent areas. 1 = Lake Baikal; 2 = Cenozoic sediments in the rift basins; 3 = Cenozoic basalts; 4 = isopachous lines of Cenozoic sediments (in km); 5 = faults; 6 = boundaries of geological provinces: SP = Siberian platform, SBU = Sayan-Baikal domal uplift, TB = Transbaikal region of moderate tectonic activity; 7 = section line (see Fig. 3).

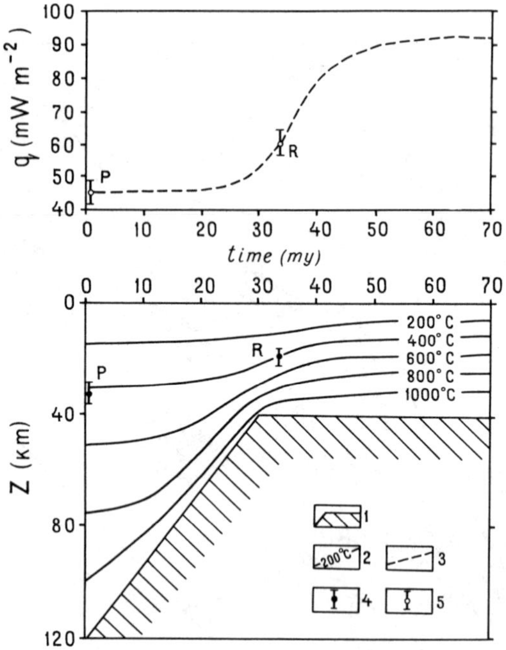

FIGURE 2. Temperatures in the lithosphere and surface heat flow versus time for the central part of the asthenospheric upwelling (version without the great dyke). 1 = asthenospheric surface; 2 = isotherms in the lithosphere; 3 = surface heat flow; 4 = confidence intervals of Curie depths (according to magnetic data); 5 = confidence intervals of average values of the measured heat flow. R = rift zone, P = Siberian platform.

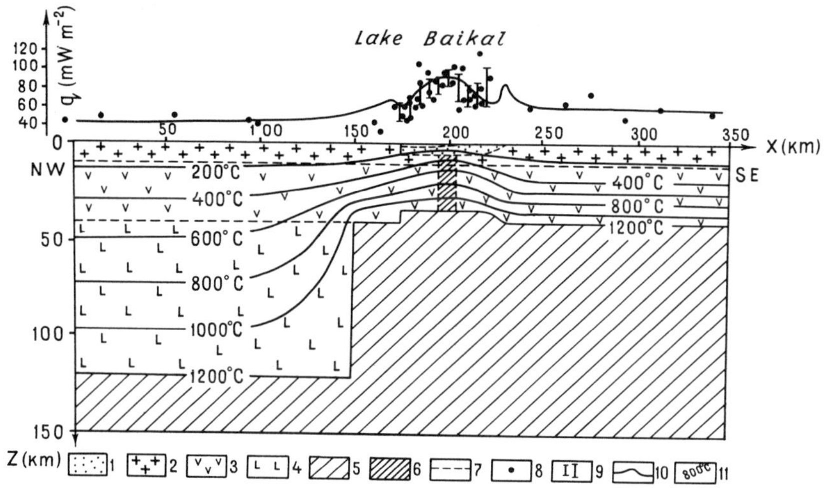

FIGURE 3. Temperature field in the lithosphere and surface heat flow for the Present along the section line (see Fig. 1). 1 = sediments in the Baikal basin; 2 = "granitic" layer; 3 = "basaltic" layer; 4 = lithosphere; 5 = asthenosphere; 6 = great dyke; 7 = layer boundaries; 8 = heat flow data projected from 150 km-wide band; 9 = confidence intervals of heat flow average values obtained from the procedure of sliding window, 10 km wide; 10 = plot of theoretically defined heat flow; 11 = theoretically defined isotherms.

THE QUANTITAVE EVALUATION OF ROCK DEFORMATION: A REVIEW

RAMSAY, JOHN G.
Geol.Inst. ETH, Zürich, Switzerland.

ABSTRACT
A general discussion of the strain state and the relationships between displacement and strain leads to a review of the practical methods whereby we are able to measure strain in naturally deformed rocks. Methods for describing strain using maps and graphical techniques are described. The geotectonic significance of making strain measurements is discussed with particular emphasis on how strain data enable a geologist to appreciate more exactly the significance of large scale and small scale tectonic features of the earth's crust.

INTRODUCTION

The measurement of the strain state of a naturally deformed rock is fundamental to an understanding of tectonic processes and tectonic structures, and strain data should always be sought out in any terrain where orogenic deformation has taken place. These measurements can be used to interpret the significance of both small scale and large scale geometric features, and the origin of planar and linear fabrics can be correctly interpreted. In deformed terrain strain determination offers a key to the correct restoration of stratal thickness and for the reconstruction of correct palinspastic profiles. From an analysis of strain state it is possible to evaluate many features of the regional displacement plan and to link large scale folds and faults with the movements that have taken place as a result of major plate displacements.

Proceedings of the 27th International
Geological Congress, Volume 7, pp. 269–293
TECTONICS

BASIC THEORY

Finite or total strain arises as a result of finite differential displacements of the particles making up a body. The movements of points in a body can be described in terms of vectors joining the initial and final positions of each point. Equations, known as coordinate transformation equations can be established which describe the relationships of all points in a body before and after displacement. Different types of equations relate to different types of body displacement and to different types of strain. In three dimensions the position of any point can be described in terms of coordinates with reference to three orthogonal coordinate axes, x, y and z: its initial positions (x,y,z) is transformed to a final position (x', y',z'). The most general coordinate transformation equations can be expressed

$$
\begin{aligned}
x' &= f_1(x,y,z) \\
y' &= f_2(x,y,z) \\
z' &= f_3(x,y,z)
\end{aligned}
\qquad (1)
$$

where $f_1(x,y,z)$ etc. are mathematical functional descriptions in three variables. Several different types of functional relationships lead to characteristic types of overall displacement plan.

1. Body translation (Figure 1B)

$$
\begin{aligned}
x' &= x + A \\
y' &= y + B \\
z' &= z + C
\end{aligned}
$$

where A,B and C are constant terms. Such a displacement plan produces displacement vectors which are the same for all initial points (x,y,z). The body is translated without rotation or internal deformation.

2. Body rotation (Figure 1C)

In three dimensions body rotation is described by special types of linear functions. For example a rotation around the z-axis through an angle of ${}_z$ is described by:

$$
\begin{aligned}
x' &= x \cos{}_z - y \sin{}_z \\
y' &= x \sin{}_z + y \cos{}_z \qquad (3) \\
z' &= z
\end{aligned}
$$

or in matrix form:

$$
\begin{vmatrix}
\cos{}_z & -\sin{}_z & 0 \\
\sin{}_z & \cos{}_z & 0 \\
0 & 0 & 1
\end{vmatrix}
\qquad (4)
$$

a completely general rotation described by rotations around the x, y and z axes can be computed from the matrix product of three matrices of the type of equation (4). Body rotation is a transformation that clearly goes on without the development of strain even though the individual displacement vectors in the body vary with position.

3. Homogeneous strain (Figure 1D)

Homogeneous strain can be defined as a condition of displacement where all initial sets of parallel straight lines change orientation and spacing distance but remain straight and parallel after displacement. This geometric condition implies that a rectangular grid of identically shaped squares (in two dimensions) or cubes (in three dimensions) before displacement become transformed into a grid of identically shaped parallelograms (in two dimensions) or parallelopipeds (in three dimensions). The material is internally deformed and, because the shape changes of like shaped initial elements is the same, the deformation is identical through the body. Coordinate transformation equations leading to homogeneous strain are linear and of the type:

$$
\begin{aligned}
x' &= ax + by + cz \\
y' &= dx + ey + fz \qquad (5)
\end{aligned}
$$

$$z' = gx + hy + iz$$

or, in matrix form:

$$\begin{vmatrix} a & b & c \\ d & e & f \\ g & h & i \end{vmatrix} \qquad (6)$$

where the nine coefficients a,b,......i are numerical constants.

4. Heterogeneous strain (Figure 1E)

This is the most general displacement plan and the coordinate transformation equations are non linear. Most geological situations in deformed rocks are of this type. The rather complex strain variations found in a heterogeneously strained body are usually analysed by considering the geometric changes taking place within small elemental units (Figure 1E). Such small units show geometric characteristics which are very close to those of homogeneous strain, and, mathematically, it can be shown that at a point this conclusion is exact. This means that any state of heterogeneous strain may be analysed in terms of a field of interrelated homogeneous strained elements. In geological terms, the strain within a large fold, for example, can be carried out by assuming that the state of strain in individual outcrops or hand specimens is effectively homogeneous (Cloos, 1947).

In a heterogeneously strained body the variations of strain that exist are tightly constrained. Because the body remains coherent, strain variation is controlled by conditions of strain compatibility. The analysis of strains within geological structures in terms of this compatibility concept has been greatly used over the past decade and has been very valuable in defining what types of strain variation are possible and what are impossible (Ramsay and Graham, 1970; Hobbs,1971; Oertel and Ernst, 1978; Cobbold, 1977,1980).

THE STRAIN ELLIPSOID CONCEPT

A particularly useful concept for the analysis of homogeneous strain can be developed by considering the shape after displacement of an initial sphere (or in two dimensions a circle) of unit radius. The sphere is deformed into an ellipsoid, the finite strain ellipsoid (in two dimensions the circle becomes the finite strain ellipse), and the geometric features of this ellipsoid enable the principal features of strain to be readily visualised. Nine components are necessary to define completely the ellipsoid and the body rotation made by the mass. These nine components (four are necessary in two dimensions to define the strain ellipse, Figure 2) are related directly to the nine terms related to displacement and the nine coefficients of the coordinate transformation equations (Equation 5). The three principal axes of the strain ellipsoid X,Y and Z are mutually perpendicular and are of lengths $1+e_1$, $1+e_2$ and $1+e_3$, where e_1,e_2 and e_3 are known as the three principal finite longitudinal strains. These axes require three orientation terms,θ_1', $\theta_2'$and θ_3' to fix the spacial orientation of the ellipsoid. These axes are derived from three other mutually perpendicular lines in the initial sphere which also require three orientation terms θ_1, θ_2 and θ_3 to fix their positions. The differences in orientation between the perpendicular lines before and after deformation can be considered as defining a three dimensional body rotation. Further discussion of the theoretical basis of the strain ellipsoid can be found in Ramsay, 1967,1969,1976; Flinn, 1979A,1979B; Ramsay and Huber, 1983.

Practical strain analysis in geology is generally carried out by making measurements of two dimensional finite strain ellipse sections of the strain ellipsoid, and constructing the three dimensional data from mathematical manipulation of the two dimensional information. The strain ellipses that produce the three dimensional result most effectively are the principal planes of the ellipsoid containing the axes XY,YZ or ZX. Generally the strain ratios are determined on any two of the three sections $R_{xy}=(1+e_1)/(1+e_2)$, $R_{yz} = (1+e_2)/(1+e_3)$ or $R_{xz} = (1 +e_1)/(1+e_3)$. This might be called the classical method of strain analysis of geological structures (Cloos, 1947). However, a knowledge of the state of strain given by <u>any three</u> strain ellipse sections even those that do not contain the principal longitudinal strain axes, is sufficient to evaluate the shape and orientation of the strain ellipsoid (Ramsay,1967, pp.142-148).

In most practical strain analyses with geological problems only 5 of the 9 components of strain are generally determinable. This is first because the three terms contained in the data for the initial orientation of the unit sphere are generally unknown. Second, because we generally do not absolute line lengths in the objects contained in a deformed rock, the strain data we acquire defines only the ratios of the semi axes of the strain ellipsoid R_{xy} and R_{yz}. Another way of appreciating the data gap is to see that, because absolute line lengths are unknown, we lack information about the volumetric or <u>dilational component</u> of the strain given by

$$v = (1+e_1)(1 + e_2)(1+e_3) - 1$$

The nine total data required to define the finite strain ellipsoid ($e_1, e_2, e_3, \theta_1, \theta_2, \theta_3, \theta_1', \theta_2', \theta_3'$) are therefore incompletely resolved into five data ($R_{xy}, R_{yz}, \theta_1', \theta_2', \theta_3'$) with, rarely, the addition of a sixth, the volume change v.

Good summaries of the theory of finite strain and its practical application to deformed rocks can be found in Ramsay,1967,1969,1976; Jaeger and Cook, 1969; and Means, 1976).

PRACTICAL METHODS OF FINITE STRAIN MEASUREMENT

The shape and orientation of the strain ellipsoid are most readily measured from embedded particles which had initially spherical or circular form. The most common objects useful for this analysis are ooids (Cloos, 1947, 1971), green diagenetic alteration spots in red haematite rich sediments shown in Figure 3 (Sorby, 1908; Wood, 1974), spherical fossils (foraminifera, radiolaria), circular fossils or circular cross sections of cylindrical shapes (crinoids, worm burrows). For the strain analysis to be accurate it is especially important that the object being measured had a rheological behaviour close to that of the rock matrix surrounding it. For example, although initially circular crinoids are sometimes very good strain markers, it may be found that the large single calcite crystal of which they are generally composed behaves very differently from that of a fine grained limestone matrix. Because of the competence contrast, the object will then give an inaccurate measure of the total deformation.

Special problems arise when determining strains from spherical objects where chemically induced deformation (pressure solution) have played a major role in the strain process. Figure 4 illustrates a deformed oolitic rock where the deformed ooids have been preferentially dissolved on the sides sub-perpendicular to the greatest shortening direction. Clearly the shape of these ooids will not exactly record the amount of rock strain. The strain determination technique most appropriate to this problem depends on an analysis of the geometric relationships between the centres of the objects. The centre to centre technique, initially developed by Ramsay (1967, pp.195-197), has been considerably refined by Fry (1969), and offers a very practical tool for many types of analytical problems posed by the redistribution of objects and grains in rocks as a result of strain (Ramsay and Huber, 1983).

Many objects suitable for strain analysis have initial forms which are sub-ellipsoidal or sub-elliptical (e.g. some ooids, conglomerate pebbles). This initial geometry poses special problems in that it is not correct to take arithmetic means of the deformed particle shapes to determine the form of the strain ellipsoid. This problem can be resolved mathematically and in a practical way by considering the change of shape of original elliptical particles (from an original ellipse ratio of R_0 to a final shape R_f) and the modification of the orientation of the particle long axis (from $\emptyset$ to $\emptyset'$). The so called $R_f/\emptyset'$ technique (Ramsay, 1967, pp.207-211) has proved of great value in practical strain analysis, especially with deformed conglomerates and clastic rocks (Dunnet, 1969; Lisle, 1977). Two other major problems arise when deformed conglomerates are used for strain determination. The first is the problem of ductility contrast so that the pebbles may not be recording the full extent of the strain, and the second is the problem posed by pressure solution along pebble-pebble contacts, Figure 5. Both features are of great interests to geologists because the heterogeneous behaviour of individual rock components in an aggregate might be of use to measure more exactly the differing ductilities of different rock types.

Other methods of strain measurement in common use in Phanerozoic rocks concern the use of the shapes of deformed fossils to establish the form of the strain ellipse. Historically, these methods were some of the first to be used by geologists (Sharpe,1847;Haughton,1856;Danbrée,1876), and the results of the application of these analytical techniques, the significance of many types of rock cleavage was established (summary in Harker, 1885). Two main groups of techniques may be noted. The first is based on an analysis of length changes in different directions in the rock, from such objects as Belemnites, or Crinoid columns: the second analyses the changes in angular elements of fossils, generally those with bilateral symmetry (Figure 6), but the changes in the complex shapes, such as the spiral forms Gastropods and Ammonites may also be used. The most detailed exposition of these techniques are set out in works by Breddin, 1956;Ramsay, 1967;Ramsay and Huber, 1983. Although fossils often offer excellent material for the computation of the form of a single strain ellipse, they are usually less good for establishing the three dimensional form of the strain ellipsoid. This is because the initial orientation of many fossils is controlled by the orientation of the bedding surfaces in the rock. The form and orientation of the bedding plane strain ellipse can be established relatively easily, but it may be very difficult to determine the forms of other strain ellipses which transect the bedding surface and which are necessary for the determination of the strain ellipsoid form.

PRACTICAL METHODS OF DETERMINING THE PROGRESSIVE STRAIN HISTORY

The techniques outlined above relate the shape of the finite strain ellipsoid to its initially spherical form. However, all rocks pass through a series of changing states of strain, known as progressive deformation, and the techniques now to be discussed review methods whereby we can measure this strain history. Progressive deformation is a more or less continuous process of shape change, but for convenience of analysis we often break up this continuity by analysing the progressive evolution in a series of small finite stepwise changes or strain increments.

Progressive deformation can only be meaured if we have some special geometric feature in the rocks which is sequentially evolving at the same time as the deformation. The main features recording such changes are first, the shapes of progressively propagating and opening vein fissures and the way these fissures become filled with new mineral growth and second, the geometric forms of minerals growing around rigid undeformable objects embedded in the rock and surrounded by pressure shadow zones.

Vein fissures developing during a deformation propogate at their tips in a direction controlled by the incremental principal strain orientations at the time of propagation. The curved forms of propagating fissure tips therefore record a change in the incremental strain directions (Figure 7). Extension vein fissures are sequentially filled with new minerals as they progressively dilate. These new crystals have a fibrous form, and any changing separation direction of the vein walls is recorded by changes in growth direction of the crystal fibres (Figure 8). Several different types of fibrous vein filling are known, and a knowledge of the fibre growth sequence is critical for a correct interpretation of the deformation increment sequence (Durney and Ramsay, 1973; Wickham, 1973). The most complete discussion of the techniques for numerically evaluating the increments are set out in Ramsay and Huber, 1983. The geometry of the fibres forming in pressure shadow zones around resistant objects (pyrite, magnetite, etc.) is closely related to the fibre form of crystals found in extension veins (Figure 9), and can be used to give exact measures of the cahnging strain increments (Ramsay and Huber, 1982). One particularly interesting and recently evolved technique concerns the determination of the progressive finite strains from measurements of the strain increments (Casey et al., 1982; Ramsay and Huber, 1983). The coordinate transformation equations for each increment are computed, and the matrices of successive incremental transformations are successively multiplied to compute the total transformation after a certain number of increments. From these successive matrix multiplications it is possible to compute the principal characteristics of successive strains. As yet this technique has only been applied to two dimensional strain problems.

RECORDING STRAIN DATA

Information regarding the shape and orientations of the principal finite strains can be recorded directly on to maps or profile sections (Cloos,1947,1971; summary in Ramsay and Huber, 1983) and such simple techniques offer especially good visual methods of showing strain variation. Variations in intensity of strain can be recorded in graphs and a number of different methods have been proposed: Flinn (1962) with R_{xy} as ordinate and R_{yz} as abscissa **Ramsay** (1967) with log R_{xy} as ordinate and log R_{yz} as abscissa, Hossack (1968) using octahedral shear strain and Lode's parameter and Elliott (1972) using a polar method. The choice of method depends upon the nature of the data and the aim of the analysis, and a discussion of choice of graphical representation is given in Ramsy and Huber, 1983. All the methods try and separate the classification of ellipsoid form (oblate versus prolate) and the amount of deformation. The Flinn and Ramsay graphs probably offer the simplest way of obtaining numerical parameters for these two features. Flinn defined a k-value to indicate form. Some idea of the amount of deformation can be obtained from the distance of the ellipsoid plot from the point (1,1) and could be termed a d-value (Figure 10A). The counterparts on the Ramsay plot are K and D respectively (Figure 10B), but there are simple relationships between these pairs of parameters. The basic theoretical problem confronting us here is that there is no generally accepted way of comparing in a single parameter the amount of deformation of a prolate strain ellipsoid with that of an oblate ellipsoid. All the graphical methods described above have a common drawback: they relate only to ratios of the three principal ellipsoid semi-axes and therefore do not take into account volumetric modifications of the rock during deformation. The interpretation of such graphs is not always clear and is subject to a certain degree of ambiguity. Ellipsoids falling into an identical position can be interpreted in several ways (e.g. flattening or constrictive) depending upon the amount of volumetric dilation (Ramsay, 1967;Ramsay and Wood, 1973).

GEOLOGICAL SIGNIFICANCE OF STRAIN

Strain measurement provides a key to the interpretation of the significance of tectonic structures on many different scales. Although much work on strain analysis has been carried out over the past twenty years this activity is not just an expression of current fashion. In fact the significance of measuring strain in deformed rocks was appreciated well before the beginning of this century by practical field geologists seeking to understand specific tectonic features. Of particular importance over the period of from about 1845 to 1910 was the use of strain measurements to determine the meaning of planar and linear fabrics in deformed rocks. In the middle of the nineteenth century slaty cleavage had been interpreted as the result of a variety of forces, magnetic, electrical and mechanical. It were the perceptive studies of Sharpe (1847), Sorby (1853,1856,1908), Phillips (1857) and Harker (1885) in England, Heim (1878) in Switzerland and Daubrée (1876) in France that led to clear realisation that penetrative cleavage, such as found in slates and schists, represented the XY or "flattening" plane of the finite strain ellipsoid. The statement made by Sorby in 1908 expresses very clearly the philosophy behind this remarkable period of research - "In the case of nearly all branches of science a great advance was made when accurate quantative measurements were used instead of more qualitative. One great advantage of this is that it necessitates more accurate thought, points out what remains to be learned, and sometimes small residual quantities, which otherwise would escape attention, indicate important facts." Harker's classic paper on cleavage (1885) gives one of the most complete summaries available today of an analysis of the significance of this common structure. However, even today, many experts in the field of fabric analysis seem unaware of this work, and proceed to rediscover many of its conclusions. In the period 1930-1945 it was unfortunate that many of those working on geological fabrics either seemed unaware of the general relevance of strain theory to tectonics, or overemphasized certain very special displacement models of pure shear and simple shear. After a period of what might term "structural geology doldrums" a sudden advance was made by the publication of the work of Cloos (1947) in which he set out the results of a very thorough study and survey of strain states in the Appalachian fold belt of Maryland. This study integrated field data at all scales from features seen in thin section through to the large scale fold structures. It was the first work to show how strain data collected at each outcrop were perfectly coherent one with another, that the small scale structures of cleavage, lineation, fold axes and vein systems had geometric forms related to the form and orientation of the finite strain ellipsoid and that the overall picture of strain variation on a regional scale showed systematic variations which clearly linked with the Apalachian fold belt as a whole. Cloos' outstanding work gave a strong research impulse in structural geology and marked the beginning of a new phase of understanding in the subject. The major topics of this readvance will be listed below.

1.ROCK FABRIC

The strain studies of Cloos using deformed oölitic limestones confirmed the work of the late nineteenth century that slaty cleavage and penetrative schistosity form parallel to the XY ($1+e_1$, $1+e_2$) plane of the finite strain ellipsoid. He showed that at about 30 percent shortening in the Z($1+e_3$) direction cleavage became observable in field outcrops, and that cleavage intensity increased with increase in shortening. This type of cleavage is therefore a unique feature of any rock. Certain late deformations can induce special types of non-penetrative planar fabric (e.g. crenulation cleavage) but these are not related to the finite strain principal planes. Especially important has been the realisation that the often complex patterns of orientation and intensity of cleavage in folded rocks (e.g. Figure 11) can be explained by different plans of finite strain variation arising from different rock rheologies. Particularly important in this respect has been the application of computer based methods of finite element analysis to make the complex mathematical computations necessary to predict an overall strain plan in a fold (Dietrich, 1969; Dietrich and Carter, 1969; Hudleston and Stephansson, 1973).

Penetrative linear fabrics can arise in several ways. Some are parallel to the axis of greatest finite extension X($1+e_1$) and represent maximum stretching directions. Others, often found obliquely oriented to the X-direction are intersections of planar XY fabric surfaces with pre-existing planar fabrics (bedding, lithological layering). Some workers have shown how some, usually weakly developed linear fabrics can result from a combination of diagenetic flattening normal to bedding combined with later tectonic elements (Ramsay and Wood, 1973; Graham, 1978). This combination can produce a total maximum stretch fabric and a <u>pencil structure</u> oblique to the maximum tectonic stretching.

2. SMALL SCALE ROCK STRUCTURES

Because of the rheological anisotropy in layered rocks, strain sets up a great variety of mechanical instabilities depending upon the nature of the anisotropy and the orientation of the anisotropic feature within the finite strain ellipsoid. The relationship of the structures termed boudinage and buckle folding to the orientation of competent layers lying in a position of positive and negative finite extension respectively has been appreciated for a long time. More recent studies have linked much more exactly the forms and orientations of these features with differing layering orientations within the strain ellipsoid, and how different types of strain ellipsoid set up a wide spectrum of combinations of boudinage and buckle folds (Ramberg, 1959, Ramsay, 1967). In many naturally deformed rocks deformation is often concentrated into zones known as shear zones. This mode of deformation appears to predominate wherever rather uniform rock materials are found (e.g. massive sedimentary layers, or plutonic igneous rocks). A study of the geometry of these shear zones, the strains and fabrics found in them and their geotectonic significance has been greatly enhanced by the use of strain measurement techniques and a discussion of geometric compatibility based on the geometric constraints of the zone walls (Ramsay and Graham, 1970).

3. LARGE SCALE TECTONIC STRUCTURES

Cloos (1947,1971) showed that strain variation was systematically related to positions in large scale folds. He also demonstrated that there is an overall increase of finite strain from NE to SW in the Appalachians, a feature that can be generally correlated with increase in metamorphic grade, and that the X-axes of the finite strain ellipsoid show rather constant trends over very large areas. Because there is a direct connection between strain and displacement, regions with constantly oriented strain X-axes imply rather constantly directed displacements. These directions may sometimes be related to the major motions between colliding continental plates. Some orogenic zones appear to show quite complex strain variations which can be separated into shearing and volumetric distortions (e.g. in the N.Apennines, Kligfield et al., 1981), while others appear to show combinations of vertical and horizontal strike slip motions (e.g. in the Pyrenees, Choukroune, 1976). Fold belts and plate collision zones with pronounced arcuate forms (e.g. the Hercynian orogenic zone of Spain, Matte, 1968) often show strain variations of great complexity with strain ellipsoid X-axes either horizontally or steeply inclined. These strain variations have been used to constrain possible geodynamic models, for example, to distinguish originally formed arcuate zones from those where an initially rectilinear zone was later bent (Ries and Shackleton, 1976). In the Armorican deformation zone of N.W.France, the strain patterns are consistent with extremely large sub-horizontal crustal motions in a steeply oriented planar zone (Jégouzo, 1980; Percevault and Cobbold, 1982).

In Precambrian terrains strain analysis is, by the nature of many characteristic gneissic and highly metamorphosed rocks, somewhat restricted. Techniques that have been productive are the measurement of Xenolith shapes in plutonic rocks and a study of the geometric features of dyke complexes using folding and boudinage relationships to determine the strain ellipsoid (Escher et al, 1975; Talbot,1970). The recognition of shear zone geometry from an analysis of the orientations and intensity of schistosity has also produced significant data (Beach, 1974; Coward, 1976). Particularly impressive have been the results best interpreted as showing that Precambrian tectonic processes did involve large horizontal and vertical displacements, not unlike those that appear to dominate currently active tectonics. Many igneous plutons are emplaced over quite a large time interval, and the earlier emplaced magmatic bodies become deformed as a result of the intrusion of younger igneous masses. These older parts often behave in a plastic manner, and strain analysis of the xenoliths contained in them can be used as a method of comparing consolidation rates of an intrusion skin with the rate of volume increase resulting from the emplacement of younger intrusions.

All geologists are aware of the usefulness of constructing geological profiles through orogenic zones. Unfortunately some still persist in reconstructing profiles by line balancing, a technique that assumes that the layers in a cross section has suffered no significant length change as a result of the deformation process. The sections produced by Hans Cloos through the large Appalachian folds show clearly that changes do go on. A knowledge of the strain state in a layer, however, enables these modifications in layer length to be fully evaluated and correction factors can then be applied to restore bed lengths and bed thicknesses to their correct original dimensions (Ramsay, 1969; Oertel, 1974; Hossack, 1978; Cobbold, 1979).

The application of incremental strain measurement techniques has produced especially interesting appreciations of the complexity of movement that has arisen in the Helvetic nappes ofthe Swiss Alps.(Durney and Ramsay, 1973; Ramsay and Huber, 1983). Remarkable differences in strain increment history are found in the lower and higher nappes of the pile and appear to be in part due to tectonic level and lower constraints on movement freedom in the highest tectonic element.

CONCLUSIONS

1. Techniques exist for the measurement of finite and incremental strains in many, if not most deformed rocks. The collection of such data should be regarded as a standard part of field and laboratory investigations.
2. Strain data enable the geologist to interpret more exactly the significance of small scale and large scale tectonic features of a deformed zone, to make correctly constructed palinspastic cross sections and to relate rock architecture to major displacement process that have taken place in the Earth's crust.

ACKNOWLEDGEMENTS

Part of this work was carried out with help from Swiss National fonds (Grants 2.895-0.77 and 2.824-0.80).

REFERENCES

Beach A. 1974. The measurement and significance of displacement on Laxfordian shear zones, North-West Scotland. Proc.Geol.Assoc.Lond. 85, 13-21.

Breddin H. 1956. Die tektonische Deformation der Fossilien im Rheinischen Schiefergebirge. Z. deutsch. geol. Ges. 106, 227-305.

Casey M., Dietrich D., Ramsay J.G.1983. Methods for determining deformation history from chocolate tablet boudinage with fibrous crystals. Tectonophysics 92, 211-239.

Choukroune P. 1976. Strain patterns in the Pyreneen chain. Phil.Trans.R.Soc.Lond. A, 283, 271-280.

Cloos H. 1947. Oolite deformation in the South Mountain Fold, Maryland. Geol.Soc.Amer.Bull. 58, 843-918.

Cloos H. 1971. Microtectonic. John Hopkins Univ. Baltimore

Cobbold P.R. 1977. Compatibility equations and the integration of finite strain in two dimensions. Tectonophysics, 39, T1-T6.

Cobbold P.R. 1979. Removal of finite deformation using strain trajectories. J.Struct.Geol. 1, 67-72.

Cobbold P.R. 1980. Compatibility of two dimensional strain and rotations along strain trajectories. J.Struct.Geol., 3, 379-382.

Coward M.P. 1976. Archaean deformation patterns in southern Africa. Phil. Trans R.Soc.Lond. A, 283, 313-331.

Daubrée G.A. 1876. Expériences sur la schistosité des roches et sur les déformations des fossiles, corrélatives de ce phénomène. Compt.Rend.Acad.Sci.Paris 82, 710-798.

Dietrich J.H. 1969. Origin of cleavage in folded rock. Amer. J.Sci. 267, 155-165.

Dietrich J.H., Carter N.L. 1969. Stress history of folding Amer.J.Sc. 267, 129-154.

Dunnett, D. 1969. A technique of finite strain analysis using elliptical particles.Tectonophysics 7, 117-136.

Durney D., Ramsay J.G. 1975. Incremental strain measured by syntectonic crystal growth. In Gravity and tectonics (De Jong and Scholten eds.), 67-96. Wiley, New York.

Elliott D. 1972. Deformation paths in structural geology. Geol.Soc.Amer.Bull. 83, 2621-2638.

Escher A., Escher J.C., Waterson J. 1975. The reorientation

of the Kangamiut dyke swarm. Canad.J.Earth Sci. 12, 158-173.
Flinn D. 1962. On folding during three dimensional progressive deformation. Quart. J.Geol.Soc.Lond. 118, 385-428.
Flinn D. 1979A. The computation of deformations. Geol.Jour. 14, 87-98.
Flinn D. 1979B. The deformation matrix and the deformation ellipsoid. J.Struct.Geol. 1, 299-307.
Fry N. 1979. Random point distributions and strain measurement in rocks. Tectonophysics 60, 89-105.
Graham R.H. 1978. Quantative deformation studies in the Permian rocks of the Alpes Maritimes. Proc.Goguel Symp. (Bur.Rech.Géol.Min.France) 220-238.
Harker A. 1885. On slaty cleavage and allied rock structures with special reference to the mechanical theories of their origin. Rep.Br.Assoc. 55 meeting.1-40.
Haughton S. 1856. On slaty cleavage, and the distortions of fossils.Phil. Mag. Ser. 4, 12, 1-13.
Heim A. 1878. Untersuchungen über den Mechanismus der Gebirgsbildung. Schwabe, Basel.
Hobbs B. 1971. The analysis of strain in folded layers. Tectonophysics 11, 239-375.
Hossack J.R. 1968. Pebble deformation and thrusting in the Bygdin area (S.Norway). Tectonophysics 5, 315-339.
Hossack J.R. 1978. The correction of stratigraphic sections for tectonic finite strain. J.Geol.Soc.Lond. 135, 229-241
Hudleston P.J., Stephansson O. 1973. Layer shortening and fold shape development in the buckling of single layers. Tectonophysics 19, 299-321.
Jaeger J.C., Cook N.G.W. 1969. Fundamentals of rock mechanics. Methuen, London.
Jégouzo P. 1980. The South Armorican shear zone. J.Struct. Geol. 2, 39-47.
Kligfield R., Carmignani L., Owens W.H. 1981. Strain analysis of a Northern Apennine shear zone using deformed marble breccias. J.Struct.Geol. 3, 421-436.
Lisle R.J. 1977. Clastic grain shape and orientation in relation to cleavage from the Aberystwyth Grits. Tectonophysics 39, 381-395.
Matte P. 1968. La structure de la virgation hercynienne de Galice (Espagne). Trav.Lab.Géol.Grenoble 44, 5-128.

Means W.D. 1976. Stress and strain. Springer Verlag, New York, Heidelberg, Berlin.
Oertel G. 1974. Unfolding of an antiform by the reversal of observed strains. Geol.Soc.Amer.Bull. 85, 445-450.
Oertel G., Ernst W.G. 1978. Strain and rotation in a multilayered fold. Tectonophysics 48, 77-106.
Percevault M.N., Cobbold P.R., 1982. Mathematical removal of regional ductile strains in Central Brittany. Tectonophysics 82, 317-328.
Phillips J. 1857, Report on cleavage and foliation in rocks. Rept.Brit. Assoc.Sci. 269, 60-61.
Ramberg H. 1959. Evolution of ptygmatic folding. Norsk. Geol.Tidsskr. 39, 99-151.
Ramsay J.G. 1967. Folding and fracturing of rocks. McGraw-Hill, New York, London.
Ramsay J.G. 1969. The measurement of strain and displacement in orogenic belts. In Kent P.E. et al. (eds.): Time and place in orogeny. Geol.Soc.Lond. 43-79.
Ramsay J.G. 1976. Displacement and strain. Phil.Trans. R. Soc. Lond. A 283, 3-25.
Ramsay J.G., Graham R.H. 1970. Strain variation in shear belts. Can.J.Earth Sci. 7, 786-813.
Ramsay J.G., Wood D.S. 1973. The geometric effects of volume change during deformation processes. Tectonophysics 16, 263-277.
Ramsay J.G., Huber M.I. 1983. The techniques of modern structural geology. Vol.1, Strain analysis, Academic Press London, New York.
Ries A.C., Shackleton R.M. 1976. Patterns of strain variation in arcuate fold belts. Phil. Trans. R. Soc. Lond. A 283, 281-288.
Sharpe D. 1847. On slatey cleavage. Quart. J. Geol. Soc. Lond. 3, 74-105.
Sorby 1853. On the origin of slaty cleavage, Edin. New Philos. J. 55, 137-148.
Sorby 1856. On the theory of the origin of slaty cleavage. Phil. Mag. 12, 127-135.
Sorby 1908. On the application of quantative methods to the study of the structure and history of rocks. Quart. J.Geol. Soc. Lond. 64, 171-232.
Talbot C.J. 1970. The minimum strain ellipsoid using deformed quartz veins. Tectonophysics, 16, 89-101.

Wickham J.S. 1973. An estimate of strain increments in a naturally deformed carbonate rock. Amer.J.Sci. 273, 23–47.
Wood D. 1974. Current views on the development of slaty cleavage. Ann.Rev. Earth Planet. Sci. 2, 369–401.

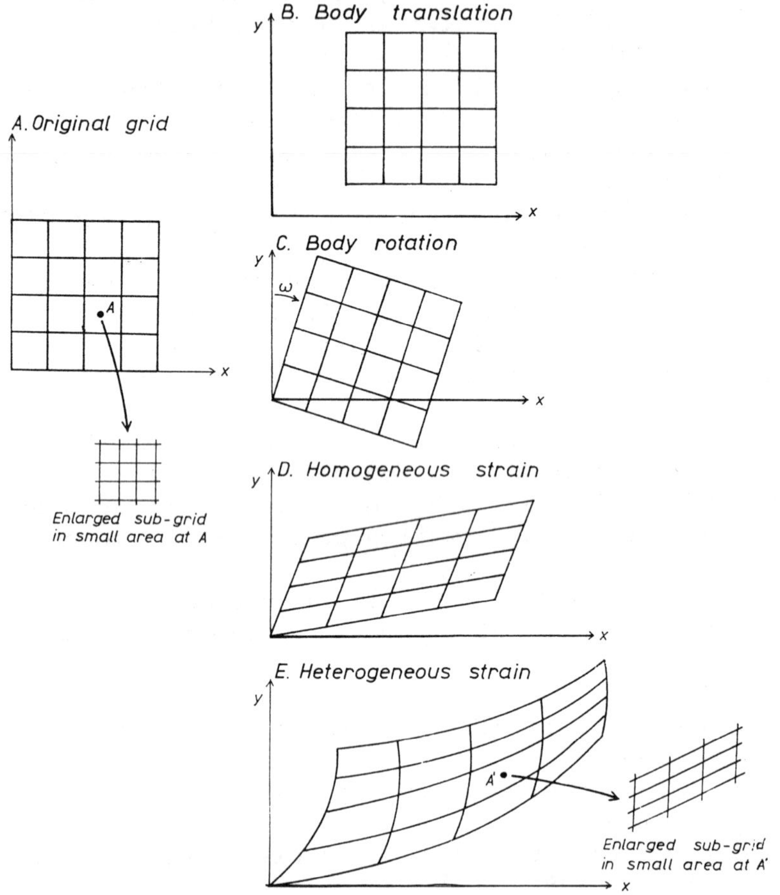

FIGURE 1. Different types of displacement plans related to the shape of an original orthogonal grid A.

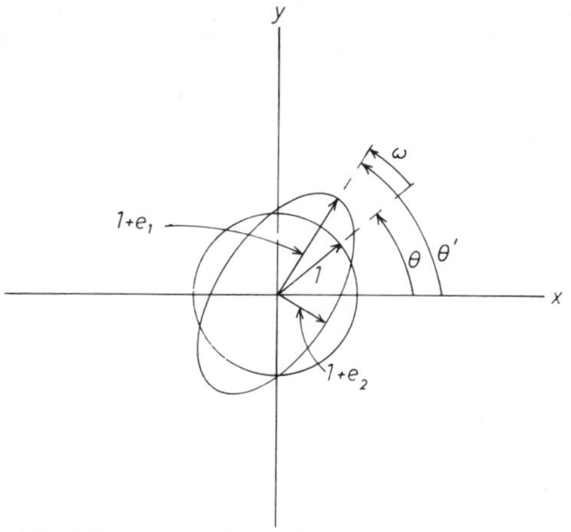

FIGURE 2. The finite strain ellipse derived from an initial circle with unit radius. The two principal strains are e_1 and e_2; the finite rotation ω is the difference in orientation between the principal ellipse axes Θ' and the positions of these lines before deformation Θ.

FIGURE 3. Elliptical forms of deformed reduction spots in a slate, N.Wales, UK.

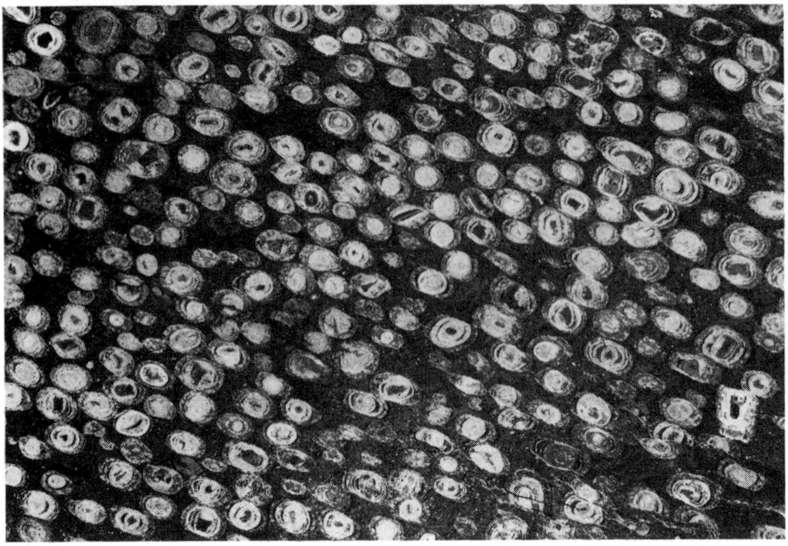

FIGURE 4. Deformed oolitic ironstone, Central Wales, UK. The ooids have been deformed by plastic flow processes and by pressure solution.

FIGURE 5. Limestone conglomerate deformed mostly by contact-contact pressure solution. Morcles nappe, Helvetic Alps, Switzerland.

FIGURE 6. Deformed trilobite Angelina Sedgwicki, N.Wales, UK.

FIGURE 7. Splay fissures developed at the tip of a vein as a result of changing incremental strain directions, Wildhorn nappe, Helvetic Alps, Switzerland.

FIGURE 8. Fibrous structure in a quartz-carbonate filled tectonic vein. Doldenhorn nappe, Switzerland.

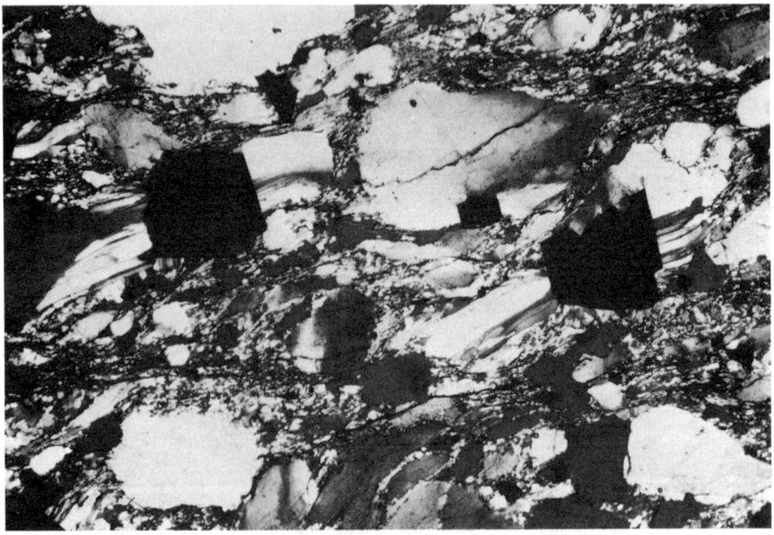

FIGURE 9. Thin section of fibrous crystalline pressure shadows around magnetite grains. S.Mountain, Baltimore, U.S.A.

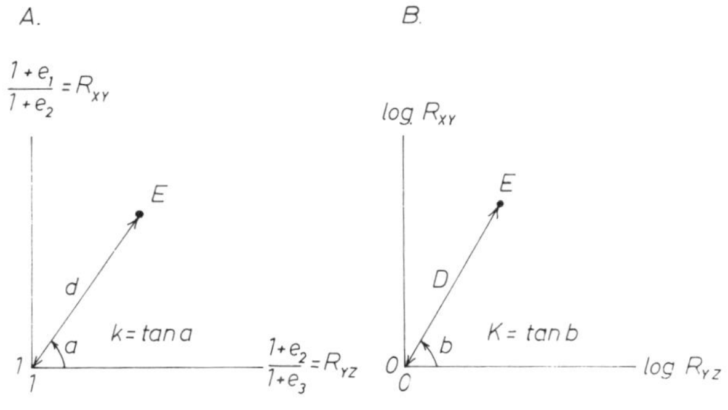

FIGURE 10. Methods of graphically representing a strain ellipsoid E: A) The Flinn plot with parameters k and d. B) The Ramsay logarithmic plot with parameters K and D.

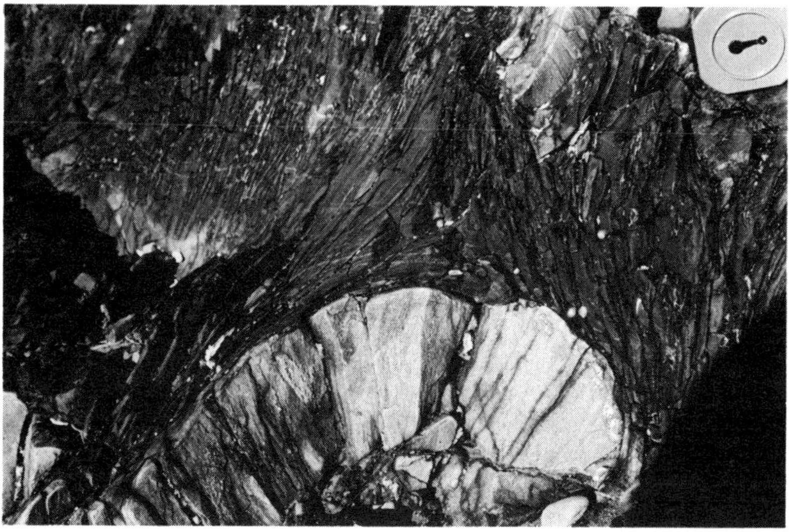

FIGURE 11. Complex convergent and divergent cleavage patterns in folded sediments, N.Wales, UK.

PROBLEMS OF TECTONIC FLOW STUDY

LUKJANOV, A.V.
Geological Institute of the USSR Academy of Sciences,
Moscow, USSR

ABSTRACT
The heterogeneity of tectonic flow causes the formation of structure patterns of various size. In order to understand the mechanism of structure formation it is necessary to study the tectonic flow quantitatively. Some problems are the most significant on this way: 1) description and quantitative analysis of structure patterns, 2) quantitative evaluation of strain and rotation, 3) study of stress-metamorphism for evaluation of deformation.

Ductile deformation and tectonic flow of rocks are of special significance in the lithosphere evolution since Precambrian till the recent time. They have been studied from different points of view. Gigantic geological bodies and lenses, composed the lithosphere, are of different composition and properties. They deform heterogeneously, exchanging the form and position, spreading, breaking, sliding of each other, and taking an active part in the whimsical movement, or better to say, flow of the lithosphere *). Tectonic layering of the lithosphere is the result of this flow to a great extent (Tectonic layering ..., 1980). Complex strain fields, magma generation, metamorphism of rocks stimulate and complicate the tectonic flow. Tectonic flow controls the localization of mineral

*) Advantage of the term "flow" is perfectly obvious on on the examination of the uppermost (ice) sphere of the lithosphere - Inland Ice Caps and glaciers. Tectonic flow of the middle sphere (the Earth crust) characterize its evolution including horizontal movements and ductile deformation also very well. The character of ultrabasic rocks deformation shows the existence of tectonic flow in the lowermost sphere of the lithosphere (the upper mantle).

Proceedings of the 27th International
Geological Congress, Volume 7, pp. 295–309
TECTONICS

deposits and seismoactive zones. That's why the practical aspect of tectonic flow study has as much significance as the theoretical one.

In this presentation I would like to attract attention to the correlation of small and extensive deformations with local intensive deformations and large translations of rock masses. Ever since, it is necessary to touch with the problems of tectonic layering of lithosphere, tectonic overthrust formation, formation of refolded folds under progressive deformation, formation of specific tectonic patterns not only in plane but in crossections also, and techniques of ductile deformation studying. It is impossible to examine all these problems in a short presentation. That is why I will touch on them only, but will speak about deformations and translations of rock masses.

We know the deformation of the crust still not very well. The main problem is the quantitative evaluation of deformation. If we knew the strain and rotation distribution in detail, we could understand mechanics of geostructures formation and maybe mechanics of crustal and subcrustal masses movement. But our knowledge of deformation distribution is not enough for such analysis. That is why the quantitative evaluation of rock deformation becomes one of the most important aims of the recent structure geology.

The basis for conclusions to be presented are my long-term works in the Pamirs, Middle Asia, Kasakhstan, Baltic Sheld and in the other regions of Europe and Asia as well as a tectonophysic modelling of geostructures. I have used also a lot of published data (Belousov, 1970, 1971, 1975, 1977; Cloos, 1958; Durney, Ramsay, 1976; Ez, 1967, 1976, 1977, 1978; Mileev, 1976; Mitra, 1978; Patalakha, 1962, 1967, 1970; Paterson, 1978; Peive, 1967; Ramberg, 1957, 1963, 1964, 1970, 1976; Ramsay, 1967; Sukrun, Segure, 1976; Sorvachev, 1978; Tokhtuev, 1967, 1972; Turner, Weiss, 1963; Zagorcev, 1981, 1982; Zwart, 1962; and others).

The crust moves, better to say - flows everywhere. Tectonic flow of the crust is considerable in the folded belts as well as on the platforms. Structure patterns of regions, deformation, magmatism and sedimentation contain information on the tectonic flow of rock-masses. But frequently we do not try to extract this information. We even do not know how to do that, or know it not perfectly.

Tectonic flow consists of strain, rotation and translation. Strain and rotation can be studied directly in the rock outcrops. Translation cannot be studied directly. It can be calculated after the analysis of structure patterns. Direct measurements of translation with paleomagnetic and paleoclimatic methods are not accurate enough, and can be used for the analysis of very large structures only. So the techniques of tectonic flow investigation splits into two methods: 1) strain and rotation measurement and 2) tectonic pattern analysis.

STRAIN AND ROTATION MEASUREMENT

I have mentioned that strain and rotation can be measured everywhere. Right theoretically it is so. But in practice we meet many difficulties on the way of the deformation measurement. Two kinds of these are the most significant. First, the small deformations (compression less than 20%) are very poorly manifested and give practically no information with ordinary methods (Lukjanov, 1980; Patalakha, 1967). On the second hand, too many measurements are necessary to evaluate the deformation. A set of not less than nine independently measured values characterize the deformation. It is necessary to find special "strain gauges" in the rock and to make very specific research to get such evaluation. We never do that during the ordinary geological mapping. But during latest decades, the geologists of many countries began to carry out special investigations and improve methodics. The data base grows, but slowly.

Strain measurement is executed in two stages: 1) evaluation of orientation of principal axes of strain (three measurements). It is easy to do that observing shistosity and lineation. 2) measurement of the magnitude of the axes (three measurements more). The latter is much more difficult. Very often we don't take these measurements and lose the information. But if we really want to study deformation, these ones are necessary. There are many strain gauges for the measurements: 1) deformed inclusions with well known original form (fossils, oolites, pebbles,pillow, and others) (Cloos, 1958; Lukjanov, 1959, 1980; Patalakha, 1962; Zagorcev, 1981, 1982), 2) inclusions with unknown original form, but originally having quite irregular position (fragments in breccia, porphyroblasts, crystalls in intrusiv rocks, and others) (Lukjanov, 1980; Patalakha, 1967), 3) newly formed minerals in cracks and at the boundaries of solid inclusions (Durney, Ramsay, 1976; Lukjanova, Lukjanov, 1979), 4) boudinage (Ramberg, 1964; Tokhtuev, 1967, 1972), 5) folds (Belousov et al., 1955; Gzovsky, 1975; Ramberg, 1963), 6) stress-metamorphysm (Lukjanov, Lukjanova, 1982a), 7) tectonofacies (Patalakha, 1967, 1970, 1978).

Rotation measurement is carried out also in two stages: 1) determination of the orientation of the axes of rotation, and 2) measurement of the rotation value. It is very difficult to measure the rotation, but in many cases it is possible (Ghosh, Ramberg, 1976, 1978; Lukjanov, 1980; Ollesen, 1978; Zwart, 1962). Geologists usually don't measure the rotation. It is a pity, because we cannot analyse the tectonic flow without determination of rotation. We cannot even differ a pure shear from a simple shear. Gauges of rotations are 1) syntectonic porphyroblasts, 2) undeformed inclusions and newly formed minerals, 3) geological bodies transsected by ductile deformation zones, 4) folds, 5) boudines. It is very important that the absence of rotation must be specially proved as well as the rotation not equal to zero.

Rock volume changes are registrated observing special forms of new mineral formation, veins, dikes. They are common in microstructure and much less frequent in macrostructure (Lukjanov, 1980). But every time it is necessary to be sure if the volume changes or not.

The systematic measurement of deformation helps to describe the structure parageneses quantitatively and to start the analysis of structure patterns. I will say about this later. Now I will enumerate some

conclusions carried out of the study of rock deformation.

1. Ductile deformation is common in the crust. Its magnitude in some locations is gigantic. The measurements made in Appalachians, Pyrenees, Alps, Kaukazus, Pamirs, Middle Asia, Kasakhstan, Kola Greenland, and many other regions lead us to this conclusion quite definitely. Sometimes the extension reaches many hundreds and thousands per cent and takes a short time to develop (for example, Paleogenic rocks on the Pamirs).

2. Ductile deformations develop in all the stages of metamorphic transformation of rocks. We have specially studied ductile deformations in the rocks of different facies and grades of metamorphism (Lukjanova, Lukjanov, 1979, 1982b). And I can state that there is no any thermodynamic threshold for the process of deformation. Ductile deformation with typical manifestations of stress-metamorphism develops even in the earliest stages of the catagenesis. The rocks suffer large deformations in all facies of metamorphism and catagenesis, under all thermodynamic conditions. But the rock plasticity successions are different in different PT-conditions (Lukjanov, 1980). The newly formed minerals are also different. Plasticity of rock or mineral association depends on the state of the system: equilibrium or nonequilibrium. The process of active metamorphic transformation of rock make its plasticity higher.

3. Surface-active liquids (solutions, fluides, melts) reduce the strength of rocks (Pertsov et al., 1982; Salnikov et al., 1982; Traskin, Scvortsova, 1982). That is why the ductility grows with the partial melting, and astenolenses can be formed near some isotherms in the crust.

4. In the geological time-scale, the lithosphere is a very mobile and easy deformed geosphere.

5. The axes of strain are oriented in many overthrusts in such a way that the axis of compression is subvertical and perpendicular to the plane of thrusting, and the axis of tension is directed along it. I even said in one of the lectures that the overthrust is a

structure of horizontal extension, and compared overthrusts with glaciers (Lukjanov, 1972). But the situation is more complex. The structure of overthrusts is inhomogeneous. It is possible to see a lack of correspondence of strain orientation in two overthrusted plates as well as in two points of a single plate.

6. Ductile deformations are distributed very inhomogeneously. Heterogeneity is one of the fundamental characteristics of a ductile deformation as a geological phenomenon. It appears in structures of different scales: platforms and folded belts, multiformity of structure parageneses and structure patterns inside folded belts, mini-heterogeneities inside parageneses, micro-heterogeneities in a texture of deformed rocks - all these features show us the multistagement of tectonic flow heterogeneity. As an illustration I refer to the structure of Kazakhstan showing on the cross-sections by Patalakha (Patalakha, 1981, fig. 83, 84) and on the map by Lukjanov (Lukjanov, 1965, fig. 62). Deformation was evaluated in the first case by method of tectonofacies, in the second case by method of structure parageneses. I had a good fortune to observe the heterogeneity of deformation and tectonic layering of series in deep canyons in the Pamirs and on the ice shelf of Greenland. These features are very much alike. I doubt whether anybody would be resolute enough to explain ice shelf leayering by the convection in mantle, so I am not sure also that it is reasonable to draw mantle convection for explanation thrusting in the Pamirs and in other places.

Apparently, the crust and tectonosphere themselves develop in such a way that specific effects arise and lead them to heterogenic deformation and tectonic layering.

STRUCTURE PATTERN ANALYSIS

The analysis of structure pattern is necessary for studying tectonic flow of the lithosphere (Burtman et al., 1963; Lukjanov, 1965; Lukjanov, Scherba, 1972; Suvorov, 1961). Many publications are devoted to the theory and practical application of the pattern analysis. I will regard only some effects which connect deformation and translation in pattern analysis.

The tectonic flow connects separated deformations in a single system. It is an integral value and contain a strain as well as rotation and translation. The structure pattern of a region is the result of the tectonic flow. Tectonophysical modelling let us better understand the mechanism of structure pattern formation. The modelling shows that heterogeneity of deformation distribution leads to some geometrical effects. These effects cause tectonic plate displacement, refolded folds formation, interchange of parallel zones of extension and compression, autonomous deformation of tectonic plates, and other features of tectonic patterns (Lukjanov, 1980).

Cline-effect, reverse-effect, press-effect, and TS-effect (effect of tectonic suturing) are the most significant for structure pattern interpretation (Lukjanov, 1982). The sense of these effects can be explained on a very simple model where a wedge is deforming into a rectangel. It is the simplest model of a heterogeneous deformation, maybe a primitive model, but it shows the pure effects. In the process of deformation the strain, rotation, and translation are distributed in the model heterogeneously, and their distribution changes in time. The role of each of them changes in time and space in all the points.

Cline-effect connects the distribution of translation and strain in different parts of the model. It is important that the value of the translation depends on a length of the model and can reach 25% of it.

Reverse-effect connects the translation with the orientation of strain in time. It is noticable that there is a zone in a model where the strain "crosses zero" during the progressive deformation. At that moment, the axes of the strain reverse sign: compression changes to tension along the same axis, tension changes to compression. This effect is especially important for the analysis of refolded folds localization in the progressive developed structure pattern. The sign changes many time in more complex models.

Press-effect is similar to cline-effect, but has the opposite signs of translation and strain. It develops in other geological structures. The pattern of Tadzhic Depression in Middle Asia can be an example of press-effect.

TS-effect determines the position of peculiar structures: the Structures of Tectonic Suturing (TS-structures). These structures are antipodes to faults. In TS-structures a peculiar type of contact is formed: the suturing of rocks being far apart before. The biggest TS-structures are "sutures" on the place of paleo-oceans. But there are many smaller TS-structures being nondescribed yet. Many of them form in axial parts of folds, in zones of boudinage, in "fault zones". Typical TS-structures form in glaciers where they flow together. Medial moraines mark the position of such TS-structures. Recognition, description, and classification of TS-structures is the task of future.

The effects of hetereogeneous deformation help tu understand many characteristic features of structure patterns. They help to correlate dissipated deformations with the large displacements along faults, to understand overthrusting, tectonic layering of the lithosphere, large horizontal movements including movements of the lithospheric plates. All these features can be explained by a heterogeneous tectonic flow of the tectonosphere without hypothetical ideas engaging deep-intratelluric processes. But for that it is necessary to know exactly the distribution of deformation in the Earth crust. Therefore the main problem of structure geology now is the quantitative evaluation of ductile deformation and study of structure parageneses and structure patterns on the new, quantitative level. It means that we need to get an enormous and new in essence scientific material from all the world around. The International Union of Geological Sciences and its Commission on Tectonics could help to collect this material with organizing special International symposiums, publishing collected articles on the subject, and giving recommendations to collect the data on quantitative evaluation of ductile deformation of rocks.

REFERENCES

Belousov, V.V., To the problem of tectonic deformations nature (in Russian). - In: H. Ramberg. Modelirovanie deformatzij zemnoj kory s primeneniem zentrifugi. M.:Mir, 1970, s. 214-223.

Belousov, V.V., 1971. Structural geology (in Russian). M.: Izdatelstvo MGU, 1971.

Belousov, V.V., 1975. The foundations of geotectonics (in Russian). M.: Nedra, 1975.

Belousov, V.V., 1977. The main features of tectonic deformation mechanism (in Russian). - In: Otcherki strukturnij geologii slozhnodislotzirovannych tolsch. M.: Nedra, 1977, s. 6-29.

Belousov, V.V., Chertkova, E.I., Ez, V.V., 1955. Modeling of folding in buckling conditions (in Russian). - Bull. MOIP. Otd. geol., 1955, t. 30, vyp. 5, s. 117-124.

Burtman, V.S., Lukjanov,A.V., Peive,A.V., Ruzhentzev, S.V., 1963. Horizontal displacements along faults and certain methods of their study (in Russian). - In: Razlomy i gorizontalnye dvizhenija zemnoj kory. M.: Izdatelstvo AN SSSR, 1963, s. 5-53.

Choukroune, P., Seguret, M., 1976. Tectonics of Pyrenees: the role of compression and gravity (in Russian). - In: Sila tjazhesti i tektonika. M.: Mir, 1976, s. 154-167.

Cloos, E., 1958. Oolite deformation in the South Mountain Fold, Maryland (in Russian). - In: Voprosy strukturnij geologii. M.: IL, 1958, s.96-156.

Durney, D., Ramsay, I., 1976. Incremental strain measured by syntectonic crystal growth (in Russian). - In: Sila tjazhesti i tektonika. M.: Mir, 1976, s. 88-115.

Ez, V.V., 1967. The isoclinal folding role in the structure of Belomorides (in Russian). - Geotektonika, 1967, No 3, s. 69-81.

Ez, V.V., 1976. Tectonics of abyssal zones of the

continental crust (in Russian). - M.: Nauka, 1976.
Ez, V.V., 1977. Main features of metamorphic complexes structure (in Russian). - In: Ocherki strukturnoj geologii slozhnodislotzirovannykh tolsch. M.: Nedra, 1977, s. 208-245.
Ez, V.V., 1978. Structural geology of metamorphic complexes (in Russian). - M.: Nedra, 1978.
Ghosh,S.K., Ramberg, H., 1976. Reorientation of inclusions by combination of pure shear and simple shear. - Tectonophysics, 1976, vol. 34, p. 1-70.
Ghosh, S.K., Ramberg, H., 1978. Reversal of the spiral direction of inclusiontrails in paratectonic porphyroblasts. - Tectonophysics, 1978, vol. 51, No 1/2, p. 83-98.
Gzovsky, M.V., 1975. The foundations of tectonophysics (in Russian). - M.: Nauka, 1975, 536 s.
Lukjanov, A.V., 1959. Ductile deformation of limestones in major fault zones (in Russian). Izv. AN SSSR. Ser. geol., 1959, No 4, s. 102-106.
Lukjanov, A.V., 1965. Structural effect of horizontal crustal movements (in Russian). - M.: Nauka, 1965, 210 s.
Lukjanov, A.V., 1972. To the problem of dynamics and energetics of tectonic thrusting (in Russian). - In: Vnutrennjaja geodinamika. L., 1972, vyp. 2, s. 22-24.
Lukjanov, A.V., 1980. Plastic deformations and tectonic flow of rocks in process of layered lithosphere formation (in Russian). - In: Tektonicheskaja rassloennost litosfery. M.: Nauka, 1980, s. 105-146.
Lukjanov, A.V., 1982. The most important effects of heterogeneous deformation (in Russian). - In: Eksperimentalnaja tektonika v reshenii zadach teoreticheskoj i prakticheskoj geologii. Novosibirsk, 1982, s. 93-95.
Lukjanov, A.V., Lukjanova, V.T., 1982a. Stress-metamorphism in Mesozoic and Cenozoic rocks of Middle Asia (synopsis of the lecture of 26 February 1981) (in Russian). - Bull. MOIP, otd. geol., 1982, t. 57, vyp. 3, s. 117.
Lukjanov, A.V., Lukjanova, V.T., 1982b. Dynamometamorphose in frühen Stadien der Katagenese. - In:

Deformation und Metamorphose von Gesteinen, v.2. Potsdam, 1982, s. 150-174.
Lukjanov, A.V., Scherba, I.G., 1972. Paragenetic analysis of structures as a basis for the middle-scale tectonic mapping in folded regions (in Russian). - In: Tektonika Sibiri, t.5, M.: Nauka, 1972.
Lukjanova, V.T., Lukjanov, A.V., 1979. Epigenetic microtexture of sedimentary rocks in stress field of different type (in Russian). - Litol. i. polezn. iskop., 1979, No 2, s. 108-123.
Mileev, V.S., 1976. Kinematic conditions of abyssal folding and the tectonics of the Proterozoic Majtjubinsky anticlinorium (in Russian). - In: Geologija i tektonika dokembrija Zantralnogo Kazakhstana. M.: Izdatelstvo MGU, 1976, s. 185-365.
Mitra, G., 1978. Ductile deformation zones and mylonites; the mechanical processes involved in the deformation of crystalline basement rocks. - Amer. I. Sci., 1978, vol. 278, No 8, p. 1057-1084.
Ollosen, N.O., 1978. Distinguishing between inter-kinematic and syn-kinematic porphyroblastesis. - Geol. Rdsch., 1978, Bd. 67, H.1, s. 278-287.
Patalakha, E.I., 1962. Deformation of concretions in Uspensky copper deposit, Central Kazakhstan (in Russian). - Izv. AN KazSSR. Ser. geol., 1962, No 3.
Patalakha, E.I., 1967. Evaluation of degree of folded series deformation (in Russian). - In: Tektonika i dinamometamorfism paleozoja Kazakhstana. Alma-Ata: Nauka, 1967, s. 48-110.
Patalakha, E.I., 1970. Mechanism of flow structure formation in crumpled zones (in Russian). Alma-Ata: Nauka, 1970.
Patalakha, E.I., 1978. About some fundamental concepts of structure geology and tectonophysics (shear laminar flow and dislocation process) (in Russian). - Izv. AN kazSSR. Ser. geol., 1978, No 3, s. 9-17.
Patalakha, E.I., 1981. Genetic basis of morphological tectonics (in Russian). - Alma-Ata: Nauka, 1981, 180 s.
Paterson, M.S., 1978. Experimental rock deformation: the brittle field. - Springer-Verl., 1978.
Peive, A.V., 1967. Faults and tectonic movements

(in Russian). - Geotektonika, 1967, No 5, s. 8-24.
Pertzov, N.V., Kogan, S.S., Lukjanov, A.V., Rastzvetaev, L.M., Traskin, V.Yu., 1982. Peculiarities of tectonic deformation of rocks in presence of surface-active liquid phase (problems of physic-chemical geomechanics) (in Russian). - In: Eksperimentalnaja tektonika v reshenii zadach teoreticheskoj i prakticheskoj geologii. Novosibirsk, 1982, s. 47-49.
Ramberg, H., 1957. Natural and experimental boudinage and pinch-and-swell structures (in Russian). - In: Voprosy eksperimentalnoj tektoniki. M.: IL, 1957, s. 160-190.
Ramberg, H., 1963. Strain distribution and geometry of folds. - Bull. Geol. Inst. Univ. Uppsala, 1963, vol. 42, No 4, p. 1-20.
Ramberg, H., 1964. Selective buckling of composite layers with contrasted rheological properties, a theory for simultaneous formation of several orders of folds. - Tectonophysics, 1964, vol. 1, No 4, p. 307-342.
Ramberg, H., 1970. Gravity, deformation and the Earth's crust as studied by centrifuged models (in Russian). - M.: Mir, 1970.
Ramberg, H., 1976. Modelling of tectonic movements produced by gravity with the help of centrifuge (in Russian). - In: Sila tjazhesti i tektonika. M.: Mir, 1976, s. 70-88.
Ramsay, J.G., 1967. Folding and fracturing of rocks. - New York: McGraw-Hill, 1967.
Salnikov, D.I., Traskin, V.Yu., Pertzov, N.V., 1982. Decrease of rocks strength under the influence of surface-active liquid phase (in Russian). - In: Eksperimentalnaja tektonika v reshenii zadach teoreticheskoj i prakticheskoj geologii. Novosibirsk, 1982, s. 110-112.
Sholpo, V.N., 1977. Complex folding in sedimentary series (in Russian). - In: Ocherki strukturnoj geologii slozhnodislotzirovannykh tolsch. M.: Nedra, 1977, s. 64-92.
Sorvachev, K.K., 1978. Plastic deformations in granite-gneiss structures (in Russian). - M.: Nauka, 1978.

Suvorov, A.I., 1961. Uspensko-Aktasskaja dynamic pair of faults in Central Kazakhstan (in Russian). - Dokl. AN SSSR, 1961, t. 137, No 4, s. 938-941.

Tectonic layering of lithosphere (A.V. Peive Ed.) (in Russian). 1980. - M.: Nauka, 1980, 216 s.

Tokhtuev, G.V., 1967. Boudinage structures and their role in localization of ore mineralisation (in Russian). - Kiev: Nauk.dumka, 1967.

Tokhtuev, G.V., 1972. Peculiarities of deformation of inhomogeneously layered medium (in Russian). - Kiev: Nauk.dumka, 1972, 122 s.

Traskin, V.Yu., Skvortzova, Z.N., 1982. Mechanisms of rock creep in condition of surface-active liquid phase presence (in Russian). - In: Eksperimentalnaja tektonika v reshenii zadach teoreticheskoj i prakticheskoj geologii. Novosibirsk, 1982, s. 112-113.

Turner, F.J., Weiss, L.E., 1963. Structural analysis of metamorphic tectonites. - New York: MacGraw-Hill, 1963.

Zagorčev, I., 1981. Early Alpine deformations in the red beds within the Poletinci-Skrino fault zone. 2. Structure and deformations in the northern parts of Vlahina Block. - Geologica Balkanica, 11.1, Sofia, Mart. 1981, p. 101-126.

Zagorčev, I., Sapundžiev, K., 1982. Early Alpinte deformations within the Poletinci-Scrin o fault zone. 3. Ujno Allochtone. - Geotect., tectonophys., and geodyn., 14, p. 3-31.

Zwart, H.J., 1962. On the deformation of polymetamorphic mineral associations, and its application to the Bosost area (Central Pyrenees). - Geol. Rdsch., 1962, Bd. 52, s. 38-65.

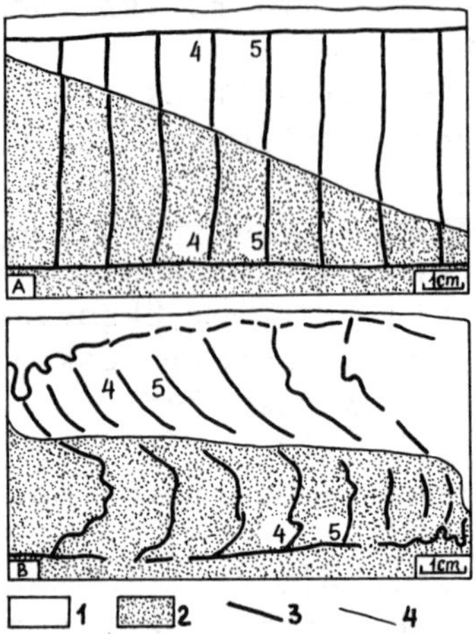

FIGURE 1. Deformation of wedge-shaped bodies in gravity field. The model consists of two silicone wedges (clines). A. - initial state. B - after centrifuging.

1 - upper cline (silicone, $\rho_1 = 2.42\ kg.dm^{-3}$,
$\eta_1 = 1.5 \cdot .10^5 \pm 20\%\ Pa \cdot s$)

2 - lower cline (silicone $\rho_2 = 1.77\ kg.dm^{-3}$
$\eta_2 = 1.0 \cdot 10^5 \pm 10\%\ Pa \cdot s$)

3 - thin plasticine interlayrs (marks) making the deformation visible

4 - clines boundary

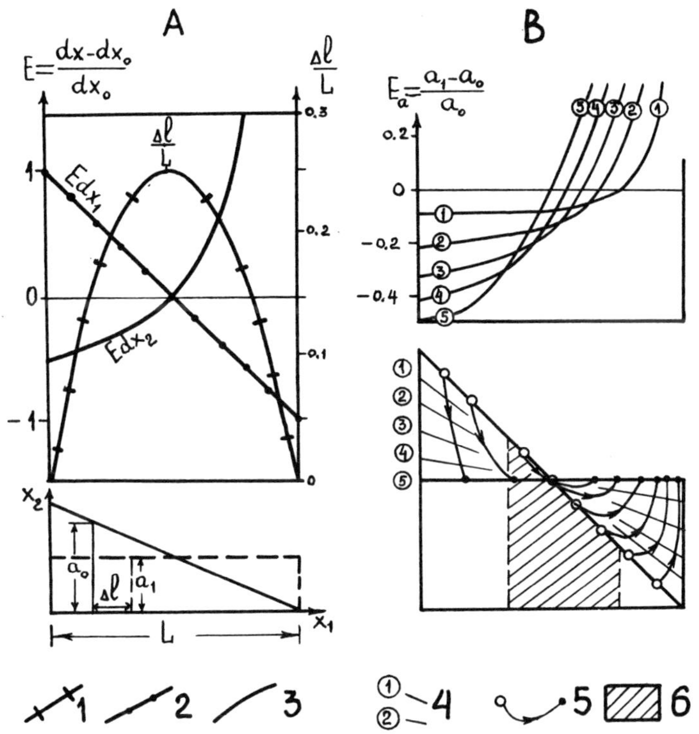

FIGURE 2. Variations of strain and translation after deformation of cline to rectangel (without friction, in Lagrangian coordinates). A - finite deformation, B - strain reversion during progressive deformation.
1 - translation (Δ1/L),
2 - elongation of elementary segments, parallel to X_1 axis (E_{dx_1}),
3 - elongation of elementary segments, parallel to X_2 axis (E_{dx_2}),
4 - stages of deformation,
5 - trajectories of particles,
6 - zone of reverse-effect.

TECTONOFACIES AND EVALUATION OF THE DEFORMEDNESS DEGREE IN FOLDED STRUCTURES

PATALAKHA, E.I.
Institute of Geological Sciences, Acad. Sci. Kazakh. SSR, Alma-Ata, USSR

In the past two decades the K.I. Stapayev Geology Institute under the Kazakh SSR Academy of Sciences has developed a peculiar method of tectonofacial analysis aimed at generalizing the great amount of information concerning the structure of deformation in the Phanerozoic period foldings which the traditional tectonic syntheses cannot cope with.

The author's works, including the book "A Tectonofacial Analysis of the Phanerozoic Folded Structures (substantiation, methods, appendix)" which is ready for publication deal with the following range of problems:

1. Thermodynamic vertical zoning of the earth's crust in folded regions.
2. Relationship of the leading dislocation mechanisms and the basic tectonic and metamorphic complexes of folded regions.
3. The model of the dislocation process.
4. The tectonofacial scale.
5. Methods and techniques of field and laboratory studies.
6. The tectonofacial analysis of folded systems.
7. The tectonofacial analysis of ore fields.

The present paper is in fact only a brief illustration to the methods of the tectonofacial analysis.

When studying folded structures in different parts of the continents, it was noticed long ago that one rock is more and the other rock is less deformed. Attempts were made to give a quantitative evaluation as regards the degree of deformedness of petrified residues of fauna and flora, rock, folded structures as a whole (by way of straightening folds - the Alps, the Appalachians, the Caucasus etc.). The investigation took a long period of time before the real degree of contrast in natural deformations and the difficulties

Proceedings of the 27th International
Geological Congress, Volume 7, pp. 311–326
TECTONICS

of the "strain-measuring" data (obtained by a dot object) at a distance became more or less clear. The mechanism of the dislocation process remaining vague, researchers often put on the same scale the observation material characterizing quite different and in fact incomparable thermodynamic levels of metamorphism. Those data characterized different in essence stages and conditions of formation of the continental earth's crust in the Phanerozoic *) folded belts. It is difficult, for example, to compare stratified and folded non-metamorphosed formations (epizone) and those touched by cleavage (mesozone) or crystallizing schistosity (katazone) as the conditions for the evolution of deformations in these zones are essentially different.

The present work elaborates the ideas set forth by the author in his book "Genetic Foundations of Morphological Tectonics" (Patalakha, 1981) which is concerned with the mechanism of the dislocation process, namely with the development of notions on tectonofacties in the cleavage mesozone of the Phanerozoic folded structures corresponding to the thermodynamic conditions of the zeolitic and green schist facies of metamorphism.

The above book gives a comprehensive substantiation of the idea relating to sutural folding as a new conception of the dislocation process (crumpling as a result of displacement) with all the far-reaching consequences. It considers in detail the main structural paragenesis of folded rock mass ("dynamometamorphic"), substantiates the unity of the general mechanism of the dislocation process in folded regions for the cleavage mesozone irrespective of the nature of the geodynamic source of disturbance. The book shows that the open Germanotype and contracted Alpinotype folds have no fundamental antipodal differences; there are only gradual intertransitions between them. This cuts the ground from under the traditional contrasting of open folding (allegedly owing to the action of vertical forces

*) Neogaen to be more precise

only) and contracted folding (as if caused by the action of lateral forces), and makes a foundation for the consideration of morphologically different foldings as consecutive phases (or stages) in the evolution of one (characteristic of the given thermodynamic regime) dislocation process of the sutural folding formation. By empirical contrastings of a big number of tectonic-type sections of Kazakhstan we distinguished ten typical stages in the formation of foldings and tectonofacies corresponding to them. For all of them we determined structural paragenesis (with due account of the lithological specificity of rock mass), sized up their differences, i.e. determined the tectonofacies boundaries. On the basis of the ideas about the dislocation process mechanism and the experimental material we found approximate kinematic characteristics for each tectonofacies (deformation components of the shift, contraction, extension etc.). By doing so we worked out the scale of tectonofacies, composed of ten types I-X joined into three tectonoformations: the lower A (I-IV), the intermediate B (V-VII) and the upper one C (VIII-X). Tectonofacies are natural rows of rock arranged according to deformedness, with a distinctive structural paragenesis.

The idea of tectonofacies did not occur unexpectedly. It had been prepared by the whole of the preceding course of morphological tectonics evolution. The positive meaning of this idea is, in the first place, that it substitutes for the tiresome and sometimes little productive study of particular structural elements (observations of fractures, cleavage, stratification etc.) by a complex-system, i.e. evolutional and paragenetic approach to the solution of problems of structural charting that can be applied to both local (prospecting too) and regional problems). It should be noted that particular structural elements are not examined separately, but as a component part, as an indicator of the structural paragenesis which is characteristic of the given specific situation. In other words it is the structural paragenesis (as a function of the deformedness degree), i.e. the structural zonation, but not merely isolated structural elements that is mapped.

So the doctrine regarding tectonofacies is based in the end on the very fruitful idea of the formational analysis as a comparative structural and paragenetic study of the inner deformation structure of foldings, which is used on a large scale in geology. Such an analysis makes it possible to reconstruct the evolution of the dislocation process as a peculiair "natural experiment" which is of interest by itself. Therefore with respect to the dislocation process the tectonofacial analysis has the same sense which the teaching about the facies of metamorphism has for the study of the problem of metamorphism.

It is important to bear in mind that the tectonofacial analysis at the present day is, perhaps, the only method which enables us to have an idea about the degree of deformedness in folded structures and their elements.

As an example we shall confine ourselves to an examination of the results of the tectonofacial analysis of the Chinghiz range in one of its cross sections *). The Chinghiz (Chinghiz-Tarbagatai) folded system consists of two structural and formational zones, Chinghiz - Chinghiz-Arkalyk proper and Akchatau-Abrala (North-East and South-West respectively) and the third - North-Balkhash zone contiguous to Akchatau-Abrala in SW and Chinghiz proper in NE (fig. 1). However, the latter zone is usually examined as part of the North-Balkhash folded system. The structural and formational zones of Chinghiz proper are distinguished rather conventionally because lying bare there are practically only different levels of one and the same thick typically eugeosynclinal volcanogenic and terrigenous profile which rather fully represents geosynclinal Cambrian, Ordovic and Silurian. That is the geosynclinal comples proper. It is composed of two structural stages.

*) The useful length of the cross section is about 140 km (about 200 km in all), the field exploration was done by A.I. Lukienko et al.

The lower one (Cambrian) includes various volcanites contaminated with terrigenous rock and with tholeiitic basalts and jaspers (ophiolitic complex) in the floor. The intermediate stage (Ordovic-Silurian) is less than half volcanogenic. The upper structural stage contains orogenic terrestrial volcanites of persilicic and medial composition of the Lower and Middle Devonian periods, Givetian pelagic basaltoids (the facies of sutural zones), Famennian and Tournaisian marine carbonaceous and terrigenous sediments, found still higher are mean-in-composition volcanites of the Lower Carboniferous period (Viséan). The pelagic Devonian and the Lower Carboniferous periods are characteristic of the North Balkhash zone. The underlying substratum of it is not yet quite clear, most likely it is the continuation of the Chinghiz structures.

Distinguished in the composition of each of the above mentioned structural and formational zones (from NE to SW) are:

Chinghiz-Arkalyk - Tundyk-Ashni synclinorium, Arkalyk anticlinorium, Chunhai synclinorium, Chinghiz anticlinoriums;

Akchatau-Abrala - Abrala synclinorium, Akchatau anticlinorium;

The North-Balkhash zone is represented by a synclinorium of the same name.

The above structural elements of lower grade reflect, as a matter of fact, the conditions of occurrence (with due account of the erosional shear) of the structural stages and suites.

Cleavage is practically not observed only in two youngest shear suites crowning the orogenic complex - the Devonian-Carboniferous and Viséan. Thus nearly all the geosynclinal tectonic complex (in the full sense of the word) which includes in this case not only two geosynclinal stages, but also (though conventionally) the lower part of the orogenic stage, may be referred to the cleavage mesozone.

One gets astonished at the abundance in the region of viscous faults, which fall steeply (up to 60°) and are very much camouflaged as usual by most diverse deformations. Here we find no "main" faults apart from the Chinghiz young displacement which to all appearances has inherited the ancient co-folded suture. In accordance with the occurrence of the faults the folds are on the whole straight, while the cleavage falls steep. So we have the basic structure of the traid which is stressed by the mutually parallel location of faults, axial planes of the folds and cleavage.

The further analysis will be given as we examine the tectonofacial composition of the following elements of the folded structure:

1. Rock of different composition
2. Rock of different age
3. Structural ages
4. Anticlinoria and synclinoria
5. Structural and formational zones
6. The folded system as a whole

This is a statistical analysis and it is based on the examination of a special table of geological conditions for the development of tectonofacties and tables of the tectonofacial composition with due account of the geological situation (being cumbersome the tables are nog given here).

1. ROCK COMPOSITION. The tectonofacial analysis of the Chinghiz range as regards the profile to be characterized has so many components that we have to analyze the rock composition at the level of suite assemblages. We studied 17 suites*), including three Cambrian, five Ordovician, three Silurian, four Devonian, one Carboniferous and one - Upper Devonian - Lower Carboniferous. As for the composition the suites are distributed in the following way. Most of them (14 out of 17) contain alternating sandstones and aleurolites (8 suites: two Cambrian, three Ordovician, two Silurian, one Devonian) and volcanogenic rock tuffs and lavas of acid and medial composition (six suites: two Ordovician, three Devonian, one Carboniferous). Individual suites

*) mapped among them are three benches of Namus, Alpeisk and Boshchekult suites apiece.

contain conglomerates (Silurian), limestones with aleurolites and sandstones (Devonian-Carboniferous), flinty aleurolites (Cambrian).

A search for any dependence of the rock tectonofacial composition on its lithology was not crowned with success. At the same time observations of the cleavage refraction show that the viscosity variations in the Chinghiz range are much lower than those found for NW Karatau, i.e. the profile of the geosynclinal complex (including the orogenic stage) is rather monotonous, unvaried. Besides, taking into consideration the considerable inner variegation of the suites that go under the heading "Sandstones and Aleurolites" we can as well refer to them "conglomerates" "limestones" and "flinty aleurolites" (which are as much internally heterogenuous) and consider in principle the whole of the investigated spectrum of the Chinghiz rock as consisting of two groups: chiefly terrigenous (sandstones in the main) and volcanogenic (11 and 6 suites respectively).

The analysis of the tectonofacial composition of all the above suites shows the following. Among the rock of the terrigenous group the highest tectonofacies C are developed only in five cases out of ten (fig. 2). Their specific significance does not overstep the limits of 10 and only in one case it exceeds 30 per cent. This maximum (about 200 metres wide), unusual for the Chinghiz range is connected with the Main Chinghiz fault. Intermediate tectonofacies B are found everywhere (in every suite without exception), and their specific significance is much higher and varies from 20 to 75 per cent. We must say that all the three tectonofacies V-VII are developed. However, a stable tendency of the decrease in the specific significance of tectonofacies from lower to higher ones can be observed. Lower tectonofacies A are developed to a greater extent (in every suite without exception). Their specific role is determined by a higher level of 30-80 and in one case it is even 100 per cent (Tundyk-Ashchisni synclinorium, Fammenian-Tournaisian imposed syncline). The observed decrease in the specific role from lower to higher tectonofacies is not found at all, while the second one is developed only in two cases. The main role is played by tectonofacies III-IV. Their correlation is not yet defined.

Similar regularities in the development of tectonofacies are also observed in the volcanogenic family of rock. The highest tectonofacies C was recorded only in one case out of six (it is significant that it is represented by tectonofacies VIII). Its specific content is about 2 per cent. Intermediate tectonofacies B is developed everywhere (in every suite). Its specific significance varies from 20 to 80 per cent. Incidentally, there is a direct evidence of a stable tendency of a decrease in the specific content in the direction from tectonofacies V to VII. Lowest tectonoformation A is developed in all parts, too. Its content varies from 20 to 83 per cent.

So if we digress from details of little importance, the tectonofacial composition of both groups rock - (terrigenous and volcanogenic) at the "suite" level is practically the same. The tectonofacial composition of both groups rock as a whole is determined by the lowest (mainly III-IV) and the middle (V-VII) tectonofacies (they are of similar specific content), the highest tectonofacies being of petty significance. No expressed tendency of dependence as regards the degree of rock deformedness on their genesis and lithologic composition was perceived. This conclusion is of extreme importance for us as it shows the lithologic peculiarities and composition, i.e. the rock viscosity properties, cannot account for the structural and tectonofacial degree of contrast, i.e. the broadest degree of contrast by the extent of rock mass deformedness recorded in the Chinghiz and other folded systems. It is explicable only from the positions of the accepted model of the dislocation process, i.e. recognition of the fracture (sutural) nature of folding.

2. ROCK AGE. Suffice it to have a look at the chart (fig. 2, left) to make a conclusion: the degree of rock deformedness practically does not depend on the age of rock. This conclusion is important in essence because from the time of contractionism there was a firm tendency to believe: the older the rock the higher

it is deformed. True, in many cases as well as in the range of very long tectonic epochs (for example, the boundaries of the base complex and the geosynclinal complex etc.) the above thesis is valid where there is angular unconformity. But is it not clear that in the places which have no such structural jumps as, say, synclinorial zones of long deflection it is impossible to specify the angular unconformities? It is for this reason that the Caledonian unconformity is often not found while the Devonian conformably lies on the Silurian.

3. STRUCTURAL STAGES. As we examine them, the conclusion we have just made acquires a special interest. From time immemorial there are three main structural stages distinguished in the Chinghiz range: the lower one (Cambrian) which is close in the general meaning to the early geosynclinal, the intermediate one (Ordovician-Silurian) which is similar to the inversion of the islandbow, and the upper one (Devonian-Carboniferous) which is close to the orogenic (fig. 2, left). In the boundaries of the mesozone (15 lower suites out of 17) we observe a rather monotonous change of tectonofacies within modal range III-VII. No jumps or contrasts in the deformation structure of the distinguished structural stages have been statistically established, though there is yet an insignificant tendency twoards a decrease in t he role of tectonofacies C and B and an increase of A with rejuvenation. However, it does not follow from this that it is not rightful to distinguish structural stages which have a more profound tectonic meaning. At the same time it is clear from the above said that we have not enough grounds to overrate and to make a fetish of the role of angular unconformities as well as the corresponding phases of folding in the evolution of folded regions (as some researchers still do). It is the character of formations, but not the phases of folding that is the basis for distinguishing the zones (laterally) and the stages (vertically).

Along with this we should take note of some contingency of angular unconformities themselves on account of not only their sutural nature of folding but also because angular unconformity (on the one hand) and the boundaries of tectonic stages (on the other) are quite different and not necessarily interrelated since for the origination of angular unconformity three elements are a compulsory condition: 1) upward motion; 2) substratum erosion; 3) submersion and sedimentogenesis. That is why angular unconformities do not always stress the boundaries of structural stages (even in the case of the major unconformity which reflects the replacement of the geosynclinal by the orogenic regime - as in the Caledonides of Kazakhstan). At the same time they rather widely appear inside the stages, including the geosynclinal ones.

4. ANTICLINORIA AND SYNCLINORIA. The cross section of the Chinghiz range under consideration is composed of four synclinoria divided by three anticlinoria (fig. 3, left). As it was noted earlier anticlinoria represent the exposure of the lower structural stage, while synclinoria - of the middle and (in two cases) the upper stages. We often register greater dislocatedness of anticlinoria, though in the given case they practically do not comprise katazone rock. The rock which we refer to the epizone (Devonian-Carboniferous and Viséan suites) is found in the North-Balkhash (locally) and Tundyk-Ashchui synclinoria where it determines the structure appearance along the profile course.

Thus, anticlinoria and synclinoria in the given case are only a somewhat different form of expressing the self same structural stages and suites. Indeed tectonofacies within the boundaries of anticlinoria and synclinoria evolve in the main (5 cases out of 7) rather uniformly as well as among the recorded rock groups of the region (North-Balkhash synclinorium, Akchatau anticlinoirum, Abrala synclinorium, Chinghiz anticlinorium, Arkalyk anticlinorium). Their tectonofacial composition is: C from 1 to 5 per cent, B from 45 to 57 per cent, and A from 40 to 53 per cent. At this tectonofacies I-II and X are practically nonexistent (with the exception of Arkalyk anticlinorium), while inside tectonofacies B we observe a tendency of the growth of the role of higher tectonofacies.

The Chunai and Tundyk-Ashchui synclinoria which have not yet been thoroughly explored present an exception from the above picture. The latter, in part, is represented along the profile course by epizone rock (Devonian and Carboniferous suite). Therefore tectonoformation A here came to 100 per cent (including 40 per cent of tectonofacies) which is a rare case for folded systems in general.

5. STRUCTURAL AND FORMATIONAL ZONES (fig. 3, left). As it was noted, there are three structural and formational zones recorded along the course of the Chinghiz profile (two Chinghiz proper: Akratau-Abrala and Chinghiz-Arkalyk, and one border zone - North-Balkhash). And the greater part of the profile course falls on the first two zones which expose mainly the rock of the geosynclinal complex proper, whereas the North-Balkhash zone exposes the rock of the orogenic complex. Despite the striking difference of the zones they are very close in their composition:

Chinghiz-Arkalyk: C-3, B-39, A-58%
Akchatau-Abrala: C-3, B-50, A-47%
North-Balkhash: C-1, B-53, A-46%

The closeness of the tectonofacial composition of different-age suites of the region, structural stages and different-type (to some extent) structural and formational zones show that the dislocation process in the region was either of one type initially (and transitional in time), or is rather young, i.e. of later age (Carboniferous or younger) or it is a combination of the above two "extreme" cases which seems to be most likely to-day. At any rate the fracture (sutural) nature of the folding seems quite evident.

6. STRIKING IN THE WHOLE of the Chinghiz folded system is a low level of the dislocation process which paradoxically contrasts with the typically eugeosynclinal nature of the system, the existence of ophiolitic sutures (that did not go into the profile course because of the narrowness of the ophiolitic sutures). The Chinghiz folded system is of the following composition: C-3, B-44, A-53%. It is natural at so a great length of the profile (140 km) and the diversity of geological situations that the composition of each

tectonofacies proved complete with the maximum of the distribution frequency obviously timed to tectonofacies III (22%).

Thus the Chinghiz eugeosynclinal folded system, strange as it may seem, corresponds in the deformedness extent not to the highest (as we think it should be), but to the lowest and intermediate tectonofacies, with the maximums of the distribution frequency falling on the lowest tectonofacies III.

7. The regularity of faults in the profile is rather stable and monotonous. In general the consistency or the step between faults in a typical case is 1-2, less often varies from 0.5 to 3 km. It is important to note that since the lowest tectonofacies I-II are practically not established in the Chinghiz, it means that individual fracture zones of crumpling join together almost everywhere usually at the level of the lowest tectonofacies due to uniform regularity and contiguity of the faults. Therefore as the first approximation the whole of the Chinghiz range may be conventionally considered as one zone of crumpling with a typical level of disturbance, corresponding to tectonofacies III.

According to the tectonofacial scale tectonofacies III is equal to components of crumpling ε_c and tension ε_a the meaning of which are respectively within the range - (17-29) and + (20-40)%. We shall assume them as equal to the order of -25 and +30%.

If the deformation structure of the Chinghiz had originated as a uniform monofracture zone of crumpling related with a directed displacement of blocks, then its origination would have required steeply horizontally sloped components of displacement A along fault ($A = L.\mathrm{tg}\,\gamma$, where L = 200 km - the width of Chinghiz, $\gamma = 25^\circ$ - angle of the shift, intermediate for tectonofacies III) commensurable with the thickness of the earth's crust under the Chinghiz. In this case we would have probably had a chance to observe the outcrop mantle in one of the fault limbs. But in reality this amplitude proved dispersed into a multitude of faults which had displacements in different directions. If we assume the step between the faults to be in the neighbourhood of 1-2 km and a similar representative width of the elementary fracture zones of crumpling, then the conventional amplitude per one fault will be on the order of hundred metres.

If the deformation structure of the Chinghiz range had originated as a result of simple moldable lateral contraction (in horizontal direction SW - NE), then the magnitude of the lateral contraction (shortened) would have been

$$(\Delta L = L_o - L = \frac{L}{\varepsilon_c - 1} - L$$

where L_o is the initial prefolded width of the Chinghiz eugeosyncline) of the order of 40 km, and the initial width of the Chinghiz zone accordingly - 240 km*).

Those are the problems that can be investigated with greater substantiation than ever before by means of a tectonofacial analysis, using, as a matter of fact very usual initial geological observations. A tectonofacial analysis is not panacea, but a real method of accounting the extent of rock deformedness, which practically found no application in tectonic research until the present day.

REFERENCES

E.I. Patalakha. Genetic Foundations of Morphological Tectonics. Nauka Publishers, Kazakh SSR, 1981, 180 pp.

*) a possible effect of spreading is not taken into account, of course

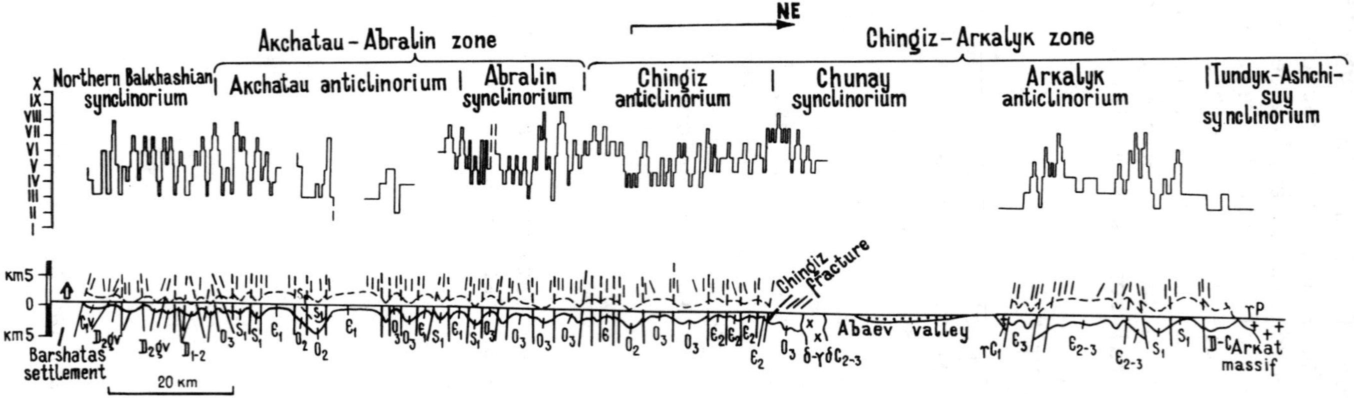

FIGURE 1. Chinghiz Tectonofacial Profile. The shading above the cross section shows the cleavage evolution, while the histogram shows tectonofacies on a ten-point scale.

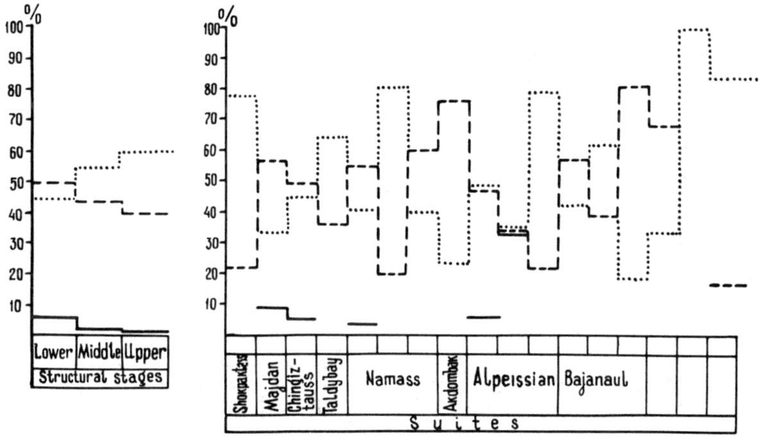

FIGURE 2. A histogram of the tectonofacial composition of suites and structural stages of the Chinghiz folded system. On the axis of ordinates are tectonofacies: A - lowest, B - intermediate, C - highest.

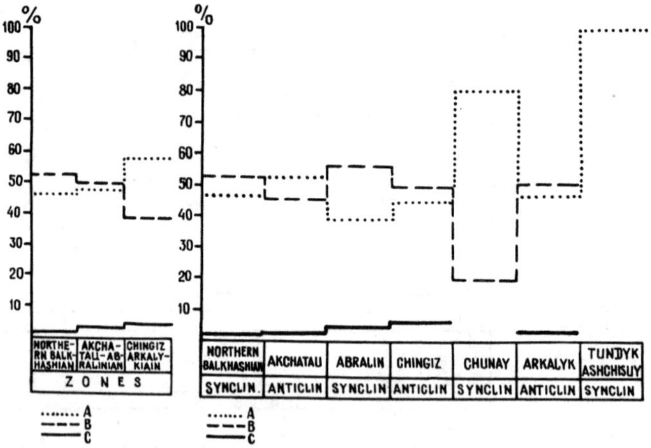

FIGURE 3. A histogram of the tectonofacial composition of anticlinoria, synclinoria and structural and formational zones of the Chinghiz folded system. y-axis is the same as in figure 2.

BRITTLE DEFORMATION WITHIN DUCTILE FAULT ZONES

WHITE, S. H.
Geology Dept., Imperial College, London SW7 2BP, England.

ABSTRACT

Ductile deformation is characterized by the following plastic deformation mechanisms: crystal plastic, mass transfer and grain boundary sliding. Each can be recognized microstructurally. Brittle deformation is by cataclasis which is marked by fracturing and fragmentation. Ductile fault zones are marked by mylonites and brittle zones bu cataclasites. This paper concentrates on cataclasis in ductile fault zones. Four types of behaviour are distinguished: 1) cataclasis of hard mineral clasts in a soft mylonite matrix, 2) strain dependent cataclasis leading to plastic flow mechanisms and subsequent mylonite development, 3) widespread initial cataclasis with concomitant retrogression followed by plastic flow and 4) cyclical cataclastic and plastic flow.

INTRODUCTION

The terms plastic, ductile, brittle and cataclastic, although frequently used in the geological literature are often confusingly used by different authors; the phenomenological being confused with the mechanistic. Strictly speaking the terms plastic, or ductile, and brittle, when referring to the behaviour of a material, describe whether or not a material is capable of undergoing appreciable permanent strains prior to macroscopic fracture (Paterson, 1978). That is, a ductile material can flow whereas a brittle does not. This simple distinction between the two classes of behaviour is largely based upon simple tensile or compression tests in metallurgical rigs designed so that the test abruptly ends if fracture occurs. However, in geological tests, performed under a confining pressure,

Proceedings of the 27th International
Geological Congress, Volume 7, pp. 327–350
TECTONICS

and in nature especially, the specimen does not gain relief from deformation upon fracturing. Murell (1977) has pointed out that under these circumstances flow can occur after fracturing. In nature the onset of large scale/regional fracturing is marked by a sudden stress drop and by seismicity. After this further flow maybe by aseismic creep. Strictly speaking, this flow is ductile. The change is also marked by a change in deformation mechanisms. Flow that does not involve massive fracturing is by crystal plastic (dislocation and twinning) mechanisms, mass transport (diffusion and pressure solution) mechanisms or by grain boundary sliding which may be accompanied by crystal plastic and/or mass transport mechanisms. In nature the first two mechanisms normally operate synchronously. Grain boundary sliding is favoured by small grain sizes and all three mechanisms may operate synchronously in fine grained rock. These mechanisms when responsible for flow will be referred to as plastic mechanisms and the resultant flow as plastic flow in the remainder of this article. Flow after massive fracturing, that is after rock fragmentation, can be solely by frictional controlled sliding and rotation of the fragments. It is then termed "cataclastic flow". This flow mode requires that neither chemical nor mineralogical changes occur due to the ingress of fluids during the dilatancy associated with the fracturing. As shall be seen later, it is more usual for the fluids to cause retrogression which produces weaker minerals that result in a mechanically softer rock and may lead to a change of deformation to plastic flow.

Essentially it is the interchange between plastic and cataclastic flow that is more evident in rock microstructures. Following on from this, a ductile shear zone is one in which crystal plastic and mass transport mechanisms dominate over cataclastic with the opposite occurring in brittle zones. In this paper the relationship between plastic and brittle deformation in ductile fault or shear zones will be treated as that between mechanisms associated with plastic flow on the one hand and cataclasis on the other hand.

RECOGNITION OF DEFORMATION MECHANISMS ASSOCIATED WITH PLASTIC AND CATACLASTIC FLOW

The pertinent microstructural features which can be used to recognize the deformation mechanisms contributing to plastic flow (crystal plastic, mass transport and grain boundary sliding) and to cataclastic flow (fracturing and fragmentation) will be briefly outlined below.

Crystal plastic

The obvious microstructural manifestation of twinning deformation is the development of deformation twins. They are distinguished from growth twins, especially in feldspars, by their needlelike morphology and their tendency to terminate in grains (Vance, 1961). Dislocation mechanisms are characterized by the presence of lattice dislocations in the grains. They are readily seen in a transmission electron microscope but their presence can also be deduced from features seen in optical micrographs. They are undulatory extinction, deformation bands, deformation lamellae, sub-grains and bent morphological features such as twins and exsolution lamellae in grains. At large strains, grains undergoing dislocation flow tend to recrystallize, particularly quartz grains, and aggregates of a given mineral have a marked preferred crystallographic fabric.

Mass transport deformation occurs by the re-arrangement of mass by diffusion through or around grains or by solution. The most common microstructural features are overgrowths, indented grains, accumulation of insoluble materials on the compressive sides of grains and the development of pressure solution stripes. Preferred crystallographic orientations do not develope in grain aggregates.

Grain boundary sliding is favoured by a small grain size with the grains retaining their original shape through large strains. Grains may become slightly elongate in the flow direction. In shear zones grains tend to align with their grain boundaries parallel to the flow direction i.e. the foliation. It is not easy to unequivocally demonstrate that grain boundary sliding has occurred as often the microstructural features of the accommodating mechanism are more obvious.

Cataclastic

Cataclastic mechanisms are characterized by the presence of fractures on all scales. These are normally infilled. Often the clasts have an irregular and angular shape although this need not always apply.

Methods used

Micro-graphs used to illustrate salient points in this contribution have been taken with an optical microscope, scanning electron microscope (SEM) using a back-scattered detector and a transmission electron microscope (TEM). The TEM was a high voltage instrument operating at 1000 KV. Mineral identification in SEM was done with energy dispersive or wave length dispersive spectrometers.

FAULT ROCKS FORMED DURING PLASTIC AND CATACLASTIC DEFORMATION

Two fault rock sequences exist, the foliated, mylonites, and the non-foliated, cataclasites (Sibson, 1977; White, 1982; White et al., 1982). There has been a recent tendency to relate mylonites solely to plastic deformation mechanisms (Sibson, 1977; White, 1976) with any cataclasis being of minor importance. It followed from this approach that cataclasis dominated in cataclasites. The macroscopic difference between the two was the development of a foliation in a mylonite and the lack of such in a cataclasite. The fault movement responsible for each were related to depth, with the cataclasis occurring at temperatures roughly below 300°C and mylonites above this temperature. The rationale for this division was based on the temperature for the brittle ductile transition during the experimental deformation of granite. Little attention had been given, until recently, to the temperature zone bordering this transition. Observation that gouges in fault zones can have a foliation and that foliated rocks can easily arise from non foliated cataclasites (Rutter & White, 1979) led White (1982) to consider the current use of the term "mylonite" and "cataclasite". So far as a mylonite was concerned, he recommended adherence tp Lapworth's (1885) original definition viz. that a mylonite is a fined grained schist that forms along a fault or shear zone and that a cataclasite is an aphanitic, structureless rock having a similar structural position as was first described by Grubemann & Niggli (1924). If these terms are used in the above sense, little or no emphasis is placed on the deformation mechanisms associated with the development of each. The role of cataclasis is emphasized by the term "cataclasite" but it is not excluded as a process in mylonotization.

At this stage it should also be remembered that, under experimental conditions, a given rock exhibits the ductile/brittle transition over a limited temperature interval which is largely pressure dependent (Paterson, 1978). However in nature where the effect of dilatancy associated with this transition is to suck water, the consequent mineralogical retrogression may dramatically lower the transition in the fault rocks, producing subsequent deformation by dominant crystal plastic and mass transport deformation mechanisms. Therefore what we can ascertain in microstructural studies of fault and shear zones is not so much the ductile/btittle transition but the inter-relationship between cataclastic and plastic flow mechanisms. Therefore the topic covered by this contribution can be further refined and it is effectively about cataclasis in mylonites zones.

CATACLASIS IN MYLONITE ZONES

Four areas can be delineated:

1) cataclasis of hard mineral clasts in a soft mylonite matrix.

2) Strain dependent cataclasis leading to the onset of plastic flow mechanisms and the subsequent development of a mylonite.

3) Initial widespread cataclasis after low strains leading to extensive retrogression and subsequent plastic flow.
4) Cyclical cataclastic and plastic flow.
Each of the above will be described separately.

Cataclasis of hard mineral clasts

It is common to find fractured or pulled apart clasts of hard minerals in mylonites which form a mechanically softer matrix. A common example is feldspar clasts in a quartz mylonite. Although the pulling apart of hard clasts is frequently observed, it is a subject that has received little attention. White et al. (1980) noted that this behaviour is similar to that seen in fibre reinforced composites and about which much had been written in the materials science literature. They described the phenomenon by the same term as used in materials science viz. "fibre loading". This has recently been taken up and extended by Lloyd et al. (1982). In this process there is a transfer of the stress from the matrix to the clast. The matrix being softer undergoes a greater displacement than the clast and shear strains form between the clast and matrix parallel to the long axis of the clast which is assumed to be parallel to the flow. The applied load (tectonic stress) becomes distributed between the clast and matrix according to the following equation (Kelly, 1973):

$$\sigma = E_f \, \varepsilon \, \frac{1 - \cosh B(L/2-x)}{\cosh BL/2} \qquad (1)$$

where

$$B = \frac{2 \pi Gm}{E_f A_f} \ln (Ro/r) \qquad (2)$$

and E_f, Young's modulus of the fibre

ε, strain in the matrix removed from the clast
L, the clast length
x, the distance along the clast from its end
Gm, shear modulus of the matrix
A_f, cross section area of the clast
Ro, the mean distance between adjacent clasts, and

r, the clast radius.

The above equations show that the tensile stress in the clast is a maximum at its centre and consequently a pull apart fracture will preferntially develop at that position. That is elongated clast will first break into halves and each half will in turn be halved as the strain increases until a constant length is developed. The spacing between the fractured segments of the clast will be greatest for the position of the initial halving and sequentially decrease along the axis of the clast i.e. parallel to the foliation. This is the characteristic microstructural feature of the fibre loading phenomenon.

Two examples of this are shown in Fig. 1. Both are taken from mylonitized pegmatites in the late Hercynian shear zones at Cap de Creus, N.E. Spain (see White at al., 1980). The pegmatites contain large tourmalines (Fig. 1a) which progressively break up with increasing shear strain. Both of the dominant matrix minerals, quartz and plagioclase, deformed by crystal plastic mechanisms. They undergo grain refinement mainly by recrystallization with minor neomineralization to form white micas in the case of the plagioclase. On the other hand, the tourmalines grain refine by progressive fibre loading (compare Fig. 1a & b). They become sliced into slivers with a height to length ratio much greater than 1. The slivers then rotate and also undergo fibre loading fracturing to form small equidimensional clasts such as shown in Fig. 1b. Garnets in the pegmatites behave a little differently. Because of internal fractures in the original clasts they tend to break into equidimensional clasts without going through a sliver stage. Characteristically, the end positions at which the initial tensile forces were a minimum remain intact (see Fig. 1c).

The ultimate size of the clasts of a given hard mineral is, according to equation 1, stress dependent. Thus the technique, at a first glance, offers a means for estimating the flow stress in ductile fault zones. However equation 1 is based on catastrophic failure theory. The termination of fractures in the tourmalines, together with mineral infills suggests a form of stress

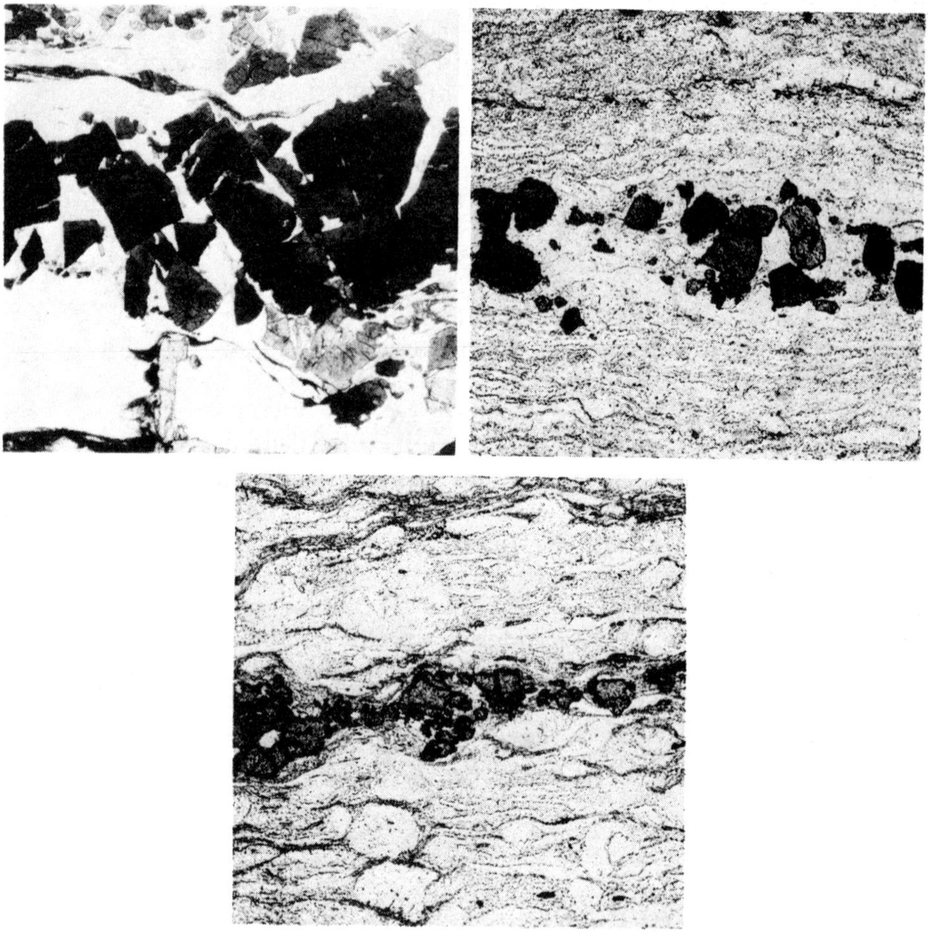

FIGURE 1. Fractured hard grains in a mylonitised pegmatite, Cap de Creus, S.E. Spain (plane polarised light).
a. Fractures in a tourmaline at low shear strains. Note that some fractures terminate within the tourmaline indicating stress corrosion form of fracturing (Mag 8x).
b. The fractured and pulled apart tourmalines at high shear strains. (Mag 20x).
c. Fractured and pulled apart garnets at high shear strains. (Mag 20x).

corrosion fracture, (Kerrich et al., 1980), that is fracture and stresses below the theoretical fracture stress of the clast (see Atkinson, 1982). The theoretical fracture stress (σ_{th}) is related to the Youngs modulus of the clast, Ec, by the following equation:

$$\sigma_{th} = \frac{Ec}{10}$$

The fibre loading theory assumes fracturing at σ_{th}.

Strain dependent cataclasis

This behaviour is illustrated by the progressive conversion of a Lewisian granite gneiss to a quartz/mica mylonite with increasing shear strain adjacent to the Arnaboll Thrust which is within the Moine Thrust Zone at Ben Arnaboll, Eriboll, N.W. Scotland. Massive gneiss crops out to within 1.5 m of the trust plane and becomes mylonitised over the next 1 m to form a narrow band of well foliated mylonite adjacent to the trust plane. The sequential development of the mylonite provides an insight into the role of both ductile and brittle processes during the mylonitisation of a granite. The sequence is illustrated in Fig. 2 with specific features illustrated in Fig. 3.

At low strains, the grain refinement processes is dominated by granular and multigranular fracturing of the plagioclase and alkali feldspars and the plastic deformation of the quartz (Fig. 2). The latter mineral tends to blunt transgranular fractures in the feldspars. Multi granular fractures only occur in areas in which feldspar grains are in contact. Some crystal plastic deformation has occurred within feldspars prior to the fracturing. Deformation twins, bent twins and undulatory extinction are common (Figs. 2a & 3a). Quartz grains exhibit well developed sub-grain structures and at this stage grain margin recrystallization is common. The quartz grains also have a pronounced c-axis fabric (White et al., 1982). At low strains, the feldspars show only slight signs of retrogression which is most noticeable in multi granular fractures. The products of retrogression, as seen in fractures, are white mica, chlorite, biotite, epidote, albite

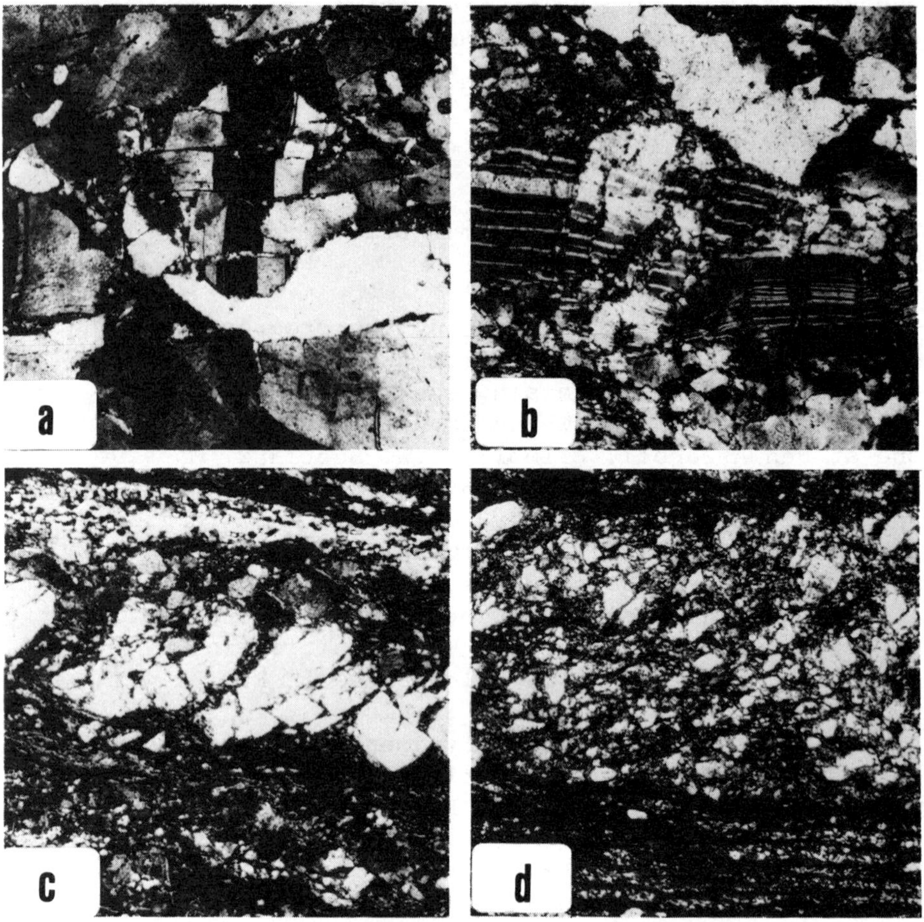

FIGURE 2. Strain dependent cataclasis leading to a plastic mylonite. Lewisian gneiss, Ben Arnaboll, N.W. Scotland. The sequence a-f is one of increasing shear strain. (Mag 20x; plx)
a. Fractured plagioclase and alkali feldspar grains.
b. A weak foliation is formed by a band of crystallized quartz.
c. Strung out fractured feldspar grains in a banded matrix. This rock is a mylonite.
d. Band of small feldspar fragments. Note the progressive reduction in fragment size from a to d.

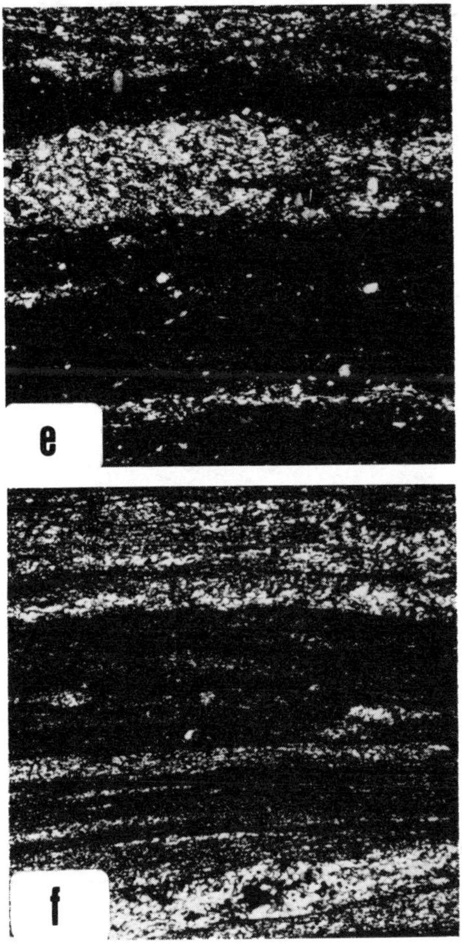

FIGURE 2 (continued). Strain dependent cataclasis.
e. At higher strains, the feldspar fragments have disappeared and are marked by the dark bands of retrogressed minerals. The light bands are of recrystallized quartz.
f. The high strain mylonite which consists of bands of quartz alternating with mica/chlorite rich bands.

and quartz. The effect of the retrogression is to seal up fractures and strengthen the rock. Permeability would decrease and the retrogression cease. However, the effect would also be to increase the pore pressure which in turn would lead to further cataclasis, a consequent additional phase of retrogression and so on. The first foliation to develop is marked by bands of recrystallized quartz in quartz rich domains and bands of intensily fractured and retrogressed feldspars in feldspar rich domains (Fig. 2b). These are equivalent of the C-planes described by Berthe et al. (1980) and are parallel to the thrust plane and eventual mylonite foliation.

The microstructural effect of an increase in shear strain is to progressively decrease the size of the clasts by tensile and shear fracturing (Fig. 2c) and to draw out the clasts into bands containing fractured fragments in a matrix of the products of retrogression. These bands alternate with quartz rich bands (Fig. 2d). At this stage the clasts are within a soft matrix and they behave as described in the foregoing section. Eventually all clasts disappear through the combined effects of fracturing and retrogression. Even the very smallest of clasts fracture (Fig. 3b & c). The rock now consists of alternating bands of quartz and the products of retrogression (Fig. 2e). As the strain increases further the latter lose albite and much of the epidote. The former feldspar rich zones are now represented by mica/chlorite rich quartz zones containing small amounts of epidote (Figs. 2f & 3d). All deformation is dominantly by plastic processes. However, some veins cut across the foliation and in cases can be seen to be sheared across the mica rich zones. This indicates that there are periods of fracturing but now the whole rock is suffering. The preferential shearing along the mica rich bands also demonstrates that the former hard feldspar bands have become even weaker than the quartz bands.

The overall effects of the retrogression of the feldspars, have been to effectively lower the temperature of the ductile brittle transition of the rock during the course of the shearing. The dominant mechanism

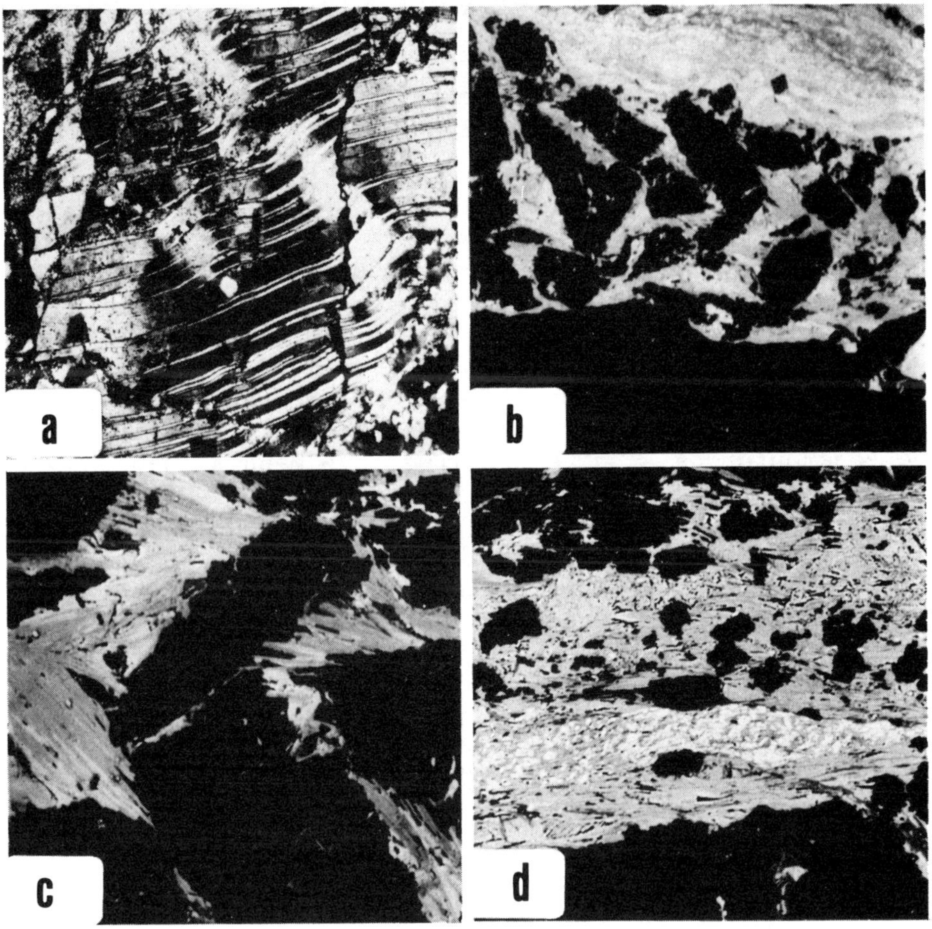

FIGURE 3. Details of the Lewisian samples (a, optical micrograph and b-d are back-scattered SEM).
a. Bent twins, undulatory extinction and fractures in a plagioclase clast. (40x).
b. Pulled apart feldspars from the dark band in Fig. 2e. Dark areas are feldspar, the light epidote and intermediate sericite. (200x).
c. Growth of oriented sericite (intermediate toning) and chlorite (lightest toning) between small feldspar fragments (600x).
d. Thin band of white mica (light) in a quartz matrix (dark) from Fig. 2f. (1000x).

at low strains was cataclasis changing to ductile processes at high shear strains. If the earlier, low strain microstructures had not been preserved, the initial dominant role of cataclasis would not be appreciated.

Initial widespread cataclasis

The Outher Hebrides Thrust in North and South Uist exemplify this type of behaviour. The structure of the thrust in this area has been described by Sibson (1977). Although several individual fault zones can be recognised, the zone is an area of large scale cataclasis and retrogression. The individual faults marked by intense cataclasis occur on the west and ones with phyllonites occur on the eastern margin of the zone outcrops. The phyllonites have been argued to represent deeper levels of the thrust zone (Sibson, 1977). However, a constant mineralogy on the cataclasite and phyllonites (quartz, albite, epidote, chlorite and white mica) argue against a depth variation. It is the theme of this section, that the phyllonites represent zones of higher strain and mark the localization of ductile deformation after cataclasis.

The fault zone occurs within Lewisian granite gneisses. The basal faults i.e. the most westerlt are marked by dark zone of cataclasites. Very little can be made of the microstructure of these zones even at their edges (Fig. 4a). Fractured fragments of quartz can be readily seen. These contain undulatory extinction and deformation bands attesting to some crystal plastic deformation. The fractured feldspars are almost totally replaced by retrogression. Occasional partially retrogressed cores can be seen but it is emphasized that these are rare (Fig. 4a). Relict twins can also be seen.

In the back scattered SEM the extent of the fracturing is revealed (Fig. 5a & b). The grains are a maze of microfractures infilled mainly with epidote but also with albite and quartz. Larger microfractures are also infilled with epidote. The fracture arrays indicate catastrophic brecciation. The back scattered micrographs show that the feldspars between the micro-

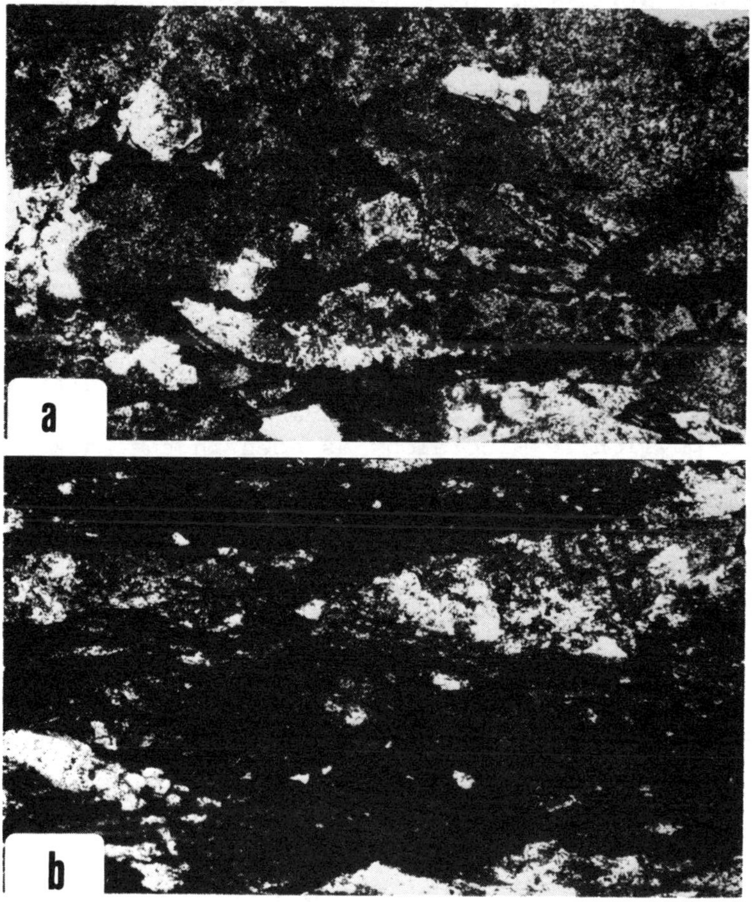

FIGURE 4. Microstructure of a cataclased granite gneiss and of a phyllonite. Outer Hebrides Thrust Zone. Uist. (ppl).
a. Cataclased gneiss. Sample consists largely of retrogressed feldspar. Relict twins arrowed. Light areas are quartz (20x).
b. Phyllonite. The light grains are quartz, the dark bands are fine grains of retrogressed minerals and the intermediate tonings are clasts of retrogressed feldspar. (20x).

fractures are extensively altered to sericite and that some feldspar still remains. The latter could not be ascertained optically. The sericite has a arndom texture indicating that no deformation followed cataclasis. The lack of disaggregation (fig. 5a & b) also suggests that little or no flow occurred after the cataclasis.

In the phyllonite zones, the products of retrogression show a dimensional alignment which marks the macroscopic foliation (Fig. 4b). The quartz grains are now elongated and have undergone grain refinement by recrystallization. There is no evidence for any random growth of sericites in the back scattered micrographs (Fig. 5d). These reveal that the epidotes have acted as hard particles and have been pulled apart. The phyllonite mineralogy as seen in the back scattered micrographs is basically similar to that in the cataclasites. There is less albite but more carbonate and pyrite.

The interpretation of the microstructures is that the initial stages of faulting were marked by widespread cataclasis and retrogression to form a mechanically soft fault rock. Subsequent deformation concentrated within the zones marked by the phyllonites and was by plastic mechanisms.

Cyclical cataclasis and plastic deformation

The Alpine Fault in the South Island of New Zealand is a transform fault marking the contact between the Australian and Pacific plates. It is a seismically active, dextral oblique reverse fault with rapid uplift on the S-E side giving rise to the Southern Alps (see Wallcott & Gresswell, 1979). The fault dips 50° S-E and is marked by a 1 km wide zone of fault rocks. The idealized fault rock sequence across the zone from footwall to hanging wall is granite - - cataclasite - green and augen mylonites - schistose mylonites - curly schists (coarser grained mylonites) and garnet oligoclase grade Alpine schists. Sibson et al. (1979) argued that the above sequence represents deformation at increasing depths from near surface conditions on the footwall side to amphibole grade

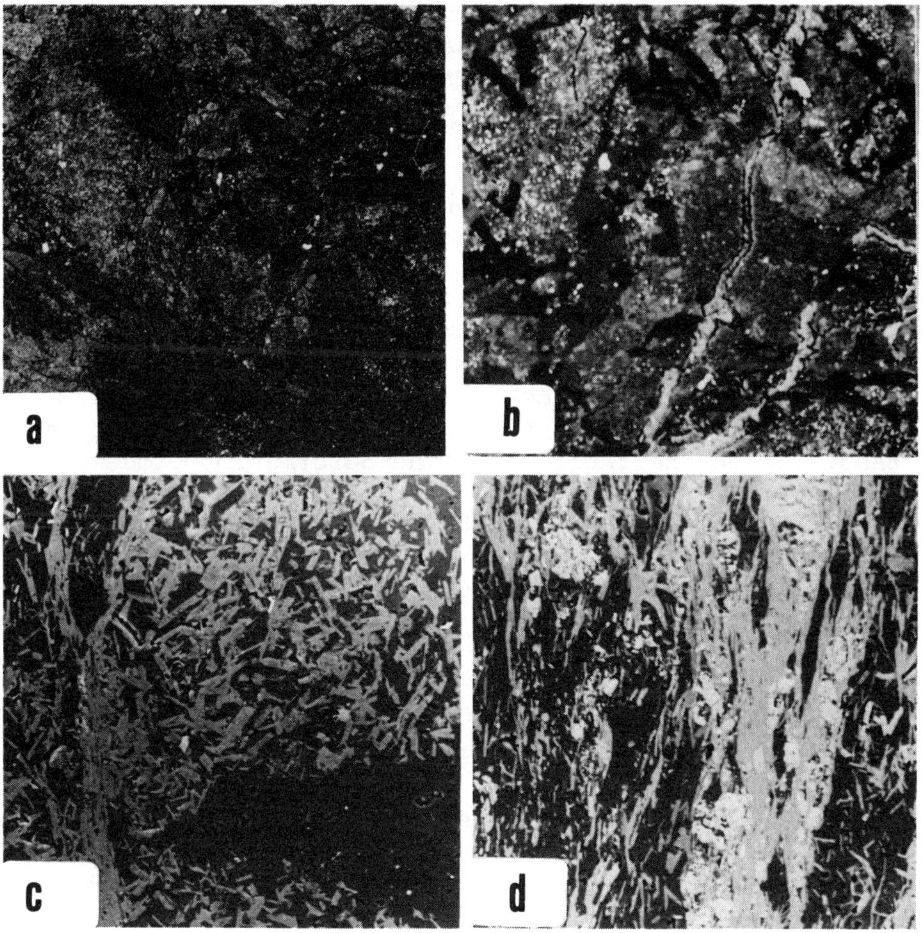

FIGURE 5. Back scattered electron micrographs of samples from the Outer Hebrides Thrust.
a. A low magnification view of a feldspar area in Fig. 4a. (60x).
b. Microfractures (vein pattern) in a cataclased feldspar grain. The dark veins are quartz the light are epidote and chlorite. (200x).
c. A sericitised cataclased alkali feldspar. (400x).
d. Microstructure of a phyllonite. The grains are quartz (dark), mica and chlorite (intermediate) and epidote (lightest toning). (400x).

conditions on the hanging wall.

The curly schists and schistose mylonites can be related to the garnet oligoclase schists and show no retrogression and no fracturing. The parentage of the green and augen mylonites is more controversial being derived both from the schists and granites. They undergone retrogression, having a mineralogy of albite, mica, chlorite, epidote and quartz and which is equivalent to low grade chlorite metamorphic facies. That is, there is a sudden decrease in grade. Calculations of temperature profiles across the fault zone indicate the likely deformation temperatures for this rock (300°C) occurs above 10 km, the depth at which microseismic activity is concentrated (Rynn & Scholtz, 1978) and which is the likely position of the ductile brittle transition in the fault zone. The deformation processes that have been operative in the green and augen mylonites has been dealt with in detail recently by White & White (1983).

Optical microscopy gives the superficial view that plastic deformation processes have been dominant. Guartz grains are elongated, in cases ribboned, and have a well developed c-axis fabric (White & White, 1983). The grains have well developed sub-grains and the quartz grain refines by recrystallization. Microcracks along with calcite has infiltrated occur around the quartz grain boundaries and across some grains.

Feldspar clasts exhibit a marked degree of fracturing; shear and pull apart fractures are common (Fig. 6a). The depth of crack penetration into the clasts is variable suggesting slow crack propagation rather than instantaneous brecciation. The small matrix albites and epidotes are also fractured. Electron microscopy revealed that fracturing was even more widespread. Healed and unhealed fractures were common in both quartz clasts and matrix (Fig. 7a). Feldspars, again both clasts and matrix grains, were commonly fragmented with sericite forming along the fractures (Fig. 7b). Transmission electron micrographs revealed that quartz and feldspars had prominent dislocation sub-structures and that intergranular fracturing was common (Fig. 7c). The backscattered micrographs

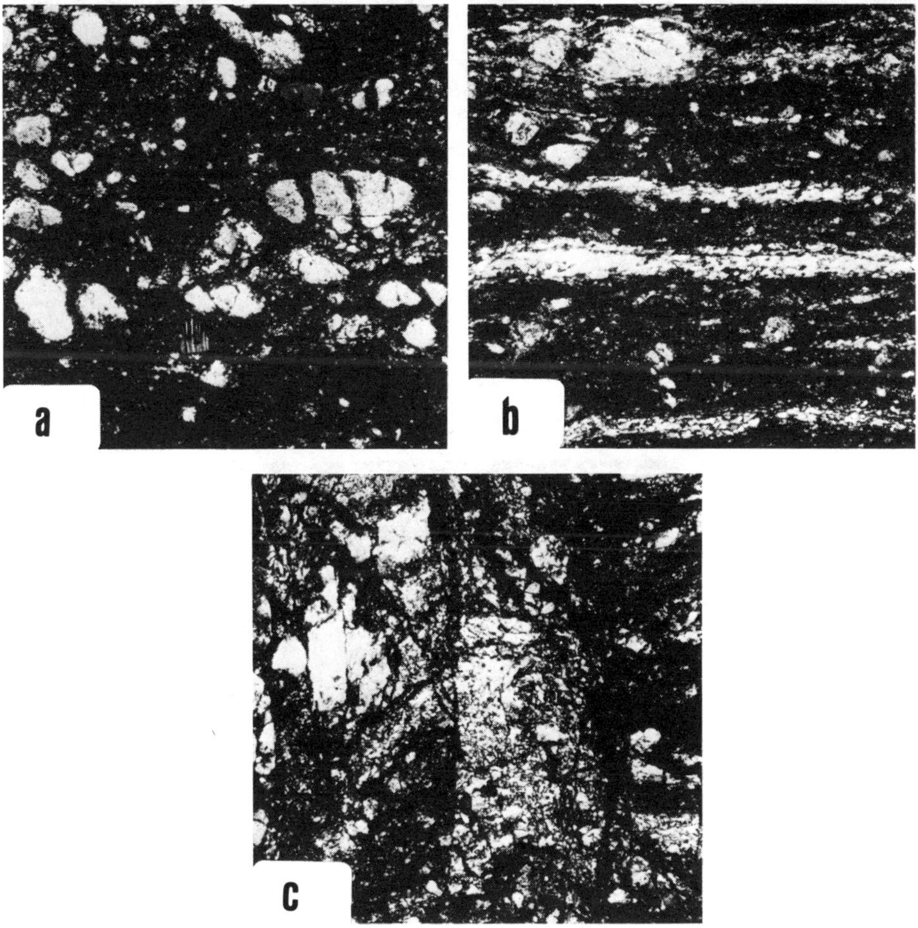

FIGURE 6. Green mylonites and cataclasite, Saddle Creek, Alpine Fault Zone, South Island, New Zealand. Crossed polarisers. (20x; plx).
a. Mylonite consists of pulled apart clasts of feld-spar (mainly albite) and quartz clasts in a matrix of epidote, white mica, chlorite, quartz and albite.
b. Quartz rich mylonite of similar overall mineralogy to the above.
c. Cataclasite. Typical microstructure of a cataclased green mylonite.

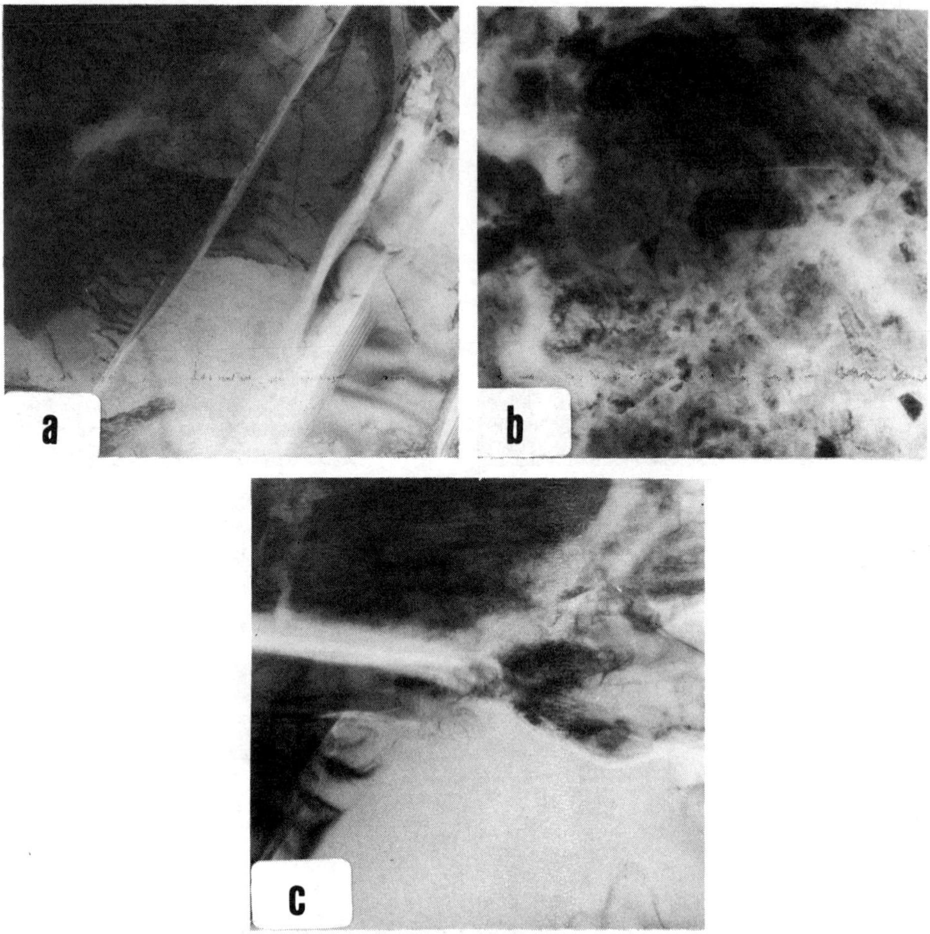

FIGURE 7. Transmission electron micrographs of the green mylonites.
a. Fractures in a quartz clast. Dislocations and sub-grains can be seen within the fragments (10,000x).
b. Fragmented plagioclase clast. Alteration has occurred in the light areas. (20,000x).
c. Fractures in white mica which lies between a quartz (bottom) and a plagioclase (top) grain. (20,000x).

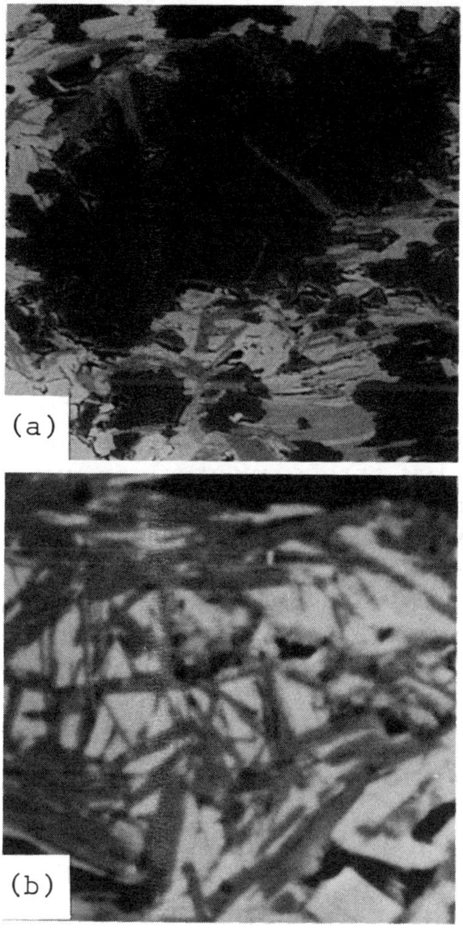

FIGURE 8. Backscattered electron micrographs showing, in detail, features in the matrix of a green mylonite.
a. A small albite clast with an irregular rim of oligoclase. Note that the rim is growing into the matrix which is made up of quartz, muscovite and chlorite (200x).
b. Randomly orientated muscovite grains growing across the mylonite foliation which is horizontal in the micrograph (1000x).

(Fig. 8a & b) revealed zones in which there was random growth of sericite grains across foliation planes. Most matrix albite grains had overgrowths of oligoclase which in cases grew across the foliation.

It can be seen from the above that, in the feldspars especially, microstructural evidence exists in the green mylonites for deformation by dislocation and pressure solution mechanisms together with cataclasis. Besides this, "post tectonic" grain growth occurred in areas. This has been interpreted as resulting from cyclical plastic deformation, cataclastic deformation and annealing during the seismic cycle. Plastic deformation occurs during stress build up leading to rupture after which there is cataclasis, dilatation and stress relief. Fluid is sucked into the cataclased zone during this low stress interval to give a phase of mineral growth which seals up the rock and leads to strength recovery and a gradual stress build up during which plastic deformation is again initiated.

The cataclasites which mark deformation at a higher crustal level (Sibson et al., 1979) often consist of brecciated green mylonites (Fig. 6b). That is, at lower temperatures, cataclasis is dominant.

CONCLUSION

The inter-relationship between cataclasis and plastic flow in shear zones and ductile faults is variable. It is strongly dependent on the extent of retrogression. The simplest situation is the cataclasis of hard clasts in a plastically flowing matrix. Here cataclasis is limited and of little importance. In other situations where cataclasis is extensive the inter-relationships with plastic flow is complex and dependent upon the degree of retrogressive neomineralization. The neomineralization is associated with fluid ingress following the dilatancy which accompanies cataclasis. The effect of this is to produce a mechanically weaker mineral assemblage and to effectively lower the ductile-brittle transition in the fault rock. This can lead to the onset of plastic flow which in cases masks evidence for earlier cataclasis.

Acknowledgements

This paper was written whilst the author was on sabbatical at Melbourne University. The hospitality of staff and students there is acknowledged. Mr R. Giddens, Imperial College, assisted with electron microscopy and Mrs M. Baker produced a typescript from a collection of scrawled pages. NERC is thanked for financial support through grant GR3/3848.

REFERENCES

Allis, R. G., Henley, R. W. and Carman, A. F. (1979). The thermal regime beneath the Southern Alps. Bull. Roy. Soc. N.Z. 18, 79-85.

Atkinson, B. K. (1982). Subcritical crack propagation in rocks. Theory, experimental results and applications. J. Struct. Geol. 4, 41-56.

Berthe, D., Choukroune, P. and Jegouzo, P. (1979). Orthogneiss, mylonite and non coaxial deformation of granites: the example of the Southern Armoucan Shear Zone. J. Struct. Geol. 1, 31-42.

Grubermann, U. and Niggli, P. (1924). Die Gesteinsmetamorphose. Berlin.

Kelly, A. (1973). Strong Solids. Clarendon Press, Oxford.

Kerrich, R., Allison, I. Barnett, R. L., Moss, S. and Starkey, J. (1980). Microstructural and chemical transformation accompanying deformation of a granite in a shear zone at Mièville, Switzerland: with implications for stress coorosion cracking and superplastic flow. Con. Min. Pet. 73, 221-242.

Lapworth, C. (1885). The Highland controversy in British geology: its causes, course and consequences. Nature Lond. 32, 558-559.

Lloyd, G. E., Ferguson, C. C. and Reading, K. (1982). A stress transfer model for the development of extension fracture boudinage. J. Struct. Geol. 4, 355-372.

Paterson, M. S. (1978). Experimental Rock Deformation: The Brittle Field. Springer-Verlag, Berlin-Heidelberg.

Rutter, E. H. and White, S. H. (1979). Microstructure and theology of fault gouges produced experimentally under wet and dry conditions. Bull. Miner. 102, 101-109.

Rynn, J. M. W. and Scholz, C. H. (1978). Seismotectonics of the Arthur's Pass region. South Island, New Zealand. Bull geol. Soc. Am. 89, 1373-1388.

Sibson, R. H. (1977). Fault rocks and fault mechanisms. J. Geol. Soc. Lond. 133, 191-213.

Sibson, R. H. White, S. H. and Atkinson, B. K. (1979). Fault rock distribution and structure in the Alpine Fault Zone. Bull. Roy. Soc. N.Z. 18, 55-65.

Vance, J. A. (1961). Polysynthetic twinning in plagioclase. Am. Miner. 46, 1097-1119.

Walcott, R. I. and Cresswell, M. M. (1979). (editors). The origin of the Southern Alps. Bull. Roy. Soc. N.Z. 18, 1-147.

White, J. C. and White, S. H. (1983). Semi ductile deformation in the Alpine Fault Zone N.Z. J. Struct. Geol. 5, 579-590.

White, S. H. (1976). The effects of strain on the microstructures, fabrics and deformation mechanisms in quartzites. Phil. Trans. Roy. Soc. Lond. A283, 69-86.

White, S. H., Burrows, S. E., Carreras, J., Shaw, N. D., and Humphreys, F. J. (1980). On mylonites in ductile shear zones. J. Struct. geol. 2, 175-189.

White, S. H., Evans, D. J. and Zhong, D. L. (1982). Fault rocks of the Moine Thrust Zone: Microstructures and Textures of selected mylonites. Textures and Microstructures. 5, 33-61.

White, S. (1982). Fault rocks of the Moine Thrust Zone: a guide to their nomenclatures. Ibid, 4, 211-221.

STRAIN VARIATION AND DEFORMATION FACIES

ZAGORCEV, I.S.
Geol. Inst., Bulg. Academy of Sciences, Sofia, Bulgaria

QUANTITATIVE STUDIES: METHODS AND TECHNIQUES

Quantitative studies on rock deformations (strain analysis) are based on fundamental laws and relationships of continuum mechanics. Rock deformations are generally heterogeneous but deformation within a restricted domain can often be approximated to homogeneous. According to the time interval between the initial (pre-deformation) and the deformed state, a rock deformation can be referred to as incremental, phasal (corresponding to a given phase or stage) or total. Each of them can be thought of as finite except for some very small incremental strains. A quantitative characterization of the homogeneous deformation can be achieved by establishing the deformation matrix (Means, 1976; Flinn, 1979). A more practical way is to distinguish between the components described as distortion, rotation, translation and dilatation on the basis of independent observations on some structural featurs (indicators).

The distortional component (sometimes not separated from the dilatational one) is most easily determined. It is given by the strain ellipsoid. The first strain analysis of that kind was performed in the midst of the last century, and the modern development followed Flinn (1962) and Ramsay (1967). Most techniques result in finding the ratios

$$a = \frac{1+e_1}{1+e_2} \quad \text{and} \quad b = \frac{1+e_2}{1+e_3}$$

from which the stretches (elongations along the principal axis) can be computed in case of no dilatation. The results are plotted either on a Flinn's plot in logarithmic version (lna vs. lnb) or on a plot containing all three stretches (nomograms - Patalakha, 1962, 1967; Lukyanov, 1980; Nadai's plot

Proceedings of the 27th International
Geological Congress, Volume 7, pp. 351–360
TECTONICS

- Owens, 1974). Ellipsoid shapes are compared on the basis of the constant

$$k = \frac{\ln a}{\ln b} .$$

The natural octahedral shear

$$E_s = (\tfrac{2}{3})^{\frac{1}{2}} \left[(\ln a)^2 + (\ln b)^2 + \ln a . \ln b\right]^{\frac{1}{2}}$$

is not a simple function of k;

$$E_s = \ln b \left[\frac{2}{3} \quad (k^2+k+1)\right]^{\frac{1}{2}} = \frac{\ln a}{k} \left[\frac{2}{3} \cdot (k^2+k+1)\right]^{\frac{1}{2}}$$

and therefore is a good measure for the strain intensity. In the case of simple and coaxial deformation path it expresses the work necessary for deforming the initial sphere into the ellipsoid (Flinn, 1978, after Nadai, 1950). The classification boundaries may be established at the value of E_s of 0.5 (lower cleavage limit), 1.0, 1.5, and 2.0 (upper cleavage limit), and of k: near 1.2, 0.8, 0.4, which correspond to the boundaries between extremely constricted, constricted, "normal" (plane strain), moderately flattened and extremely flattened ellipsoids.

A considerable number of strain (distortion) indicators have been studied since the first systematic grouping by Ramsay (1967). Most of them deform inhomogeneously with their matrix thus yielding smaller (in case of greater competence) strain ratios. Direct measurements on strain indicators in three dimensions rarely give sufficient grounds for unambiguous strain determination (Patalakha, 1962; Ramsay, 1967). Most reliable data are obtained in determining two-dimensional distortion on several planes and computing from them the three-dimensional state (Ramsay, 1967; Oertel, 1978; Milton, 1980). The two-dimensional distortion is determined through a number of different techniques in accordance to the indicators:

- initially subspherical indicators: harmonic mean (Lisle, 1977);
- initially ellipsoidal markers: most reliable are the R_f/ϕ method (Ramsay, 1967; Dunnet, 1969) and the algebraic method of Shimamoto & Ikeda (1976); the

method of Dunnet & Siddans (1971) as well as the algebraic method of Matthews et al. (1974) take account for the pre-deformational preferred orientation;

- any pre-deformational shape: Robin (1977);
- known symmetry or angular markers of shear (strain rosettes): Ramsay (1967); Sanderson (1979) etc.
- linear randomly oriented indicators: Lisle (1974); Lukyanov (1980) etc.

The confidence limits for the measured and computed strain ratios are variable. Strain ratios less than 2.5 - 2.0 are subject to larger errors due to variations of the initial shape and density distribution. For ratios less than ca. 10.0 errors are estimated at about 10%, and for ratios greater than 10.0 the errors become considerable because of needed extreme accuracy in measuring angles (less than 1° in some cases) which is geologically impossible.

The dislational component can be isolated from the distortional one in several specific cases (Roberts & Siddans, 1971; Barr & Coward, 1974). In other cases the geologically sound procedure is to compare the behaviour of a given rock type containing a specific indicator, in different deformation environments: from non-deformed to highly deformed rocks. It is also preferable to compare the behaviour of several different indicators which possess different sensibility to dilatation. Some indicators (reduction spot, oolites) show different sensibility to compactional dilatation as a function of lithology, time of formation, etc. Thus for reduction spots Graham (1978) reported compactional ratios as large as 1.7; recent studies in boreholes within the Moesian platform (North Bulgaria) on non-deformed reduction spots from Lower Triassic red shales, siltstones and sandstones showed axial ratios not greater than 1.1 - 1.2 down to 3776 m.

The rotational component is very important but is evaluation is enhanced by a number of factors. A number of viable techniques are applicable to heterogeneous simple shear zones (Ramsay & Graham, 1970; Patalakha, 1970; Ramsay, 1967, 1980). The amount of rotation may be also estimated using specific indicators as syntectonic garnets, albites etc. (Powell & Treagus, 1970; Ghosh & Ramberg, 1976; Lukjanov, 1980, etc.), boudins, some inclusions, fossils, etc. It should be noted that rotation of some inhomogeneously deformed markers may also happen during irrotational deformation of the corresponding domain; this fact imposes additional caution and restriction to interpretations.

SCALE AND STRAIN VARIATION

A practical problem of some importance is the observation scale and the size of the deformation domains. Strain in different fold mechanism models has been largely discussed (e.g. Ramsay, 1967; Patalakha, 1967, 1981 ; Ramberg, 1961, 1963, 1976, etc.) and apparently strain may vary considerably within a very small area (minor centimetric folds) each deformation domain being restricted to a single or a few millimetric indicators. On the other hand, deformation may vary insignificantly over comparatively large areas (e.g. Wood & Oertel, 1980) in the kilometric scale. In most frequent cases strain variation is recorded over distances of several tens of meters (e.g. Zagorcev et al., 1976; Patalakha, 1981) and is dependent on decametric to hectometric folds or onto large thrust sheets with almost parallel lying rock sequences.

Consequently the strain variation measured in small almost homogeneous domains is strongly dependent on the variation of heterogeneous deformation (expressed in fold and thrust development) over large regions. The strain variation (the "internal" rock deformation) should be compared to the "visible" mes- to macrodeformation expressed in minor folds, boudinage etc., and to the "megadeformation" expressed in large folds, thrusts and nappes.

In the transition from mesoscopic to macro- and megascopic scale, the attention is drawn to the importance of cohesive and non-cohesive (Means, 1976) geologic boundaries. Regions of low gradient strain variation with transitional boundaries between deformational domains are sharply cut by domains of intensive shear and/or translation. The latter are very narrow (faults, thrusts) in the upper parts of the crust (brittle behaviour), or sufficiently large zones of heterogeneous shear with transition boundaries in the lower parts (ductile behaviour) (Ramsay, 1967, 1980; Patalakha, 1970). Another type of such boundaries separate lenticular bodies (including cores of isoclynal folds and inliers) with high competence contrast. They are characterized by very high strain concentration but often without visible angular discordance. Superimposed or coeval metamorphic parageneses often obliterate the discordant character. Such boundaries are often described as slides or intreaformational glides, and they may transform a tectonic unit into a pseudostratigraphic unit (personal communications by A.V. Lukjanov, Yu. Morozov, S. Kuprin and others; publications by Hutton, Zagorcev and others).

DEFORMATION FACIES

The notion of facies is widely used in geological science and is based on a complex of descriptive features related to genetic particularities. In structural geology the deformation facies should describe particular structural parageneses (assemblages) corresponding to a given type of deformation (deformation matrix and deformation path). A similar idea has been introduced by Hansen (1971) as strain facies: "Physical properties of a structural assemblage express the conditions of its formation". However, the practical application of the concept is not yet well developed, and the three type facies described in the Trollheimen region refers to fluid dynamics considerations about stream line orientation rather than to real strain distribution and history.

The full description of a deformed facies should take account for the poly- or monocyclic development of the structures present as well as for poly- or monophasal history of each cycle. Monocyclic deformation is usually developed in two or three major phases, the interphasal time being characterized by quantitatively insignificant deformations. The quantitative approach to each phase is most frequently hindered by the lack of specific indicators: the indicators present usually allow us to characterize the total strain as well as to evaluate incremental strains during one later phase of the same cycle when the rock or some markers have had already brittle behaviour.

The second group of classification features covers the predominant type of "internal" rock deformation, with distinction of two dominant types (irrotational and rotational), and of the important type of predominant simple shear deformations, within the latter. Presence of dilatation component may be significant for some early diagenetic phases (compaction) or coeval to migmatization or diaphthoresis deformations (volume increase) in high-grade metamorphics. The predominant fabric (non-tectonic - NT; linear - L; planar + S; combined linear & planar - LS) and its deformation history is another fundamental feature.

The third group of classification features refers to the distribution of heterogeneous deformation (and the corresponding structures) on a larger scale. Types and frequence of non-cohesive boundaries are also taken into account, as well as fold styles and predominant folding mechanism).

A very important classification feature concerns the behaviour of deformed rocks due to different deformational environment, rock lithology, strain rate, etc. Three fundamental groups of brittle, semi-ductile (brittle-ductile) and ductile formations (Ramsay, 1980) are easily distinguished in most deformation studies.

A few facies belonging to a complete facies classification could therefore be expressed as follows (only for deformations arisen during a single deformation phase):

- facies of predominant block tectonics, with subfacies of extensional (normal faults), compressional (upthrusts) and shear (wrench faults) tectonics;
- facies of block - fold deformations;

- facies of low-intensity fold deformations: low strain (E_s 0.5), upright folds formed by buckling, NT fabric);
- facies of intensive fold deformations: moderate to high strains (E_s 0.5 - 1.5); upright formed by buckling or by buckling and flow (shear), or inclined folds with pronounced vergence; L, S or LS fabrics;
- facies of fold-thrust deformations: moderate to high strains (E_s 0.5 - 1.5): vergent to isoclinal folds, often recumbent, with transition to nappes; S or LS fabrics;
- facies of complex thrust deformations: very high strains, isoclinal folds, development of thrust and nappe complexes; S or LS fabrics;
- facies of very intensive deformations with formation of secondary laminated (lenticular) structure: verhy high strains (E_s about and more than 2.0); glide boundaries and secondary (superimposed) foliation with transposition of earlier S-structures; S or LS fabric. Facies of other type and subfacies will be further developed in implementation of the above-exposed principles.

REFERENCES

Barr M., Coward M.P. 1974. A method for the measurement of volume change. - Geol. Mag., 111; 293-296.

Dunnet D. 1969. A technique of finite strain analysis using elliptical particles. - Tectonophysics, 7; 117-136.

Dunnet D., Siddans A.W.B. 1971. Non-random sedimentary fabrics and their modification by strain. - Tectonophysics, 12; 307-325.

Durney D.W., Ramsay J.G.R. 1973. Incremental strains measured by syntectonic crystal growths. - In: De Jong, Scholten (Eds.), Gravity and tectonics, Wiley, London, 67-96.

Flinn D. 1962. On folding during three-dimensional progressive deformation. - Quart.J.Geol. Soc. Lond., 118; 385-433.

Flinn D. 1979. The deformation matrix and the deformation ellipsoid. - J. Struct. Geol., 1; 299-307.

Ghosh S.K., Ramberg H. 1976. Rotation of inclusions by combination of pure shear and simple shear. - Tectonophysics, 34; 1-70.
Graham R.H. 1978. Quantitative deformation studies in the Permian rocks of Alps-Maritimes. - Men. BRGM, 91; 219-238.
Hansen E. 1971. Strain facies. - Springer-Verlag, Berlin; 208 pl.
Lisle R.J. 1974. Deformed lineations as finite-strain structures. - Tectonophysics, 21; 165-179.
Lisle R.J. 1979. Strain analysis using deformed pebbles; the influence of initial pebble shape. - Tectonophysics, 60; 263-277.
Lukjanov A.V. 1959. Plasticheskie deformacii izvestnyakov v zonakh droblenya soprovozhdayushchikh krupnie razlomi. - Izv. AN SSSR, ser. geol., 4; 102-106.
Lukjanov A.V. 1980. Plasticheskie deformacii i tektonicheskoe techenie gornykh porod litosfery. - In: Peyve A.V. (ed.) Tectonicheskaya rassloennost' litosfery; Trudy, vip. 343, Nauka, Moscow, 105-146.
Matthews P.W., Bond R.A.B., van den Berg J.J. 1974. An algebraic method of strain analysis using elliptical markers. - Tectonophysics, 24; 31-67.
Means W.D. 1976. Stress and strain. - Springer Verlag, N.Y., 339 p.
Milton N.J. 1980. Determination of the strain ellipsoid from measurements on any three sections. - Tectonophysics, 64; T19-T27.
Oertel G. 1978. Strain determination from the measurement of pebble shapes. - Tectonophysics, 50; T1-T7.
Owens W.H. 1974. Representation of finite strain state by three-axis planar diagrams. - Geol. Soc. Am. Bull., 85; 307-310.
Patalakha E.I. 1962. Analiz odnorodnoy plasticheskoy deformacii v svyazi s izucheniem vnutrennikh tectonicheskikh deformaciy gornykh porod. - Izv. AN Kaz. SSR, ser. geol. 6 (51); 42-52.
Patalakha E.I. 1967. Ocenka stepeni deformirovannosti skladchatikh tolshch. - Trudy IGN AN Kaz. SSR, v. 19; 48-110.
Patalakha E.I. 1970. Mekhanizm vozniknoveniya struktur

techeniya v zonakh smyatiya. - Nauka, Alma-Ata; 216 p.
Patalakha E.I. 1981. Geneticheskie osnovy morfologicheskoy tektoniki. - Nauka, Alma-Ata; 180 p.
Powell D., Treagus J.E. 1970. Rotational fabrics in metamorphic minerals. - Mineralog.Mag.,37; 801-814.
Ramberg H. 1976. Superposition of homogeneous strain and progressive deformation in rocks. - Bull. Geol. Inst. Univ. Uppsala, 6; 35-67.
Ramsay J.G. 1967. Folding and fracturing of rocks. - McGraw Hill, N.Y.; 560 p.
Ramsay J.G. 1969. The measurement of strain and displacement in orogenic belts. - In: Kent, P.E. et al. (eds.) Time and place in orogeny; Geol. Soc. Lond., 43-79.
Ramsay J.G. 1976. Displacement and strain. - Phil. Trans. R. Soc. Lond. A, 283; 3-25.
Ramsay J.G. 1980. Shear zone geometry: a review. - J. Struct. Geol., 2; 83-99.
Ramsay J.G., Graham R.H. 1970. Strain variation in shear belts. - Can. J. Earth., 7; 786-813.
Ramsay J.G., Wood D.S. 1973. The geometric effects of volume change during deformation processes. - Tectonophysics, 16; 263-278.
Robberts B., Siddans A.W.B. 1971. Fabric studies in the Llwyd Mawr ignimbrite, Carnarvonshire, North Wales. - Tectonophysics, 4; 283-306.
Robin P.-Y. 1977. Determination of geologic strains using randomly oriented strain markers of any shape. - Tectonophysics, 42; T7-T16.
Sanderson D.J. 1977. The algebraic evaluation of two-dimensional finite strain rosettes. - Math. Geol. 9; 483-496.
Shimamoto T., Ikeda Y. 1976. A simple algebraic method for strain estimation from deformed ellipsoidal objects. L. Basic theory. - Tectonophysics, 36; 315-338.
Siddans A.W.B., 1980. Analysis of three-dimensional homogeneous, finite strain using ellipsoidal objects. - Tectonophysics, 64; 1-16.
Wood D.S. Oertel G. 1980. Deformation in the Cambrian slate belt of Wales. - J. Geol., 88; 309-326.

Zagorcev I.S.1981. Rannealpiiskie deformacii v krasnocvetnikh otlozheniyakh Poletinsko-Skrinskoy razlomnoy zony. 2. Stroenie i deformacii severnoy chasti Vlakhinskogo bloka. - Geologica Balc., 11; 101-126.

Zagorcev I.S., Tzankov Tz.V.,Sapundziev K.M. 1976 Strain variation in a part of the Skrino anticline, District of Kjustendil, determined from deformed reduction spots. - C.-r. Acad. bulg. sci., 29; 1333-1336.

DEEP INHOMOGENEITIES IN THE UPPER MANTLE AND THEIR INFLUENCE ON THE STRUCTURE AND DYNAMICS OF THE EARTH's CRUST

V.B. SOLLUGUB, A.V. CHEKUNOV and Yu.K. SHCHUKIN
Institute of Geophysics, Acad. Sci. Ukrainian SSR, Kiev, USSR

In the times of the earliest geological studies, pioneers and ore explorers were, undoubtedly, aware of the inhomogeneous structure of the rock mass. They observed crumpled and fractured sedimentary layers varying in composition and structure, unconformable with huge intricately shaped bodies of crystalline rocks cut by ore-bearing veins.

The invention of a microscope and other tools for the fine structure and composition analysis made it possible to recognize the internal inhomogeneities of geologic formations to the point of discreteness that was beyond the direct perception of the man.

At last, recent large-scale studies, especially geophysical investigations, revealed the existence of inhomogeneities so large that they could be compared not only with vast regions, but with the size of the planet, on the whole. Inhomogeneities in structure, composition and phsycial properties at all the hierarchic levels of classification and subordination is, thus, an inevitable and important feature of the earth's interior.

In the present paper we consider deep inhomogeneities in the lithosphere only. It is clear from the title of the paper that we believe the mantle to be a supplier of matter and the major generator of motion within the earth's crust, whatever are the ways of their transformations. As will be seen, the crustal inhomogeneities and dynamics are interrelated with deeper inhomogeneities within the mantle and are strongly dependent on them. On the other hand, it should be emphasized that, in many cases, no direct correlation is found, which implies the existence of new and very interesting phenomena. Furthermore, we stipulate that the earth's crust is neither simply

Proceedings of the 27th International
Geological Congress, Volume 7, pp. 361–383
TECTONICS

some 'slag' of the passive mantle, nor is a sclerotic shell, but a rather dynamic system with inherent internal processes that, to a certain extent, has a feedback effect on the mantle.

Lithospheric inhomogeneities may be classified into two important varieties - sharp inhomogeneities and distributed ones. The former, naturally, mean sharp, clearly expressed borders. These are layers, blocks, massives, intrusions, etc. The cutlines of the latter are blurred and indistinct and, sometimes, can only be approximately recognized. These are the areas of gradual petrographic and lithological transition from one rock to another, zones of partial melting and regions of anomalous gradients of the physical property variations, etc. The pattern is, of course, very general, since it cannot cover all the varieties of natural models. For example, the top and bottom of the layer are, indeed, often indistinguishable, while within the layer, facies replace each other laterally. The clearness of the blocks' borders depends on the width and expressiveness of the bordering faults, the blocks themselves being almost always layered with physical properties varying within the layers; intrusive massives are always stratified, etc. The complexity of the structure of lithospheric inhomogeneities reflects the diversity of the causes that have produced these inhomogeneities. However, the pattern is far from being chaotic, its regularity can be revealed when specific geologic environment is analyzed.

The inhomogeneities are generally caused by the following reasons:

a) differences in composition and physical properties of structure-petrographic complexes ("layers") or other lithospheric formations; the differences arise either historically - along the vertical - or in the course of the subsequent lateral movements;

b) differences in physical properties of structure-petrographical complexes ("layers") or other lithospheric formations of the same composition; these result from the later thermodynamic effects;

c) differences in physical properties and composition

of various lithospheric zones due to the operating processes of matter transformation, metamorphism, phase transformation, magmatism, partial melting, seismic stress release, etc.;
d) differences in physical properties and composition of intensely fractured and other anomalous zones in the lithosphere due to their possible water saturation or saturation with other fluids, ore concentration, etc.

Layers are one of the major types of inhomogeneities. The thickness of the upper lithospheric layer - the earth's crust - varies from 5 - 6 km to 10 - 15 km below the oceans and from 20 - 25 km to 60 - 75 km below the continents. The thickest crust (of 50 km and thicker) is characteristic of orogenes of various age, mainly, of young orogenes, but, sometimes, of rather old ones, even the early Proterozoic. A thin crust is found below deep continental depressions, especially, under oceanic throughs.

The earth's crust (1-3, etc.) is composed of the sedimentary layer and consolidated part. On the continents, the latter is divided into the "granitic" and "basaltic" complexes, the transition between the two being, generally, smooth and gradual; below the oceans it is represented only by the "basaltic" complex. Each complex is built of rocks with specific petrographic composition and physical parameters. Within the earth's crust, the thickness of a single layer or complex may range widely: the sedimentary thickness ranges from 0 to 25 km, while the thickness of the "granitic" complex varies from 0 to 40 km and that of the "basaltic" complex from 0 to 20 km. These layers and complexes may occur in various combinations depending on the geologic history of the region and the type of geodynamic processes that have operated there. This, naturally, leads to formation of inhomogeneities at various depth.

In the earth's crust of the regions which have been covered by sufficiently detailed and comprehensive geological and geophysical studies we, sometimes, cannot neither identify fine layers and complexes, nor determine their age. For example, below the well exposed regions of the old pre-Ryphean consolidation in the south of the European part of the USSR (the Ukrainian Shield), the following units are recognized (4): 1) metamorphosed, intensely dislocated early Proterozoic sediments and volcanic rocks with V = 6.0 - 6.2 km/s; 2) the upper Archean formation of the same appearance with V = 6.2 - 6.4 km/s; metamorphosed Lower Archean sedimentary and volcanic rocks with V = 6.3 - 6.5 km/s. Then follow formations with V = 6.5 - 6.8 km/s which, at a greater depth, are gradually replaced by the "basaltic" complex with V = 6.8 - 7.4 km/s.

In many regions of the globe, at various depth levels within the lithosphere, low velocity layers are found. These may result from the following reason: different petrographic composition of the rocks composing different crustal complexes; temperature anomalies responsible for the appreciable degradation of elastic properties of rocks to the point of partial melting and alteration of thermodynamic conditions; the mechanical intrusion or overthrusting of high velocity deposits on the low velocity ones during tectonic displacement; hydrothermal springs providing water to saturate rocks in large volume, especially at a depth of a few dozens of kilometers; intense fracture of the medium due to earthquakes that occur at the depth of focal zones (Tashkent, Gazli in Armenia), etc.

We believe that low velocity layers within the earth's crust (mostly in the intermediate portion and at the base) may, possibly, have two important effects:

1) subhorizontal redistribution of the structure-material complexes of rocks at different levels not only within geosynclines and orogenes, but also, probably, on platforms;

2) strong deformation of the medium, including its intense softening with formation of deep zones highly permeable to fluids of various kind and gaseous components.

In the consolidated crust, apart from low velocity layers, high velocity layers may also be observed.

Combinations of both form peculiair geologic units of the "puff-pastry" type.

Below the tectonically active regions (continental rifts, epigeosynclinal orogenes), between the earth's crust and upper mantle, a thick (to 10 - 15 km and more) layer of the "crust-mantle mix" is found with elastic velocities of 7.4 to 7.7 km/s that are too high for the crust and too low for the mantle. By physical properties, the "mix" is similar to the crust-mantle, or "serpentine-basaltic", complex of mid-oceanic rifts (5).

It should be noted that the layers, as well as other lithospheric inhomogeneities, are, in general, unstable. Many of them are being transformed, migrating, changing their physical habit. This is, especially, is the case in the areas where active processes go on. The erosion of crystalline rocks, both acidic and basic, leads to accumulation of sedimentary landmasses (Fig. 1). As a result of metamorphism, these masses, as well as crystalline basic rocks, can be granitized, with the thickness of the "granitic" complex growing and that of the "basaltic" complex decreasing. The granite, in turn, undergoes reduction and basification. As is shown by the analysis of the evolutional series composed of real geologic units, the crust and mantle convert one into the other through the "crust-mantle mix" depending on which direction - the constructive or destructive one - evolution takes (5). It is evident, that all these processes, as well as others (magmatism, tectonics, etc.), lead to the corresponding changes in the structure and properties of lithospheric inhomogeneities. These processes are accompanied with expansion or compression of material, redistribution of stresses, deformations, etc., which are different at different levels of depth. These differences are so significant that the sign of deformations (compression or extension) occurring simultaneously may repeatedly change on the same vertical section of the lithosphere.

The layering of the lithosphere differs from stage to stage both vertically and over the area. The differences between the structural stages of the upper crust horizons, composed of sediments and volcanic rocks of the Phanerozoic and late Precambrian complexes metamorphosed to a various degree, are due, in the first place to the different nature of deformation during different tectonic cycles. These stages provide different reaction to tectonic stress of any kind, transform the stress, the material layering being complicated by the separation and displacement of one layer about another.

The above mentioned processes operate, also, in deeper lithospheric layers. The deformation of these layers, their topography and disharmonic structure revealed by the modern geophysical methods are the results, to a significant degree, if not mostly, of recent processes.

Analysis of seismological phenomena also reveals the difference between structure at different depths. The intensity of these phenomena in the lithosphere varies from layer to layer. Extended foci and focal zones form seismogenic structures which are subordinate to the structural-dynamic inhomogeneities and concentrated at a depth varying from several kilometres to several hundreds of kilometres. The regularities of the regional distribution of earthquake foci at depth indicate the presence of subhorizontal seismogenic structures.

Essential inhomogeneities within the earth's crust and upper mantle are due to deep-seated faults crossing the entire lithosphere from the asthenospheric layer up to the ground surface. Within the fault zones, individual blocks and plates move both in the vertical and horizontal direction. Here, areas of the partial melting of the upper mantle rocks can be encountered, as well as variations of physical properties of rocks, elastic velocity, density and seismic energy absorption, etc. Weakened areas may control the construction and properties of the medium in separate blocks: in some of them, to a depth of several hundreds of kilometres, earthquake foci are located. The fault zones cut the lithosphere into separate blocks and, to a considerable degree, are responsible for the stage-by-stage structure of the earth's crust and upper mantle, the stages differing in energy capacity, density, velocity characteristics of rocks, etc.

Large lithospheric inhomogeneities are produced by the temperature regime (6) which conditions, in particular, the origination and location of the chambers of partial melting of the mantle matter and is dependent on the depth to the base of the magneto-active block (Curie isotherm), position of the top of the electricity conducting layer, etc.

In studying the lithosphere it is very important to consider its dynamic features. The entire area of the continental Earth can be divided into two types - the passive and the active. In the areas of the passive lithosphere, no tectonic processes are observed at present. Here, the thermal field is, generally, rather quiet, seismic activity is weak, the recent movements of the earth's crust have amplitudes as low as several tens of millimetres per year, etc. The other type is the regions with the active lithosphere. In these regions, active tectonic movement is observed along with intrusions, anomalous heat flow and numerous earthquake foci. The lithosphere of both the types includes a number of inhomogeneities. These are especially abundant in areas of the junction of the active and the passive lithosphere, which form the zones of the most complex structure where especially active tectogenesis is found.

Geologic structures of the earth's crust and deep inhomogeneities of the upper mantle do not always correlate directly. This invites special consideration of obliquely dipping deep faults, mildly sloping tectonic contacts, geodynamic interaction of large lithospheric blocks with different regimes of their evolution and different areas of their junction. All these problems are also related to the prediction of earthquakes which is still the problem that is not solved.

The existence of the deep mantle inhomogeneities where mineral deposits are, probably, concentrated to be channelled to the ground surface through implicit zones of permeability in geological units, requires a change in the strategy of the comprehensive studies of deep structure in the direction of a further increase of the depth studied by using independent methods of survey, such as magnetotelluric sounding (MTS), heat flow studies, etc.

All the said is illustrated on the following specific examples. Fig. 2 shows a lithospheric cross-section along the line Isle of Crete - Vrancea. Here, both the crustal and upper mantle inhomogeneities are discernible. The crustal seismogenic areas occupy mainly the region of the Carpathians - Balkans and Crete. In the upper mantle, only two weakened zones of intense earthquake foci concentration - Carpathian and Aegean - are observed. In these sites, the greatest depths of the earthquake foci range to 200 m, whereas in the area of the Balkanides they are only 40 - 50 km. The number of earthquake foci also varies along the vertical. In the area of the Hellinides, most of the foci are concentrated at depths of 60 to 80 km, while in the Carpathians, they are about 150 km deep. The regions considered are characterized by a specific law of the vertical propagation of elastic waves (curves A and B). Fig. 3 presents a lithospheric cross-section along the line Kerch - Tarkhankut Peninsula - Vrancea (the junction of the Eastern and Southern Carpathians) - Apuseni Mts. (the Transylvanian Basin) (7). The geotraverse has been covered by comprehensive geological and geophysical studies; the seismic data on the geotraverse let us distinguish the young basement surface, horizon C_o, pre-Ryphean basement surface C_1, horizon C-M. i.e. the top of the crust-mantle layer, the M-discontinuity, seismic horizons P_1, P_2 in the upper mantle, the conducting horizon as derived from MTS data and the top of the asthenospheric layer according to heat flow data.

On the territory of the Crimean Peninsula and within the water area of the Odessa Gulf, horizon C_o with V_b = 5.8 - 6.2 km/s forms a basin with maximum depth values west of the Tarkhankut Peninsula.

The pre-Ryphean basement surface (C_1) with V_b = 6.2 - 6.5 km/s is traced all along the geotraverse.

The horizon forms a deep depression in the eastern section of the line. Its greatest depth is observed in the area of Jankoy (12 km) and west of the Tarkhankut Peninsula (10 km). In the area of Galats (northern Dobruja), horizon C_1 is 5 km deep. Further west, in the area of the Ciscarpathian Foredeep, the horizon plunges along the fault to a depth of 15 - 20 km to form a narrow trough. In the area of the Carpathians (in the region of Vrancea), the pre-Ryphean basement surface rises to a depth of 5 - 10 km.

Below the pre-Ryphean basement surface, in the Odessa-Jankoj Rift Zone 2, the seismic interface C-M with V_b = 7.5 km/s is found to lie practically flat at a depth of 28 to 30 km.

Another seismic boundary traced along the geotraverse considered (Fig. 3) is the M-discontinuity. Its hypsometry is quite different from that of the aforementioned interfaces. In the easternmost section of the line, the M-discontinuity occurs at a depth of 55 km and is the major feature of the Orekhov-Pavlograd zone of a thick crust bounded by fault V (4) in the west. The area between faults V and IV^1 is the southward continuation of the Zaporozhye -Sumy Median Massif. The deep faults IV - IV^1 border the Krivoj Rog - Krupets early Proterozoic geosynclinal area traced from the Ukrainian Shield to the Crimean Lowlands and further south, below the Crimean Mts. and the Black Sea water area, where its south-north strike changes to the southeasternly.

Further west along the geotraverse, the southward continuation of the Kirovograd Massif is found where the M-discontinuity occurs at a depth of 40 km and plunges to 50 km further west which is characteristic of the Odessa-Yadlov early Proterozoic Geosyncline bordered by two deep faults P and P^1. In the Northern Dobruja, the M-surface rises gradually to 40 km. Then follows another thick zone that characterizes the junction of the Eastern and Southern Carpathians and represents mountain "roots" of the Alpine age. West of this zone, the M-discontinuity is uplifted as high as 30 - 32 km, the depth being typical of the

Transylvanian Basin.

Below the M-discontinuity, under the Odessa Gulf, two seismic horizons P_1 and P_2 are established at depths of 85 and 100 km; the layer that occurs between them is characterized by velocities $V_{int.}$ = 7.9 - 8.0 km/s that are lower compared to those of the over- and underlying layers.

A similar low velocity layer is revealed in the upper mantle in several areas of the Ukrainian Shield - on geotraverse VI near the towns of Odessa (50 - 70 km) and Rovno (110 km) and on geotraverse III north of the towns of Kharkov (50 km) and Ukhta (70 km). The contour map compiled for the top of this layer (4, Fig. 2, p. 4) shows the existence of a long north-eastward striking swell whose axis is traced along the line Bucharest - Kishenyov - Kursk. As is seen in the contour map, the top of this layer as located on geotraverse V, form an anticlinal bend with minimum depth of 50 km in Dobruja. The bend is shown on the geotraverse by the dashed line P_1.

Important information about deep structure is contained in the heat flow and MTS data (6). Fig. 3 shows temperature distribution within the earth's crust and upper mantle to the depth of the top of the asthenospheric layer as derived from heat flow data.

The highest temperatures and the position of the asthenospheric layer within the geotraverse considered correspond to zones of relatively young tectonic and magmatic activity. In the west, in the area of the Pannonian Massive and Transylvanian Basin, the top of the asthenosphere is 50 km deep (according to explosion seismology data) (8), then it plunges eastward to a depth of 80 km. In the Volcanic Carpathians, the asthenospheric layer rises locally to 70 - 75 km ($t = 1300 - 1400^{\circ}C$), then it sinks to depths of about 200 km due to the presence of a vertically inhomogeneous layer where numerous earthquake foci are concentrated (9). Further east, a second zone is observed where the top of the asthenosphere is uplifted to 100 km - the Odessa - Jankoy Rift Zone.

Thus, the depth to the asthenosphere as derived from the geotraverse varies from 50 km in the Pannonian Massive to 200 km in the Odessa Gulf.

The reported data on the behaviour of the top of the asthenosphere along the geotraverse considered are confirmed by MTS data in the Crimean Lowlands and Eastern Carpathians.

As is seen, variations of the depth to the top of the asthenosphere amount 5o 50 - 60 km. The areas of a sharp asthenospheric depth variation are, evidently, coincide with faults of the mantle origin. The westernmost fault (G^{I}, G^{II}) passes through the seismo-active zone of Vrancea to include all the earthquake foci of this zone. Note that no foci exist in the asthenospheric layer, they all are concentrated in the lithosphere.

The deep mantle faults (G^{I}, G^{II}) are very wide and related to the following features of the earth's crust: the Carpathian Foredeep with thick sediments (to 15 - 20 km), the thick crust (to 55 km) below the Carpathians (the Alpine mountain roots and Zavaritsky-Benioff zone as assumed by some scientists).

The second mantle fault (D^{I}, D^{II}) where significant asthenospheric depth variations are also observed (amounting to 70 km or more) bounds a very thick (to 200 km) litospheric block in the east. This fault runs somewhat further south of Odessa and coincides with a deep fault bordering the Odessa - Yadlov geosynclinal zone in the west. Here, the top of the asthenospheric layer is uplifted to the level of 100 - 110 km. Below the second block with the least lithospheric thickness, some inhomogeneities are recognized within the earth's crust and upper mantle, such as low velocity layer at a depth of 80 - 90 km, etc.

The third mantle fault E^{I} E^{II} bounds an elevated asthenospheric block in the east and is revealed on the geotraverse considered in the eastern part of the Crimean Lowlands. Here, the depth variations for the top of the asthenosphere reach 50 km.

Fig. 4 shows the location of the major faults in the Ukraine, including the extraordinarily deep mantle faults $G^I G^{II}$, $D^I D^{II}$ and $E^I E^{II}$. In the main, two fault systems are outlined - northward mantle faults originating from the asthenosphere and deep (crustal) faults striking south-north and changing their trend to the south-easternly in the south of the territory.

Mantle faults cut the tectonic units of the territory considered at different angles and, as will be seen later, affect them in a different manner. They cut, naturally, deep (crustal) faults as well, but do not change their orientation. It seems that the mantle faults were the first to be formed, while all the other faults were formed later. As is known, the deep crustal faults striking south-northward (4) bound the early Proterozoic protogeosynclinal zones; therefore, the age of these faults is thought to be early Proterozoic. Consequently, the north-eastward striking faults are Archean or older. Mantle faults have been traced far beyond the Ukraine. So, according to Yu.K. Shchukin (10), the zone $G^I G^{II}$ passing through the seismoactive region of Vrancea is traced in the south-westward direction as far as the Tyrrhenian seismoactive zone. Deep structures in the form of seismogenic anomalies (11) are also observed in the north-eastern direction. It should be noted that in the north-western part of the Ukrainian Shield where the consolidated crust is known to have been subjected to deep erosion, the northeasternly faults have been established by geological evidence as well. In the rest of the Ukrainian Shield, faults of this orientation are identified only locally. All these facts prove the deep nature of the mantle faults described. As mentioned above, the mantle faults contributed to the formation of certain tectonic units. For example, the fault zone $G^I G^{II}$ mentioned above is traced in the area of the Tyrrhenian Sea (Fig. 5D) and is represented here as a typical Benioff-Zavaritsky zone. In Vrancea, the lithosphere is also very complex, as it was mentioned at the beginning of the paper. At the intersection of the fault $G^I G^{II}$ with the Odessa early Proterozoic zone, two anomalous phenomena are observed. First, the Odessa Zone is forked into two branches -

the Odessa - Yadlov Zone (P^a) and Odessa - Vinnitsa Zone (P^b) (4). Second, in this area a so-called Golovanev gravity high is established which is interpreted as an intrusion of basic and ultrabasic rocks from the mantle into the upper consolidated crust (Fig. 5C).

Further north-east, the fault zone $G^{I}G^{II}$ cuts the Dnieper-Donets Paleorift. In this area, a Ryphean graben terminates (or wedges out) within the paleorift, the former being traced from the east (fig. 5B) (4). Finally, in the central part of the East-European Platform, the zone $G^{I}G^{II}$ coincides with the abnormally elevated Curie surface (12) and the areas of the most productive oil and gas province. All the listed features of the mantle fault $G^{I}G^{II}$ are expressed as anomalous geophysical fields within passive regions and in a stronger seismicity of the crust and mantle in active regions.

The second mantle zone $D^{I}D^{II}$ is manifested within the consolidated crust in the change of the strike of the ancient platform border and some deep sublatitudinal faults within the Shield.

The third zone $E^{I}E^{II}$ played an important role in the formation of the western part of the Crimean Mts., the eastern termination of the Odessa - Jankoy Rift and western termination of Donbass.

These examples illustrate clearly the variety of structural inhomogeneities in the lithosphere caused by the processes operating deep in the lithosphere and asthenosphere.

REFERENCES

1. Belousov V.V. Principles of Geotectonics. "Nedra", Moscow, 1975, p. 262.
2. Muratov M.V. Origin of Continents and Oceanic Basins. "Nauka", Moscow, 1975, p. 176.
3. Khain V.E. Regional Geotectonics: Extraalpine Europe and West Asia. "Nedra", Moscow, 1977, p. 359.
4. Sollugub U.B. Crustal Structure of Ukraine. "Geofiz. Zurnal", 1982, 4, p. 3-25.

5. Chekunov A.V. Evolution of the Crust-Mantle Interface. "Geofiz. Zhurn., 1980, 6, p. 18-25.
6. Kutas R.I. Thermal field and model of the earth's crust. "Naukova Dumka", Kiev, 1978, p. 140.
7. Sollugub V.B., Chekunov A.V., Sollogub N.V. Structure of the Lithosphere below Carpathian Crimean Segment of Mediterranean Folded Belt. "Geofiz. Zhurn.", 1983, 1, p. 3-14.
8. Poshgay K. Crustal and Upper Mantle Structure from selected wave data. In Crustal and Upper Mantle Structure of the Central and East Europe. "Naukova Dumka", Kiev, 1978, p. 43-44.
9. Kondorskaya N.V., Slavina L.B. Shchukin Y.K. Structure and Dynamics of the Lithosphere of Carpathian region and Deep earthquake Region of Vrancea from seismological and Geophysical Data. In: "Geophysics, Geology and Disastrous Natural Phenomena". Moscow, 1980, 10, p. 89.
10. Shchukin Y.K. Geodynamics and Seismicity. Scientific and Technical Results. "Obshchaya Geologia", 14, Moscow, 1981, p. 86.
11. Shchukin Y.K. Seismogenic Structures. In "Crustal Structure of Central and East Europe from Geophysical data". "Naukova Dumka", Kiev, 1980, p. 126-129.
12. Borisov A.A., Volodarsky P.F. Kruglyakova G.I., Lopatina N.P. Analysis of the Nature of Anomalous Magnetic Field on the Territory of the USSR. In "Relation between sub-surface and deep-seated crustal structures", "Naukova Dumka", Kiev, 1971, p. 136-143.

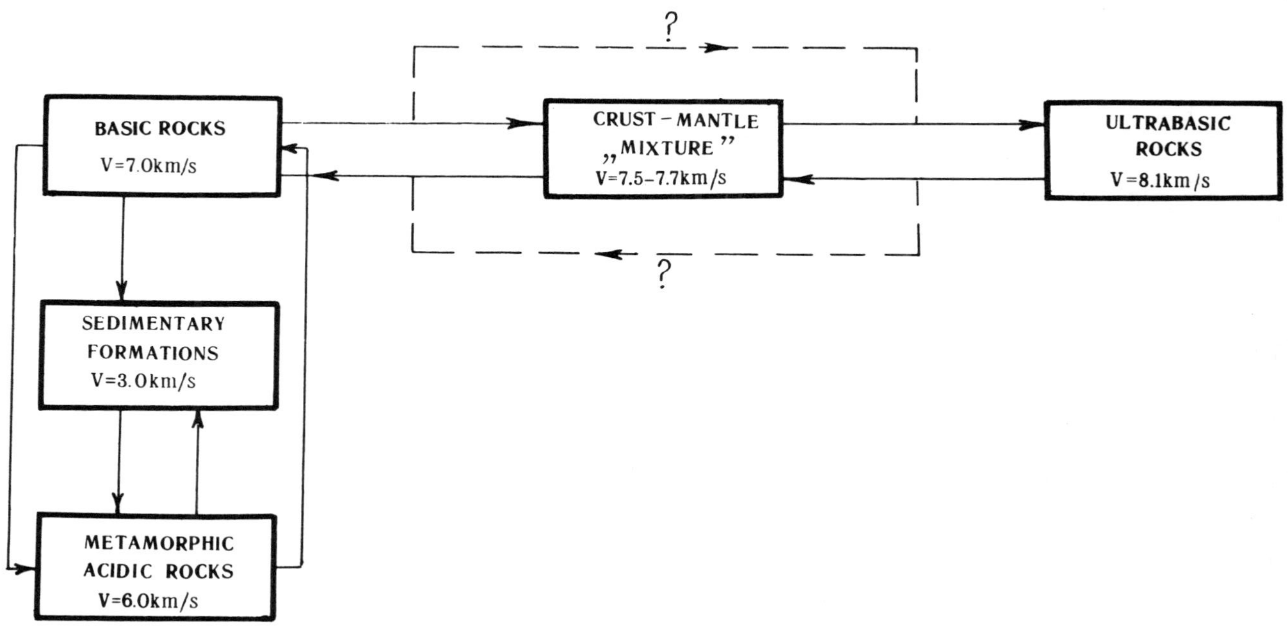

FIGURE 1. Block diagram of rock transformation in the course of geological evolution.

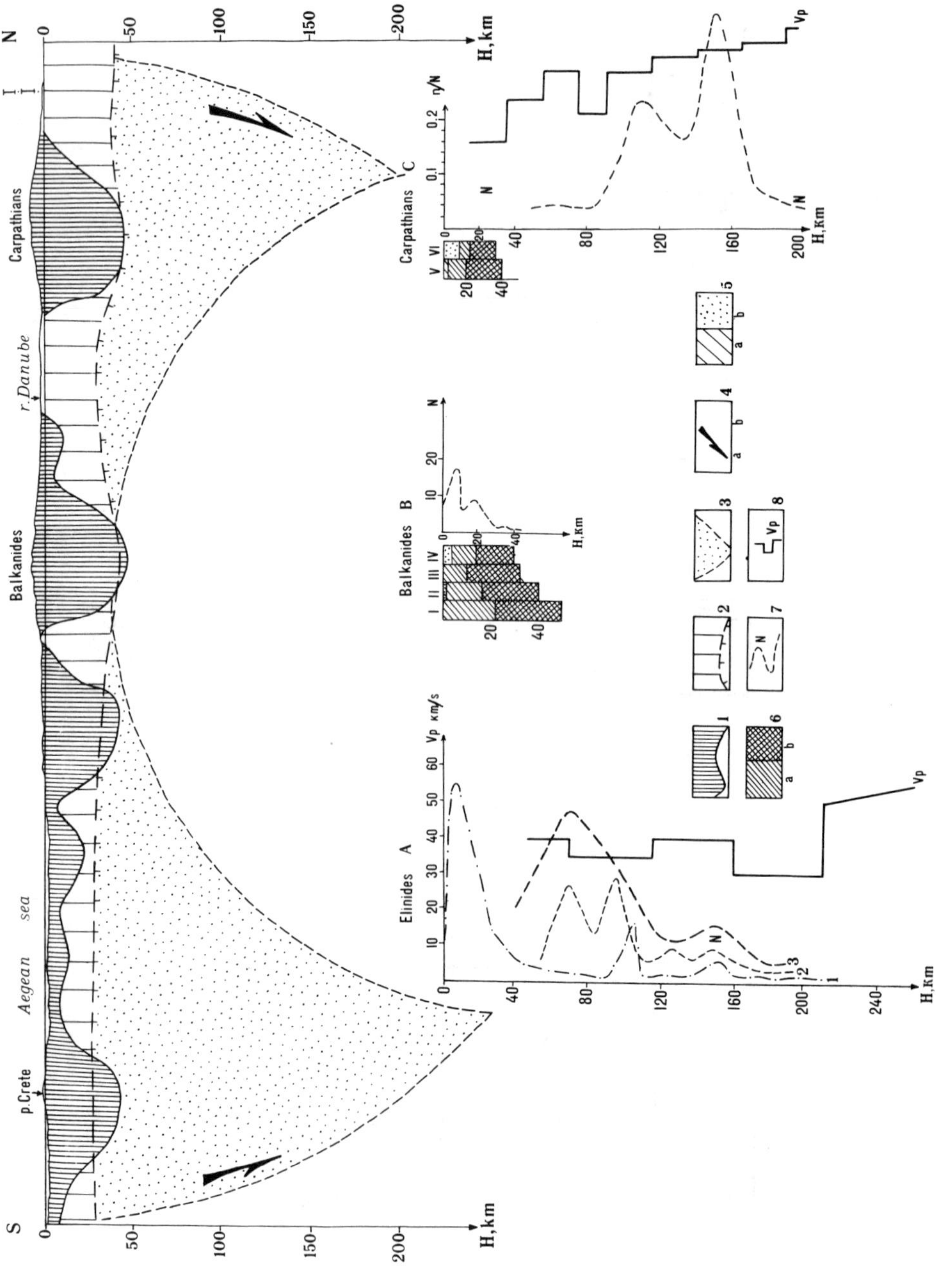
S
N
p.Crete
Aegean sea
Balkanides
r.Danube
Carpathians
H, km
C
Elinides A
Balkanides B
Carpathians C
Vp km/s
Vp
H, Km
I II III IV
V VI
1
2
3
4
5
6
7
8

FIGURE 2. Schematic vertical section of the lithosphere along line Isle of Crete/Mediterranean Sea/ -Vrancea/ East Carpathians (after Y.K. Shchukin) Numbers on the line 1 - seismogenic crustal layer rich in earthquake foci with M 5-6; 2 - aseismic portion of the earth's crust and M-discontinuity; 3 - weakened part of the upper mantle - a zone of concentration and energetic influence of strong subcrustal earthquake foci; 4 - dominant trends of dipping of deep earthquake focal zones in the Aegean and Carpathian regions.
Geological sequence in Balkanides ("B) and Carpathians ("C") 5a - Alpine folded rocks, b - sedimentary cover; 6a - granitic layer, b - basaltic layer. Geological sequence in Balkanides ("B") I - Rhodopian Massif, II - Mid-mountaineous area; III - Stranga; IV - Moesian plate. Geological sequence in the Carpathian region ("C") V - folded structures; VI - foredeep; 7 - earthquake foci-depth distribution graphs (graph "A"): 1 - for Europe, 2 and 3 - for Aegean region with different averaging window; 8 - curves of longitudinal velocity distribution with depth: for South-East Europe including Hellinides (curve "A") and East Carpathians (curve "B").

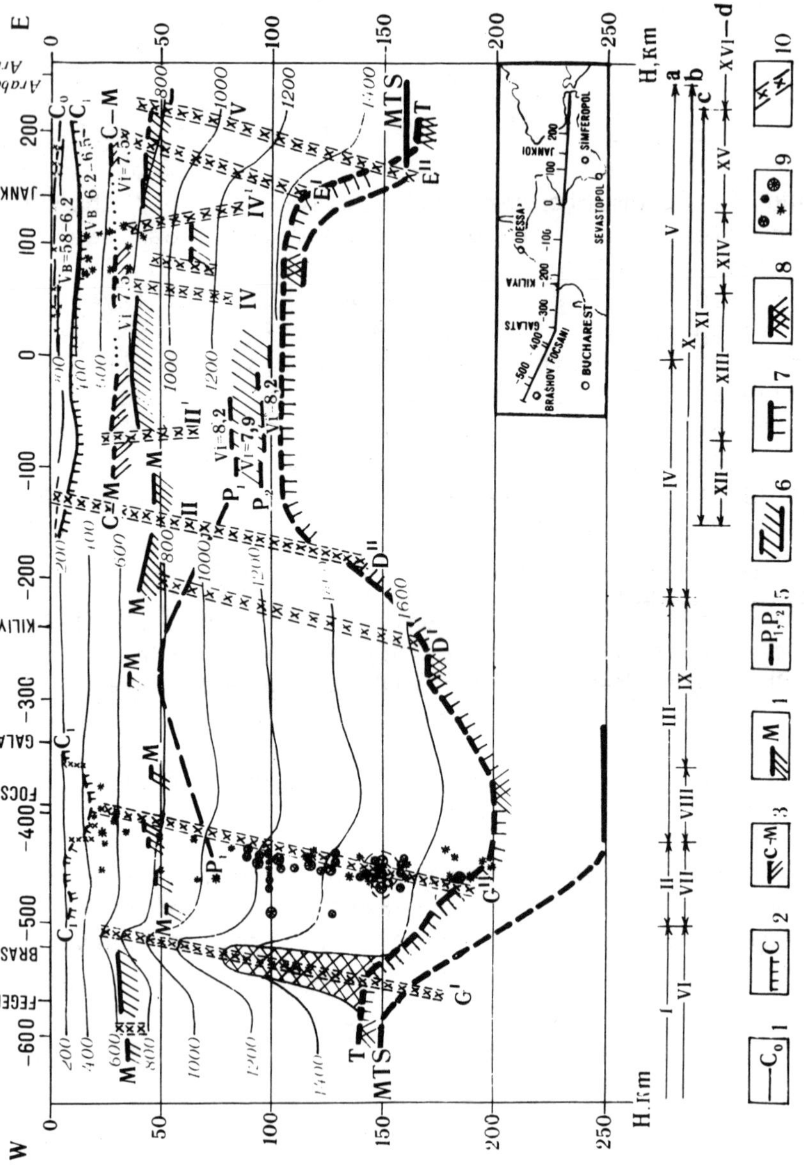
W
E
H, Km
FEGERASH
BRASHOV
FOCSANI
GALATS
KILIYA
JANKOI
Arabatskaya Arrow
MTS
ODESSA
SIMFEROPOL
SEVASTOPOL
BUCHAREST

FIGURE 3. Lithospheric cross section along Geotraverse Kerch-Vrancea Transylvanian Plain. I - young basement (Co) surface with V_b=5, 8-6, 2km/s; 2 - Preryphean basement (C_1) surface with V_b=6, 2-6, 5km/s; 3 - crustal mantle layer (C-M) surface with V_b=7,5 km/s; 4 - M-surface with V_b=8,1 km/s; 5 - seismic horizons P_1-P_2 in the upper mantle; 6 - low velocity layer in the upper mantle as derived from DSS; 7 - surface of the asthenospheric layer calculated from heat flow data; 8 - asthenospheric layer surface from MTS data; 9 - location of the asthenospheric layer surface from combined data; 10 - earthquake foci; 11 - deep-seated faults; 12 - deep faults in the mantle; 13 - faults; 14 - isotherms; 15 - zones of possible rock melting; 16 - zoning: a - by geographic criteria (I - Transylvanian Plain, II - East Carpathians, III - Peri-Dahubian Plain, IV - Gulf of Odessa, V - Crimean Peninsula);b-d - by geological criteria (VI - Transylvanian Basin, VII - East Carpathian folded structures, VIII - Carpathian Foredeep, IX - Dobruya folded structure, X - Near-Black Sea depression, XI - Odessa-Jankoy Rift Zone, XII - Odessa-Yadlov Early Proterozoic Geosyncline, XIII - Kirovograd Massif, XIV - Krivoy Rog Krupyat Early Proterozoic Geosyncline, XV - Zaporozhye Median Massif, XVI - Orekhov Pavlograd Early Proterozoic Geosyncline).

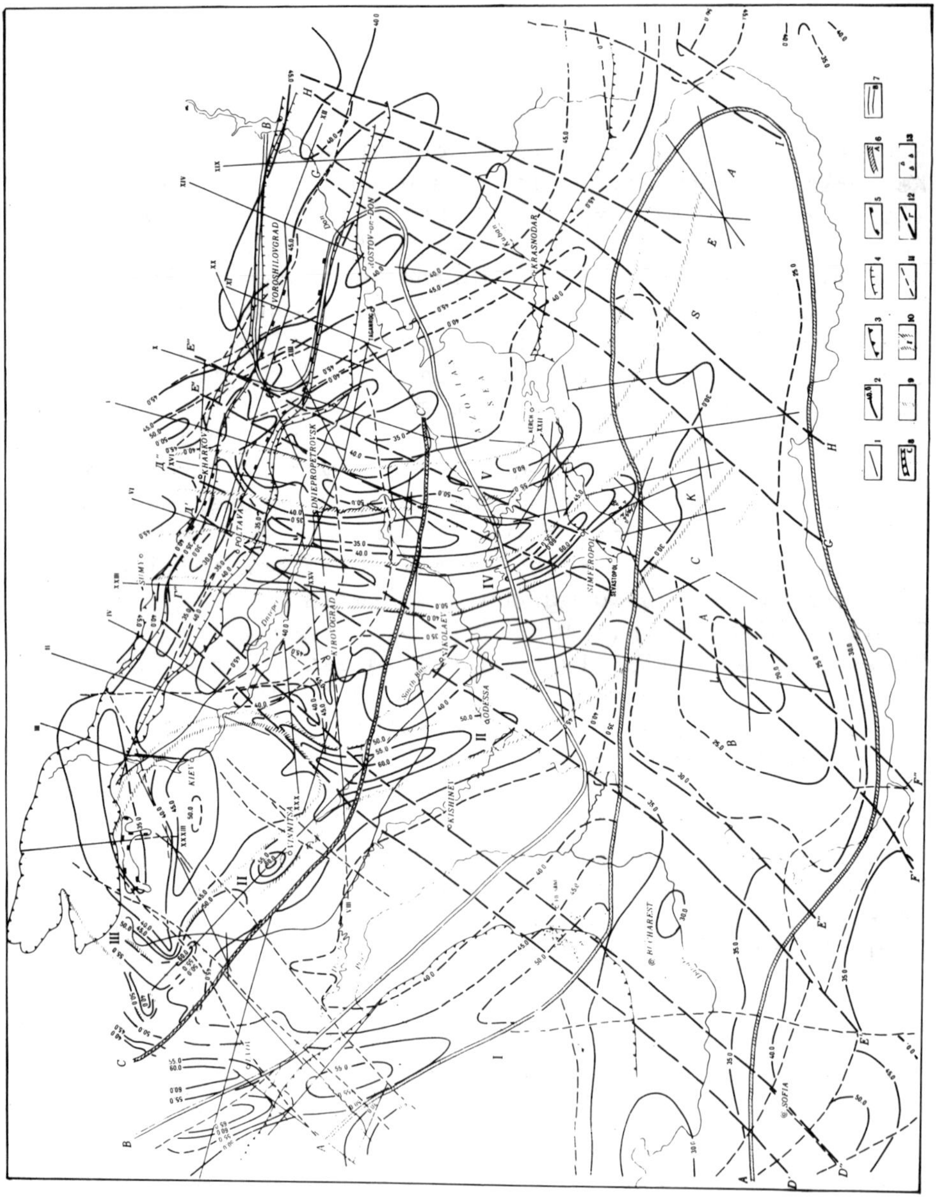
KIEV
KHARKOV
POLTAVA
DNIEPROPETROVSK
VOROSHILOVGRAD
ROSTOV-on-DON
KRASNODAR
ODESSA
KISHINEV
VINNITSA
SIMFEROPOL
SEVASTOPOL
KERCH
BUCHAREST
SOFIA

FIGURE 4. Scheme of deep tectonics of the Ukraine (After V.B. Sollugub) 1 - DSS lines; 2 - Misohypses; 3 - Dnieper-Donets Paleorift contour line from geologic data; 4 - Dnieper-Donets Paleorift contour from DSS data. 5 - Riphean Graben contour within Dnieper-Donets Paleorift; 6 - Contour lines of other tectonic regions; 7 - Oder-Caucasian Lineament; 8 - border of East European Platform; 9 - Sovietsk-Melitopol Lineament; 10 - deep Early Proterozoic faults; 11 - deep faults; 12 - especially deep north-west striking fault zones in the mantle (D'D'' - Sofia-Vrancea Sumy zone, E'E'' - Sevastopol-Slavyansk zone); 13 - earthquake foci.

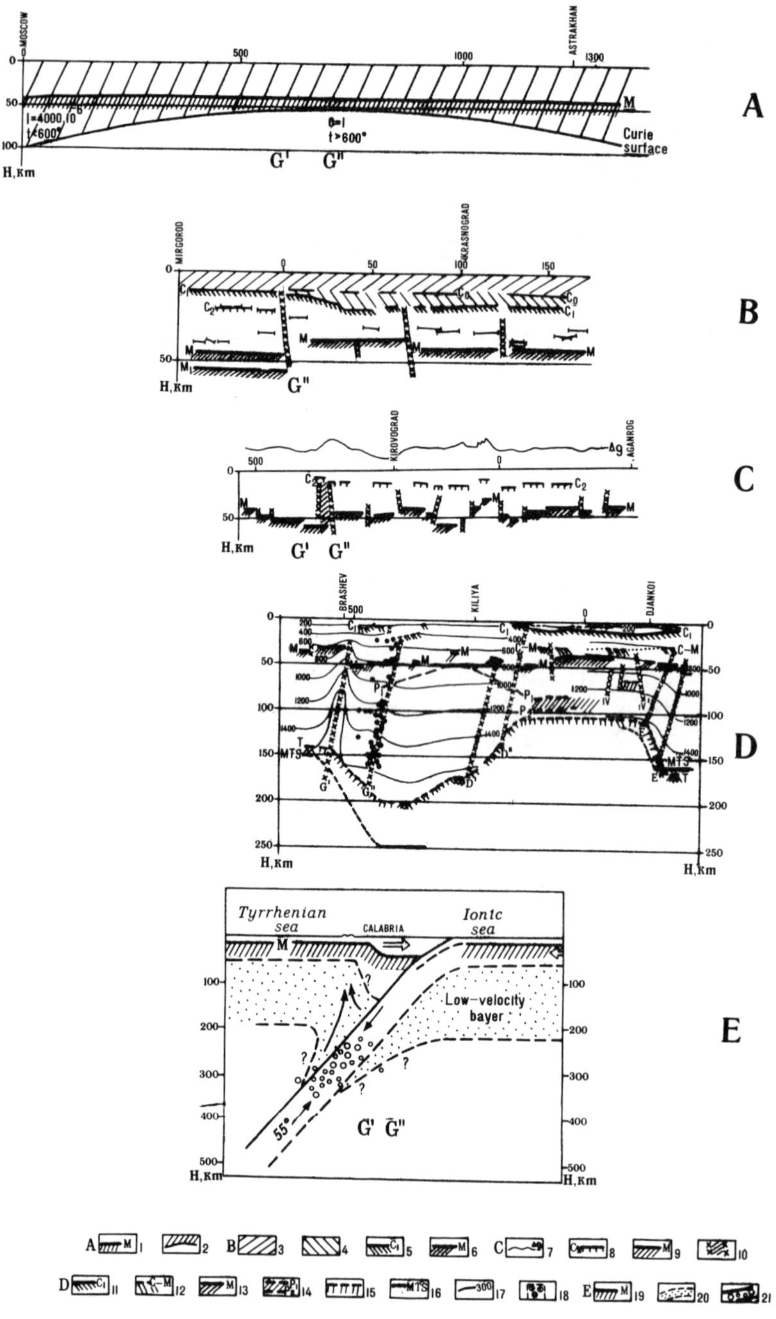
A
MOSCOW
ASTRAKHAN
500
1000
1300
M
Curie surface
t>600°
G' G"
H, Km
B
MIRGOROD
KRASNOGRAD
50
100
150
G"
C
KIROVOGRAD
TAGANROG
Δg
G' G"
D
BRASHEV
KILIYA
DJANKOI
MTS
H, Km
E
Tyrrhenian sea
CALABRIA
Ionic sea
Low-velocity layer
55°
G' G"
H, Km

FIGURE 5. Lithospheric cross sections showing inhomogeneities in the especially deep mantle fault zone. A - lithospheric cross section in central part of East-European platform; 1 - M-discontinuity, 2 - Curie surface; B - Crustal cross section of Dnieper-Donets Paleorift; 3 - Mezo-Paleozoic sediments, 4 - Ryphean sediments, 5 - Pre-ryphean Basement surface, 6 - M-discontinuity; C - crustal cross section of Ukrainian Shield; 7 - gravity profile, 8 - protobasement surface, 9 - M-discontinuity, 10 - intrusion of basic or ultrabasic rocks; D - lithospheric cross section along the line Kerch-Vrancea; 11 - preryphean basement surface, 12 - crust mantle layer surface, 13 - M-discontinuity, 14 - low velocity layer, 15 - asthenospheric layer surface from geothermal data, 16 - asthenospheric layer surface by MTS data, 17 - isotherms, 18 - earthquake foci; E - lithospheric cross section along the line Tyrrhenian Sea-Calbria; 19 - M-discontinuity, 20 - low velocity layer (the asthenosphere), 21 - Benioff Zavaritsky zone.